WATER POLICY IN SPAIN

Water Policy in Spain

Editors

Alberto Garrido

*Department of Agricultural Economics and Social Sciences,
Universidad Politécnica de Madrid, Madrid, Spain*

M. Ramón Llamas

Department of Geodynamics, Complutense University, Madrid, Spain

 CRC Press
Taylor & Francis Group
Boca Raton London New York

CRC Press is an imprint of the
Taylor & Francis Group, an **informa** business

CRC Press
Taylor & Francis Group
6000 Broken Sound Parkway NW, Suite 300
Boca Raton, FL 33487-2742

First issued in paperback 2017

Typeset by Vikatan Publishing Solutions (P) Ltd., Chennai, India

Library of Congress Cataloging-in-Publication Data

Water policy in Spain / editors, Albert Garrido, M. Ramón Llamas.
 p. cm.
 Includes bibliographical references and index.
 ISBN 978-0-415-55411-4 (hardcover : alk. paper) 1. Water resources development–Spain. 2. Water–Government policy–Spain. 3. Water-supply–Spain. I. Garrido, Albert. II. Llamas, Manuel Ramón. III. Title.

HD1697.S7W38 2009

333.9100946—dc22

2009021770

ISBN-13: 978-0-415-55411-4 (hbk)
ISBN-13: 978-1-138-11443-2 (pbk)

**Visit the Taylor Francis Web site at
http://www.taylorandfrancis.com**

**and the CRC Press Web site at
http://www.crcpress.com**

Dedications

To my parents, Ginés and Amparo.

Alberto Garrido

*To all the secretaries who have suffered my awful handwriting
with almost infinite patience and understanding.*

M. Ramón Llamas

Table of contents

V – *Conclusions*

Foreword

When we started working on this volume, we had a clear idea about the topics that should be covered to provide a fair view about Spain's water policy. Although a generation stands between us, we were both convinced that Spain was undergoing a profound transition from an almost exclusive supply-side type of policy to the much more complex and conflicting context in which most mature water economies typically evolve. We seized the opportunity to assemble a collection of essays about water policy in Spain convinced that the volume would need to contain a clear vision on how to face the most pressing needs and challenges. In 2004 the change of Government brought to power new ideas and seemed ready to embark in a new era, after a decade of discussions and two failed attempts to pass a National Hydrological Plan. The book, we thought, would go to print by the time the new Government should have had the time to put forward new water policies. These, we surmised, would be more targetted to address the most pressing environmental problems, to facilitate more efficient water allocation, and to solve or mitigate the current water political conflicts.

Unfortunately, the swinging regime of Spanish water policies did not stop, and in fact one would come to the conclusion that the pendulum now moves more erratically than before, and more heavily influenced by policies that are drafted and approved at the European Union level, such as the upcoming reform of the Common Agriculture Policy (the famous 'CAP' health check') or the application of the Water Framework Directive. Yet, as editors, we felt the obligation to ensure that the most updated and forward-looking strategic thinking had a voice in the volume. Second thoughts, however, suggested otherwise. For, in the absence of structured and clear strategy, it was far more valuable to let the reader build his or her own conclusions about the future. We are convinced that the analysis of the successes and failures of the Spanish water policy may be useful for water decision-makers in arid and semiarid countries where irrigation is the main water use.

Our purpose, then, was to provide all the needed background as well as cover the present and past policies, together with a description of the state of water resources in Spain in sufficient detail. In writing this foreword, we are tempted to entertain the hypothesis that drafting 20-year national water plans and getting them approved and accepted is similar to Penelope's weaving of Laertes' shroud by day and undoing it by night. Perhaps it is impossible to have such a robust and detailed plan, a conjecture that would be followed by the conclusion that Spain has lost 15 years searching for such chimeric goal. On the other hand, technological and political advances, generally positive, are so rapid and relevant that the adaptation to them may be one of the most important aspects for any water policy. Fortunately, despite such never ending discussions about what best water planning is or should not be, the country learned a lot and both the lay public and experts became more educated on complex water problems. Ultimately, this is what this book is about.

Despite the fact that the reader will run into that so-called and demised National Hydrologic Plan in many chapters, the objectives and the constraints too have changed as a result of a better understanding of what is economic and politically feasible. Since problems related to water resources have been aggravated as a result of economic growth and increasing water demand, policies below the national level (regional or local) have proved to be far more superior, efficient and expedite in many instances. In addition, the European Union's Water Framework Directive (WFD) passed in 2000 has added more constraints and put the environmental objectives just below ensuring a sustainable and reliable water supply to the people. However, more than in any other European country, the main water consumptive use is for irrigation (75%). Therefore the issues or conflicts with farmers are far more relevant than in most European countries. In numerous respects, Spain began the WFD race ahead of many other EU Member States because it has fairly well developed and resilient water institutions. So the elements for terminating the transition period and for implementing a national strategy for water are now in place. It is the policy delivery part what most volume's contributors seem to be missing, and we are not in a position to make that up.

Nevertheless, our final perspective is rather optimistic. There are already now some available technological improvements that hopefully will contribute significantly to solve the current problems. For instance, communication technologies (internet) is going to be a great help in order to improve education of farmers and the public at large, and to facilitate stakeholders participation thanks to a greater transparency in data information. Another relevant aspect is that seawater desalination technology and a cheaper and faster transport of food is going to change the concepts of water and food security.

We formulated the objectives and broad contents of the book after extensive thinking and numerous conversations. But we have benefited to no measure by the clarity with which the whole project was conceived from Dr. Ariel Dinar, Professor at the University of California, Irvine. We can only express words of gratitude and professional respect for his advise, generosity and clear ideas.

We are also indebted to Javier Herrero (Iberdrola), José María Fluxá (Foro del Agua) and Guido Schmidt (WWF-Spain, Madrid) for going through Chapter 20 and providing valuable suggestions. Alberto Garrido must personally acknowledge the Department of Agricultural and Resource Economics of the University of California, Berkeley, for hosting his 2005–06 sabbatical, coinciding with part of the volume's work. The economic support of the Spanish Ministry of Education and the Universidad Politécnica de Madrid for spending one year at Berkeley is also acknowledged. We are heartily grateful to Beatriz Salgado, Alberto Garrido's wife, for helping with the tedious but necessary task of formatting the chapters and making a thorough review of the whole material. Last but not least this book could not have been edited without the chapters' authors. We are indebted for their work, their friendship and their patience.

The Editors
Alberto Garrido & M. Ramón Llamas

Note:
All contributors use the following exchange rate €1 = $US1.3 when making conversions to the American currency, as this was the exchange rate in July 2008. In terms of measuremente units, the volume uses the metric system (1 hectare = 2.5 acres; 1233 cubic meter = 1 acre-foot; 1 cubic hectometer = 1 million cubic meters (Mm³) = 811 acre-feet = 1 gigalitre).

List of acronyms

AC	Autonomous Community – Spanish Regional Governments (*Comunidad Autónoma*)
AEAS	Spanish Association of the Water and Wastewater Treatment Companies (*Asociación Empresas de Abastecimiento y Saneamiento*)
AGUA	Water Management and Use Programme (*Actuaciones para la Gestión y la Utilización del Agua*)
ATLL	Waters of Ter and Llobregat (*Aguas del Ter y Llobregat, Catalonia*)
CAP	Common Agricultural Policy—The EU farm programs
CA	Albufeira Convention (*Convención de Albufeira*)
CADC	Commission for the Development and Implementation of the Convention (*Comisión para el Desarrollo y Aplicación del Convenio*)
CIP	Common Implementation Strategy of the Water Framework Directive
CofP	Conference of Parties in the context of the Albufeira Convention
CYII	Madrid Water Company (*Canal de Isabel II*)
EC	European Commission
EPTI	Provisional Scheme of Important Matters (referred to the Basin plans, in Spanish: *Esquema Provisional de Temas Importantes*)
EU	European Union
FCR	Full cost recovery rates
GNP	Gross National Product
GVA	Gross Value Added
ID	Irrigation Districts (*Comunidades de Regantes*)
ICG	General Quality Index (*Índice de Calidad General*)
IGME	Spanish Technical Institute of Geology and Mining (*Instituto Geológico Minero de España*)
INAG	Portuguese National Water Institute (Instituto da Água, I.P.)
INE	National Statistics Institute of Spain (*Instituto Nacional de Estadística*)
IWA	International Water Association
MAPA	Ministry of Agriculture, Fisheries and Food of Spain (*Ministerio de Agricultura, Pesca y Alimentación*), now called the Ministry of Environment and Rural and Marine Affairs, after merging with the MIMAM in 2008.
MERMA	Ministry of Environment and Rural and Marine Affairs (Ministerio de Medio Ambiente, y Medio Rural y Marino)
MIMAM	Environment Ministry of Spain (Ministerio de Medio Ambiente)
NHP	National Hydrological Plan (the Law of ..., *Ley de Plan Hidrológico Nacional*)
NIP	National Irrigation Plan
OECD	Organisation for the Economic Cooperation and Development
OPA	Offer of Public Purchase (*Oferta pública de adquisición de derechos*)
O & M	Operation and Management
PEAG	Especial Plan of the Upper Guadiana (*Plan Especial del Alto Guadiana*)
PG	Plan Gassett
RBA	River Basin Authority (*Organismo de Cuenca Hidrográfica*)
SC	Spanish Constitution
TST	Tagus-Segura (Water) Inter-Basin Transfer
WFD	Water Framework Directive (see the list of laws and statutes)
WWF	World Wide Fund for Nature, also known as World Wildlife Fund (USA)
yr	year

Official and shortened names of laws and statutes

Official denomination	Shortened name
○ Water Act 1866	1866 Water Act
○ Water Act 1879 (Act of 13th June, 1879)	1879 Water Act
○ Public Infrastructure and Water Works Act, 1911	1911 Water Works Act
○ Spanish Constitution of 27th December 1978	Spanish Constitution or SC
○ Water Act 29/1985 of 2nd August 1985	1985 Water Act
○ Water Act 46/1999 of 13th December 1999	1999 Water Act
○ Directive 2000/60/EC of the European Parliament and of the Council of 23rd October, establishing a framework for Community action in the field of water policy	Water Framework Directive or WFD
○ Consolidated text of the Water Act, 2001 (Royal Decree 1/2001, 20th July 2001)	2001 Consolidated Water Act
○ National Hydrological Plan Act 10/2001 5th July 2001	NHP Act 2001
○ Transpostion of the Directive 2000/60/EC has been translated into Spanish Act through Act 62/2003 of 20th December, modifying the Consolidated Water Act, passed by Executive Order 1/2001 of 20th July	Transposition of the WFD
○ Amendment of the National Hydrological Plan Act (Royal Decree 2/2004, 18th June 2004)	NHP Act 2004
○ Concessions Act of May 23, 2003 for the regulation of the Concession Contract for Public Works	Concessions Act 2003
○ Law 62/2003 Title V	2003 Consolidated-WFD Water Act
○ Act by which the National Hydrological Plan Act 2001 is amended (Law 11/2005 of 22nd June)	NHP Act 2005

List of contributors

Maite M. ALDAYA, Ph.D. in Ecology and MSc from the London School of Economics, is a researcher on the water footprint at the University of Twente (The Netherlands). She has worked in several international organizations such as the European Union or the Food and Agriculture Organization of the United Nations.

Gaspar ARIÑO ORTIZ is a Professor of Public Law at the Autonomous University of Madrid, and Managing Partner of the Legal Firm "Ariño and Associates". He has worked for the United Nations Organization, Ford Foundation, French Government, Brookings Institution, and the Rand Corporation. Author of numerous publications including eleven books, dozens of articles and multiple papers on the fields of his speciality.

Pedro ARROJO has a Ph.D. in Physics and is an economics Professor at the University of Zaragoza. He was the Managing Director of the New Culture of Water Foundation (FNCA) during its beginning years. In 2003 he was awarded the Goldman Price. He has authored tens of books and journal articles on many issues related to the economics of water projects and water allocation.

Ana BARREIRA has a degree in Law holds two LL.Ms degrees in Environmental Law from London University and in International Legal Studies from New York University. President of the Spanish environmental law center *Instituto Internacional de Derecho y Medio Ambiente*. She received the British Council European Young Lawyer Award in 1993. She has published extensively on environmental and water law topics.

Enrique CABRERA is a Professor of Fluid Mechanics at the Polytechnic University of Valencia. He is the Director of the Institute for Water Technology. His main fields are Water Resources Planning and Management with especial emphasis on water conservation (from leakage control to water economics) and Urban Hydraulics (mainly pressure systems). He has published extensively in the urban hydraulic field, with more than 150 publications, including 23 edited books.

Enrique CABRERA ROCHERA is an Industrial Engineer, Ph.D., MSc., and Associate professor of fluid mechanics in the Polytechnic University of Valencia His work focuses on water management, performance indicators and benchmarking. He is secretary of IWA's Specialist Group on Efficient Operation and Management of Urban Water Systems, and leader of IWA's Group on benchmarking.

Javier CALATRAVA has a Ph.D. in Agricultural Economics (Technical University of Madrid) and is an Associate Professor of Agricultural Economics at the Technical University of Cartagena since 2000. He has conducted research in the economics of water resources and soil conservation, with a particular focus on agricultural water markets, and participated in 10 public research projects funded by the European Union or the Spanish Government. He has authored 20 publications, including 13 articles in JCR referred journals and chapters in international books, and conducted consultancy work on water economics and agri-environmental policy for several Spanish administrations, the European Commission, and the OECD.

Ricardo COBACHO has a degree in Engineering and Ph.D. in Urban Hydraulics. He is an Assistant Professor at the Polytechnic University of Valencia on Fluid Mechanics and Hydraulics. His research focuses on efficiency in management and use of water, and has developed projects for the Regional and National Governments. Co-editor of 2 books published by Balkema.

Emilio CUSTODIO is an Engineering Professor at Technical University of Catalonia and known international expert in Hydrogeology. Between 1996 and 2007 he was the President of the Spanish Technical Institute of Geology and Mining, leading numerous research projects related to technological and groundwater-related topics. He has authored numerous papers in leading journals and monographies edited by international publishers.

Leandro DEL MORAL is a Professor at the Department of Human Geography at the University of Seville. He has leaded a number of EU Research Projects on hydrological risk management, and has authored diverse journal articles and books on water institutions and social risk perception. He is the Managing Director of the New Culture of Water Foundation (FNCA).

Shlomi DINAR is Assistant Professor in the Department of Politics and International Relations at Florida International University in Miami. His research spans the fields of international environmental politics, security studies, and negotiation with particular interest in conflict and cooperation over transboundary freshwater. His current work investigates the role of water scarcity in promoting international cooperation and the relationship between climate change and treaty compliance. He has authored two recent books on transboundary rivers and treates.

Antonio EMBID IRUJO is a Professor of Administrative Law in the University of Zaragoza. He had a Research grant in the March Foundation and the Alexander von Humboldt Foundation, Worked in the Max Planck Institut für ausländisches öffentliches Recht und Völkerrecht (Heidelberg, Germany). He was President of *Cortes Aragón* (*Parlamento de la Comunidad Autónoma de Aragón, Spain*). He has authored fifteen books and numerous articles about water law, educational law and federal law.

Francisco FLORES has a Ph.D. in Hydraulics Engineering and was the head of the Planning Office of the Tagus River Basin Authority in Spain. His recognized professional experience in the integrated planning of water resources and conflict resolution among water users is reflected in his numerous contributions to both academic and professional publications.

Francisco GARCÍA NOVO is Professor of Ecology at the University of Seville (Spain). He has 30 years of research experience in the area of plant-water relationships and the role of water in ecosystems. He has been studying the relationships of aquifer discharge in the diversity of wetlands, in special on Doñana National Park (SW Spain). He has authored more than one hundred journal references.

Alberto GARRIDO is an Associate Professor of Agricultural and Resource Economics at the Technical University of Madrid. His work focuses on natural resource and water economics and policy. Has been consultant for OECD, IADB, European Parliament, European Commission, FAO, and various Spanish Ministries and Autonomous Communities. He is the author of 100 academic references.

Luis GARROTE is a Professor of Civil Engineering: Hydraulics and Energetics at the Technical University of Madrid and received formal training at the Massachusetts Institute of Technology. His work focuses on planning and management of hydrological systems, flood forecasting, reservoir management, and intelligent decision support systems. He has developed widely applied projects for private companies, the Ministry of Education, Science and Technology, and the European Union.

Carlos Mario GOMEZ is a Professor of Economics at the University of Alcala de Henares. He has a long experience of applied research on issues related to natural resources (mainly water and energy uses) in tourism and other sectors. His work focuses on user-friendly models to analyse water demand, environmental costs and benefits of energy uses and to implement cost effectiveness analysis.

Carlos GRANADO-LORENCIO is Professor of Ecology at the University of Seville. He has worked on 28 national and international projects. He is a Fisheries Scientist (American Fishery Society, USA) and a technical adviser of Fondo para la Investigación Científica y Tecnológica, Agencia Nacional Promoción Científica, Argentina, Colombia and Spain. He is the author of more than 80 academic references.

Nuria HERNÁNDEZ-MORA holds MS degrees in Natural Resources Policy and Management from Cornell University and in water resources management from the University of Wisconsin-Madison. She has worked with environmental organizations in the US, and as a researcher and consultant for private and educational organizations in Spain. She has co-authored several articles, monographs and a book.

Ana IGLESIAS is a Professor of Agricultural Economics at the Technical University of Madrid. Her research focuses in understanding the interactions of global change with agriculture and water resources. She has contributed to programs of the United Nations, European and National environmental programmes. Her collaborative work has been published in over one hundred research papers.

Manuel Ramón LLAMAS is Emeritus Professor of Hydrogeology at the Complutense University of Madrid. Fellow of Spain's Royal Academy of Sciences, where he chairs the Section of Natural Sciences and the International Relations Committee. Author of one hundred books or monographs and almost two hundred scientific papers. President of the International Association of Hydrogeologists (1984–1989). Vice-president of the International Water Resources Association (2001–2003). Fellow of the European Academy of Sciences and Arts (2004).

Elena LÓPEZ-GUNN holds a BSc (Econ) (University of Wales), Mphil and Ph.D. (London School of Economics), and is currently a Tutorial Fellow at the London School of Economics and Political Science. She has carried out comparative research on water rights, public participation, and paradigm shifts in Spanish water policy. She has a Postdoc position at the Complutense University in Madrid.

Esperanza LUQUE holds a degree in Agricultural Engineering by the Technical University of Madrid (1989). Has developed most of her career in Consultancy companies in water and environmental area. She has contributed from the consultancy area to the different Hydrological Plans adopted in Spain during the 90's. Currently she works in research studies on management of shared river basins and manages projects on agricultural and environmental risks.

Josefina MAESTU is an economist with wide experience of work in water economics, structural funds, water institutions and policy at National, European and International level. She is currently coordinating the implementation of economic analysis in the Water Framework Directive in Spain. She is an Associate Lecturer on Economic Management of Natural Resources, at the University of Alcala de Henares.

Pedro MARTINEZ-SANTOS holds an Honours BE (Civil) and a Masters of Technology Management (University of New South Wales, Australia), Master in Civil engineering of the Politechnical University of Madrid, and a Ph.D. in Geology of the Complutense University. His research interests relate to groundwater hydrology and aquifer resources. He just finished his Ph.D. at the

Universidad Complutense (Madrid) on the uncertainties of integrated water resources management in the Upper Guadiana Basin, Spain.

Luis MARTÍNEZ CORTINA has Ph.D. in Civil Engineering. He is a senior researcher at Spanish Technical Institute of Geology and Mine, and has carried out some hydrogeological studies and groundwater flow numerical models. He has worked as a researcher of several European Union Research Projects, and also of the Groundwater Project, launched in 1999 by the Spanish Foundation Marcelino Botín. He is co-author of four books and monographs, and is the author or co-author of about 40 scientific articles.

Manuel MENÉNDEZ PRIETO holds a degree in Civil Engineering and has been a Lecturer in Hydrology in the Technical University of Madrid. He was in charge of the Hydrology area at CEDEX (Experimental Centre. Ministry of the Environment). He currently works as a Technical Director in the Cabinet of the Secretary of State for Rural Development and Water in the Spanish Ministry of Environment and Rural and Marine Affairs.

Marta MONEO is a Post-doc Researcher in the Potsdam Institute for Climate Impact Research in Germany. She graduated in Environmental Sciences at the Universidad de Alcalá de Henares, Spain and has a MSc on Rural Integrated Planning at the Mediterranean Agronomic Institute of Zaragoza (IAMZ-CIHEAM) and a Ph.D. on Agricultural and Natural Resources Economics at the Technical University of Madrid. She continues to contribute to several national, European and international projects on climate change.

Paula NOVO, Agricultural Engineer from the Technical University of Madrid and MSc in International Development Studies from Wageningen University, is currently a junior researcher on virtual water and Ph.D. candidate at the Technical University of Madrid.

Roberto RODRÍGUEZ CASADO, Agricultural Engineer from the Technical University of Madrid, is currently a junior researcher on the water footprint at the Technical University of Madrid.

Mónica SASTRE BECEIRO is an Associate barrister at Ariño & Asociados specialised in water law. She has a Master in Environmental Law by the University of the Basque Country (Spain) and Doctoral Degree in Law. She is author of several publications about water law and water market in Spain, including "Leyes de aguas y Política Hidráulica en España" [Water Law and Hidrological Policy in Spain] written with Professor Gaspar Ariño.

Guido SCHMIDT has Ph.D. in Landscape Planning by the University of Hannover, Germany (1994). He is environmental consultant, focused on environmental planning, public participation and freshwater ecosystems. For a decade, he participated in Spain's water policy being the head of the World Wide Fund for Nature (WWF) Freshwater Programme in Spain and leading their conservation activities in the Doñana wetlands.

Lucia DE STEFANO holds a degree in Geological Sciences from the University of Pavia, Italy (1995), and a Ph.D. in Hydrogeology from the Complutense University of Madrid (2005). She has worked on water resources management both in the private sector and at the university. From 2002 to 2008 she worked for the World Wide Fund for Nature (WWF) on water policy issues: she has been the coordinator of a pan-European assessment of water policy and has been actively involved in the implementation of the WFD in Spain and in the EU. Currently she holds a postdoc research position in the Program in Water Conflict Management & Transformation of Oregon State University (USA).

Julia TOJA SANTILLANA is Professor of Ecology at the University of Seville. She has been lecturing on Limnology since 1978. Her research has focused on plankton studies in reservoirs and ponds. She has authored 65 scientific papers and a monography on limnology of reservoirs and a book on Doñana Park.

Consuelo VARELA ORTEGA is an Associate Professor in the Department of Agricultural Economics at the Technical University of Madrid, with research interest in agricultural policy and the environment, water economics and polices, land market and agricultural institutions. She has collaborated with a numerous international organizations (FAO, IDB, WB) and has published extensively in scientific journals and books. She serves in the advisory boards of CIRAD, CGIAR and IFPRI.

I
Introduction

CHAPTER 1

Scope and objectives

M. Ramón Llamas
Department of Geodynamics, Complutense University, Madrid, Spain

Alberto Garrido
Department of Agricultural Economics, Technical University of Madrid, Spain

1 INTRODUCTION

While the modern Spanish State was formed in the 15th century, historical records of works, statutes, and water utilization date back to at least two centuries BC. Its ecological wealth and its mild climate, plus the relatively abundant watercourses flowing from inland Spain to the Mediterranean and Atlantic coasts, favored long-term human settlements, which soon traded with other Mediterranean cultures.

Roman engineers built outstanding hydraulic structures in Spain, many of which can be still admired today. The Muslim occupation of most of the Iberian Peninsula in the 8th century brought new methods of water supply and management. Until 1858, for example, the city of Madrid was supplied with water mainly through infiltration galleries (khanats), a technology imported by the Arabs from Iran. They also set up bottom-up institutions to manage the scarce irrigation water and avoid social conflicts. The well-known *Tribunal de las Aguas de Valencia* (Valencian Water court) conserves over nine centuries of records.

Surface irrigation systems were also developed in many other regions of Spain, but almost exclusively on the flood plains of many rivers. Their stream flow was diverted by small dams. A third of Spain's present irrigated acreage was already cultivated at the end of the 19th century.

The main concern of Spanish politicians and social philosophers during the 17th and 18th centuries was to make the country's rivers navigable. The attempts to emulate the situation in other European countries failed because of Spain's rugged topography, and the low flows during the dry season. In the 19th century private entrepreneurs made several attempts to develop hydraulic systems, mainly for irrigation, in a similar way to the development of railroads. Most of them were economic failures.

In 1898 Spain lost Cuba and the Philippines, its last colonies. This was one of the lowest points in Spain's history. A group of scholars and politicians, known as the *regenerationists*, tried to explain and overcome this depression in Spain. One of the most pervasive mottos was that the Spanish people needed *Escuela y Despensa* (education and abundant food). As this book compellingly claims, the regeneration of Spain laid the foundations for a century on irrigation and water policies. Water projects and land reclamation were the means of dragging Spain out of poverty and illiteracy. However, this development could not be achieved by private initiative. It had to be a public or government-led action, and many politicians and intellectuals supported these ideas.

In 1902 the Ministry for Development prepared the "Gasset Plan" (GP), named after the engineer commissioned to draft this National Waterworks Plan. This first comprehensive national plan was first and foremost an inventory of dams and canals, primarily meant to provide water for irrigation. Incidentally, the GP was passed in the same year that the US Bureau of Reclamation, which shared similar aims, was set up. Both were preceded, during the 19th century by similar plans drafted by British engineers across the British Empire, mainly in India. This is what Allan (1999) describes as the "hydraulic mission".

Hardly any of the infrastructures listed in the GP were built in the following four decades, probably because of political unrest and Spain's economic difficulties. Nevertheless, it helped to raise public awareness about the indisputable importance of water management and irrigation. In 1926 the Government set up the *Confederacion Hidrográfica del Ebro* (Ebro River Basin Authority) for the integrated management of water in the Spanish basin. In less than two decades all surface water in Spain was managed by such path-breaking institutions. Present-day water institutions are still founded on the basin authorities.

In 1933 the Ministry of Public Works prepared a draft National Water Plan. This included the transfer of water from northern (Ebro river) to southeastern Spain. The Government never formally approved this National Water Plan, but most hydraulic engineers considered it to be a good solution. It was basically the same model as designed and later implemented in California to solve the scarcity of water in the southern part of the State.

Spain passed its first Water Act in 1866 (amended in 1879), which all experts agree to be a monument of fine legal drafting. While the 1866 Act was to remain in force until the 1985 Water Act was passed, which attests to its value and adequacy, it was actually meant to provide the foundations for modern governmental water policy and planning initiatives. Surface water rights for irrigation, urban water supply and hydropower could be soundly established in the framework of the 1879 Water Act. In Spain groundwater was under private ownership, as it still is today in other countries like the USA (California and Texas), India or Chile.

Between the end of the Civil War in 1939 and the enactment of the Spanish Constitution in 1978 that restored democracy in Spain, the country experienced an intense rate of waterworks construction. Spain almost doubled its surface water irrigated area, reaching 2.5 million hectares by 1975. The paradigms that drove the whole process remained unquestioned for five decades. Supplying water to the fields, controlling rivers and installing more hydropower capacity were undisputed objectives.

As a result of a steady pace of construction from 1950 to 2000 (during which about 20 dams a year were put into operation), Spain has about 1,300 large dams today. In terms of dams per capita, Spain is fourth in the world.

By the end of the 20th century, as this book details, Spain had a grand scheme of waterworks, and a whole institutional edifice had been erected. Even so, water problems peaked and came to the notice of most Spaniards. The policy thrust to start addressing the water problems of the 21st century was to come from the European Union.

While many aspects of the institutional foundations for beginning to think seriously about the most pressing water problems were already in place, the 1985 Water Act maintained the principles that secured very cheap water for hundreds of thousands of farmers, many of whom were also given houses, tractors and other capital goods to settle in semi-arid terrains and depopulated areas. Water allocation was governed by engineering constraints and, in times of droughts, rationed through strict administrative rulings.

The process of forming interdisciplinary teams involved in the preparation of the National Water Plans was similar to the one described by Dooge (1999), albeit a couple of decades later. Dooge divides the design and implementation of hydroprojects into several periods. Up until the 1950s engineers were the only decision makers. During the 1960s, economists began to play a part. In the 1970s environmentalists also started to participate. In the 1980s, the people affected by the project began to have a say. Finally, NGOs now play a significant role.

In Spain pre-1960 plans were prepared exclusively by civil and agricultural engineers. In the 1960s project documentation included cost-benefit analyses for the first time, but this had little or no impact at all on decision makers.

With its accent on water planning, the 1985 Water Act ushered in a new phase. Nevertheless, the 1993 draft of the National Water Plan attached little importance to economic analyses, and environmental impact assessment was practically nonexistent. The National Water Plan approved by the Spanish Parliament in 2001 did contain economic and environmental assessments. However, these assessments were strongly criticized by many scholars and conservation groups, especially members of the New Water Culture Foundation, as discussed in several chapters of this book.

These differences of opinion, together with the strong opposition to the Ebro water transfer by the regions of Aragon and Catalonia, a cause that had also been championed by the Government elected in 2004, put an abrupt end to the more than 100-year era of huge water projects.

Meanwhile a 'revolution' within groundwater uses in the late 1960s and early 1970s silently came to be the most intractable problem facing Spanish water policy in the mid-1990s. Looking back, it is ironic that the most productive agricultural water uses were those initiated by private individuals tapping groundwater resources and not those served by irrigation projects developed to make the *regenerationists'* dreams come true.

By the time the 1985 Water Act was passed, the Spanish Constitution already provided the umbrella for governments to administer the hydraulic public domain (all surface and ground waters) and to intervene in cases of groundwater 'overexploitation'. Yet, the enforcement of the 1985 Water Act provisions on groundwater resources failed on most accounts.

In 1986 Spain turned its back for good on a history of political unrest, isolation, conflict and swinging regimes when it became a member of the European Economic Community, now the European Union (EU). The Water Framework Directive enacted in 2000 by the EU is meant to deliver important ecological improvement on most EU water bodies by 2015. Most authors of this volume agree on the importance of two landmarks that has brought traditional Spanish water management to an end. The first is the passing of the European Union's Water Framework Directive, which all 27 member states must enforce. The second is the above demise of the grand Ebro inter-basin transfer, proposed formally in 2001 but conceived since 1998, and stopped in 2004. The two also spelled the end of the univocal definition of *common interest* and, with that, the broad consensus on water policies. Post-2004 water policies have to be diverse, less centralized and more focused on the environment. On these accounts, the period 2004–08 can be characterized by four simultaneous developments: (1) the devolution to the Autonomous Communities (as regions are called in Spain) of significant water competencies, (2) the failure to develop sufficient desalination capacity along the Mediterranean coast to substitute for the water supply that would have been serviced from the Ebro transfer, (3) the increasing use of water market exchanges and water banks, and (4) the recognition that implementing the WFD would entail enormous difficulties because of the severe environmental deterioration of many water bodies. These four ideas run across many of the books' chapters and are dealt in detail from diverse perspectives.

2   OBJECTIVES AND SCOPE

Against this turbulent history, this volume aims to offer a self-contained overview of water policy in Spain. This book aspires to bridge a gap that only the work of Maas and Anderson (1978), still a key reference on traditional Spanish water institutions, has been partially filled for English readers.

Specifically, the book intends to:

- provide a detailed description of Spanish geographical, climatic, and hydrological features;
- review the last hundred years of Spanish history to give an understanding of its water policy achievements and failures;
- identify the major water challenges that Spain needs to face;
- based on a detailed analysis, speculate about the country's potential to look after its water resources in an integrated manner and rely on demand management to meet its more pressing needs; and, lastly,
- list the lessons that are potentially applicable to countries and regions evolving similarly.

The book's structure is intended to offer a comprehensive overview of Spanish geographical and political diversity during four historical periods. The first ends with the beginning of the Spanish Civil War in 1936, when the grand water policy lines were traced and detailed, though not developed. The second ends with the death of General Franco in 1975, which paved the way for a political transition that culminated in the Spanish Constitution 1978. The third period came to a gradual close starting in December 2000 with the passing of the WFD and the inception of preparatory works, and ending abruptly after the change of government in the 2004 election. During this third stage Spain formulated new and innovative policies that never materialized in significant accomplishments. The last and fourth stage began in March 2004 and led to a political agenda, which—being part of the EU WFD—is outside direct Spanish influence and for which the country is accountable to the European Commission and other member states. Recent experience and compliance with the WFD has shown that agreements and alliances must now be forged to approve, never impose a course of action. These four stages cross-cut most chapters of the book, as they have been the cause and effect of most water-related topics. Apart from having a decidedly historical dimension, the volume is meant to help readers grapple with Spanish geographical and political diversity. Furthermore, the approval of the 1978 Spanish Constitution led to a signifi-cant process of political decentralization. Seventeen Autonomous Communities (ACs), most with boundaries crossing the territories of the main river basins, took up significant responsibilities in the area of environmental and water policies. This certainly added another factor of complexity, unprecedented in all Spain's water policy history.

3   THE BOOK'S STRUCTURE

In addition to this introductory and a concluding chapter, the book is divided into five parts. Part II, covering Chapters 2 through 6, reviews Spain's physical, economic, environmental and climatic conditions. Part III (Chapters 7 through 9) provides an overview of the constraints, opportunities and social perceptions of water policy and water issues. Part IV, including ten chapters, develops Spanish water policies in detail, covering an array of topics. Part V is the closing section with **Chapter 20** containing the conclusions and the editors' speculations about the future, identifying Spain's strengths and weaknesses in response to the main challenges reviewed in the volume, as well as lessons other countries can learn.

Part II begins with **Chapter 2** (Martínez-Cortina), which provides a comprehensive descrip-tion of the physical characteristics of all water resources. **Chapter 2** reviews major man-made infrastructures describing their regulation and storage capacities at the basin level. **Chapter 2** is the geographical guide and source of maps for all the other chapters. The volume's two chapters on major environmental and pollution processes attest to the severity and magnitude of the chal-lenges ahead of Spain if it is to fulfill the WFD's objectives. First, **Chapter 3** (García Novo *et al.*) reviews the massive process of reclaiming wetlands and impounding water bodies, summarizing some of the main ecological implications in order to define what 'good ecological status' means in the context of the WFD. Second, we learn from **Chapter 4** (Schmidt & de Stefano) that the primary causes of ecological problems are man-made impacts on water bodies, and the political unwillingness to enforce statutes and legislation.

The economic and social dimension of water uses are the subject of **Chapter 5** (Maestu & Gómez) and **Chapter 6** (Aldaya *et al.*). These chapters portray a clear picture of the importance of irrigation as the major water user, and the increasing importance of urban, and industrial uses. Sectorial use trends reflect the changes that have taken place over the last fifty years, showing the differences across regions and major basins. In terms of water uses, Spain is still a predominantly agricultural country, where 75% of the available resources are used for irrigation. Yet, booming sectors, such as tourism, second-home developments and golfing, and the expansion of cities are pushing for a redefinition of water allocations and further liberalization. **Chapter 6** reports recent evaluations of Spain's water footprint, discussing the role of virtual water trade examined in light of the increasing globalization of water.

Part III covers a number of topics that explain why inherited policies failed or succeeded, and influence the choices that are available for the future. **Chapter 7** (Iglesias *et al.*) focuses on drought and climate risks. The chapter first gives an overview of the indicators that can be associ-ated with the risk of drought and water scarcity. Second, it discusses the concept of risk integrat-ing drought sensitivity and vulnerability to exposure. Finally, the chapter examines issues related to global environmental change. Technological gaps, in addition to increasing water demands and climatic vulnerability, are factors conducive to the overall environmental deterioration of water bodies and ecosystems, as reviewed in **Chapters 3 and 4. Part III** includes two further chapters about urban and sociological issues. **Chapter 8** (Cabrera *et al.*) reviews the situation of the water utilities in Spain and identifies the major challenges in the area of urban water supply and sanitation. **Chapter 9** (del Moral) reviews the changing discourses in Spanish society throughout the 20th century, which led to the breakdown of consensus. As a result of the tremendous social upheaval the country experienced as of 1975, water policy became a fertile ground for regional, social and ideological disputes.

Part IV covers all relevant policy issues. It begins by critically reviewing water laws and water regulation in **Chapters 10** (Ariño & Sastre) and **11** (Embid Irujo). **Chapter 10** provides an overview of water legislation, beginning with 1866 Water Act and ending with the provisions enacted in 2003. It focuses on water sector regulation and liberalization, reviewing the way water markets were defined in the 2001 Consolidated Water Act. It also describes the rules and statutes that facilitate the private sector's participation. **Chapter 11** builds on **Chapter 10**, and focuses on the foundations of the 1985 Water Act. Although **Chapters 10** and **11** highlight the prevalence of rule of public law on regulating and managing water matters, their conclusions about the role of private initiative and water markets are markedly different.

Considering that water institutions are widely known to be as important as legal statutes, **Chapter 12** (Varela-Ortega & Hernández-Mora) focuses on water institutions and institutional reform. This chapter shows that traditional institutions are still remarkably present in the institutional framework currently in place. Irrigators, and their associations (one major edifice of Spanish water institutions), play a leading role in the basin agencies (the other major institutional component). The chapter provides an updated description the institutional set-up as it stands in 2009, after the merger of the Ministries of Environment and Agriculture in 2008 (Ministry of the Environment and Rural and Marine Affairs).

The economics of water resources in Spain is the consequence of the social prevalence of irrigation, the traditional view of the State as the major water supplier and the disregard of externalities and intrinsic environmental values. **Chapter 13** (Garrido & Calatrava) describes how water prices are set in Spain, reviews their levels in the multi-tier system established for most users, and portrays a picture marked by the heterogeneity across regions and users. In the final section the chapter reviews the most recent experience with water trading mechanisms put in place by the 1999 amendment of the Water Act.

Two chapters deal with one of the major challenges of water policy: the management of intensively exploited groundwater resources. A demand-driven process with significant policy implications began in the early 1970s as a result of the use of groundwater resources, as discussed in **Chapter 14** (Custodio *et al.*). **Chapter 14** shows that the 'silent revolution' driven by thousands of modest farmers led to a very productive and export-oriented agricultural sector. This whole process took shape behind the scenes, and governments and policies either ignored it or failed to harness these developments, as **Chapter 15** (Lopez-Gunn) claims. **Chapter 15** draws on the public policy literature to identify the most significant factors underlying the failed attempts at tackling some of the most pressing and intractable problems associated with groundwater resources since the 1985 Water Act was passed. This is possibly due to the fact that intensive groundwater use is a recent phenomenon in Spain as indeed it is elsewhere.

Although all book's chapters touch in passing on a number of issues related to the Water Framework Directive (WFD), **Chapter 16** (Menéndez Prieto) is the volume's reference for this important piece of European Union legislation. Strict compliance with the word and spirit of the WFD is both a major source of uncertainty and difficulties—because even the EU is learning by doing—and a major turning point in 25 years of water policy history. For one thing, restoring the ecological quality of water resources, habitats and ecosystems is now a major priority, second only to meeting direct human needs. We know from **Chapters 3** and **4** that implementing this change of priorities will come up against tremendous technical and conceptual problems, as well as practical difficulties. Water projects, basin plans, and the criteria behind water pricing schemes must be redefined to meet the goals of the Directive.

Furthermore, the strict rules described in **Chapter 12** for facilitating the participation of users in designing policies and in managing water projects have also led to much more open processes of public participation. **Chapter 17** (Barreira) sheds light on the differences between a closed and hierarchically structured format of participation, aimed at resources sharing and water projects design, and a much more open process of public participation, targeting the generation of agreed actions and river basin plans. **Chapter 18** (Garrido *et al.*) puts Spain in the Iberian context to review the cooperation with Portugal about four main rivers they share (Minho, Limia, Douro, Tagus and Guadiana). A historical review of the treaties and legal texts that lead to the Albufeira Convention (CA) signed in 1998 confirms a vision of Spain's past water policies that almost completely disregard the interests of the neighbor country. The amendment of the CA in February 2008 poses significant challenges for Spain, as the chapter explains, but the fulfillment of WFD in terms of the basin plans due by the end of 2009 is partly a joint responsibility of Spain and Portugal for their shared rivers.

Even though most other EU countries are also experiencing serious difficulties, the Spanish case is perhaps more telling, because, as **Parts II** and **III** describe in detail, pressures and demands for water are still growing and all global change projections indicate that the climate will become less favorable. Not surprisingly, water conflicts have multiplied on many small and large-scale fronts, as **Chapter 19** (Arrojo) details. This chapter confirms what common sense would dictate in view of the contents of the previous chapters. A democracy, which grants liberal powers to the regions (Autonomous Communities) and encourages the identification of issues that the lay public can easily grasp, is highly vulnerable to conflicts related to the management of scarce natural. As **Chapters 12** and **17** details, EC environmental law and the WFD makes information widely accessible to the general public, and formal participatory mechanisms must be applied to reach agreement on the best actions and initiatives.

We learn from **Chapter 19** (Arrojo) that opacity presided over the drafting the National Hydrological Plan approved in 2001, and heavily amended in 2004. One conclusion is that not only have tensions, based on the spatial and regional dimensions, grown and become more poignant, but also water problems have come to be used as ammunition in the political debate. These tensions are increased by the predominance of emotional factors and the absence of transparent hydrological and economic data. Yet technological progress in the form of accessible desalted water has partially eased the tensions in many water-scarce Mediterranean regions.

The volume suggests that there has been a great deal of policy inconsistencies, marked political swings and little bipartisan continuity across subsequent national administrations. In view of such an incomplete or constantly evolving process, one wonders what it takes to find water policy and plans that enjoy sufficient support. In the concluding part (**Chapter 20**, by Garrido & Llamas), we further speculate on Spain's major strengths and weaknesses and summarize the most important lessons to be learned from the Spanish experience.

REFERENCES

Allan, J.A. (1999). The Nile Basin: evolving approaches to Nile water management, Paper presented at the *Conference on Water Quality*, convened by the International Institute of Water Quality Sciences, Jerusalem.

Dooge, J.C.I. (1999). Hydrological science and social problems. *Arbor* 646 (October), 191–202.

Maass, A. and R.L. Anderson (1978) *... and the Desert Shall Rejoice. Conflict, Growth, and Justice in Arid Environments*. Cambridge, Massachusetts: MIT Press.

II
The natural resource base,
the environment and the economy

CHAPTER 2

Physical and hydrological characteristics

Luis Martínez-Cortina
Spanish Technical Institute of Geology and Mining, Madrid, Spain

1 MAIN GEOGRAPHICAL FEATURES OF SPAIN

Spain occupies about four-fifths of the Iberian Peninsula, and Portugal, the remainder. The Balearic Islands in the Mediterranean Sea, the Canary Islands in the Atlantic Ocean, and the cities of Ceuta and Melilla on the Northern coast of Africa, make up the rest of the country.

Spain has a total surface area of about 506,000 km², with an estimated population of 43 million (INE, 2004). This is a population density of 85 inhabitants per km². Approximately 78% of the population lives in towns.

Politically, the country is divided into 17 Autonomous Communities and two Autonomous Cities. For the most part, administrative divisions do not coincide with hydrographic boundaries. This raises a question of water resources management competencies.

While institutional issues are beyond the scope of this chapter, an overview of Spain's administrative framework is in order. Figure 1 illustrates Spain's administrative division into Autonomous Communities. It also shows Spain's ten biggest cities in terms of population size: Madrid (3 million), Barcelona (2 million), Valencia (1 million), Seville (700,000), Zaragoza (550,000), Málaga (500,000), Murcia (425,000), Las Palmas (375,000), Palma de Mallorca (350,000) and Bilbao (350,000). Figure 2 shows the boundaries of the main river basins. Figures 1 and 2 are intended as a reference for this and other chapters of the book.

2 PHYSICAL SETTING

2.1 *Orography*

The most outstanding physical feature of the Iberian Peninsula is a large central plateau (called the *Meseta*). This area is generally flat, although the *Sistema Central* (Central Range) (shale and granite) runs across it like a dorsal spine. The Meseta has an average altitude of 600 m above sea level and occupies roughly half of the peninsula. The southern half of the *Meseta* is slightly lower in elevation. The main features of this southern Meseta are the *Montes de Toledo* (Toledo Mountains), which distribute water to the north (Tagus basin) and the south (Guadiana basin).

The *Meseta* is surrounded by mountain ranges. These ranges condition the peninsula's hydrographic network, at the same time isolating the central area somewhat and assuring a well-defined continental climate. The *Meseta* then is enclosed by the *Cordillera Cantábrica* (Cantabrian Mountains) to the north, the *Sistema Ibérica* (Iberian Mountains) to the East, and Sierra Morena to the south. The western extreme is open to the Atlantic Ocean.

There are two triangular-shaped depressions, filled with tertiary materials, outside the central plateau: the Ebro valley, towards the northeast, which extends all the way to the Pyrenees Mountains and the Coastal Catalonian Range; and the Guadalquivir valley, to the south, limited by the *Cordilleras Béticas* (Baetic mountains). These ranges are a product of alpine folding, and boast the highest mountain peaks of the Iberian Peninsula. Narrow coastal plains complete the orographic scenario of peninsular Spain.

Figure 1.   Spain's political map: the Autonomous Communities.

Figure 2.   Spain's water basins.

The orography of the Balearic Islands is a continuation of the Baetic mountain range, while the Canary Islands' relief is the result of volcanic phenomena. Both are characterized by the absence of permanent water courses.

2.2   *Climate*

The climate of the Iberian Peninsula, located between two large water bodies (the Atlantic Ocean and the Mediterranean Sea) and two large continental masses (Europe and Africa), is extremely diverse. Although differences can be observed at a regional or even local level, Spain can be said to have the following climatic zones:

- The northern area of the peninsula (Galicia, the Cantabrian Mountains and the Pyrenees) has an Atlantic climate, with mild temperatures throughout the year and high rainfall.
- The Mediterranean coast and the Guadalquivir basin experience long dry summers and mild winters. Rainfall is generally low, although the Eastern Mediterranean coast is sometimes subject to sporadic heavy precipitation phenomena in autumn and winter.
- The rest of the peninsula features a more or less continental climate, with hot dry summers and cold winters. Temperatures range widely: maxima often rise to over 35°C in summer and minima fall below 5°C in winter.

The Canary Islands have a very dry climate, with hot summers and very mild winters. Rainfall is very low. This is also typical of the southeastern part of the Mediterranean coast: Almería (East Andalusia) and Murcia.

The average rainfall for the whole country is 684 mm (equivalent to a total of 346,000 million cubic meters per year, hereafter, Mm^3/yr). Winter tends to be the wettest time of the year (average rainfall in December is about 80 mm), while the summer months are drier (20 mm in July) (MIMAM, 2000).

Evapotranspiration estimates were calculated for the Spanish Water White Paper (MIMAM, 2000). Potential evapotranspiration (PET) for the whole country is about 862 mm/yr, while real evapotranspiration (RET) is 464 mm/yr. PET values are highest in the southern half of the peninsula, in the Canary Islands and in the Ebro valley. Obviously, the differences between PET and RET are greater in drier areas and lower in wetter regions.

Table 1 gives an idea of the spatial variability of these values, indicating PET and RET (columns 4–5) for the different hydrological planning areas (see Figure 2). Table 1 also includes estimates of *useable rainfall* (column 6), that is, rainfall minus RET, as defined by the Water White Paper (MIMAM, 2000). This term refers to the volume of surface water or groundwater that contributes to overall runoff and is therefore useable. The average total volume of generated water resources (column 7) can be estimated from the ratio between *useable rainfall* and land surface area.

Note that Table 1 lists average values. There is a great spatial and time variability for the whole country, and especially intra-annually and across provinces.

The Canary Islands, for instance, have the lowest average rainfall (302 mm/yr). In this region, maximum and minimum precipitation for the 1940/41–1995/96 period was 574 and 119 mm/yr, respectively. During the same period, the North III area (highest average rainfall) featured an absolute minimum of 958 mm/yr and a maximum of 2,282 mm/yr. Average rainfall for the whole country was 684 mm/yr, the maximum and minimum average values being 970 and 469 mm/yr, respectively.

3   WATER RESOURCES

Spain's physical environment is characterized by diversity, not only in terms of orography or climate, but also with regard to factors such soil types and vegetation. These features affect the hydrological response and are a factor accentuating the spatial and temporal irregularity of Spain's water regime. This irregularity is an important focus for water planners, as it has often been identified as the country's main water problem.

Table 1.   Columns 3–8: Average annual precipitation, evapotranspiration, useful rain and total volume of generated water resources. Period 1940/41–1995/96. Columns 9–10: Storage capacity and theoretical availability of surface water per River Basin.

Areas of hydrological planning	Surface (km²)	Precipitation (Mm/yr)	PET (mm/yr)	RET (mm/yr)	Useable rain (mm/yr)	Volume (Mm³/yr)		Storage capacity (Mm³/yr)	Theoretical availability (Mm³/yr)
						Total	Groundwater		
Galicia Coast	13,130	1,577	737	644	933	12,250	2,234	3,655	7,383
North I	17,600	1,284	709	563	721	12,689	2,745		
North II	17,330	1,405	653	604	801	13,881	5,077		
North III	5,720	1,606	695	673	933	5,337	894		
Douro	78,960	625	759	452	173	13,660	3,000	7,463	6,095
Tagus	55,810	655	898	460	195	10,883	2,393	10,974	5,845
Guadiana I	53,180	521	977	438	83	4,414	687	9,659	2,150
Guadiana II	7,030	662	1,075	511	151	1,061	63		
Guadalquivir	63,240	591	991	455	136	8,601	2,343	8,782	2,819
Mediterranean B.A.	17,950	530	969	399	131	2,351	680	1,041	359
Segura	19,120	383	963	341	42	803	588	1,129	626
Júcar	42,900	504	881	424	80	3,432	2,492	3,346	2,095
Ebro	85,560	682	792	472	210	17,967	4,614	6,504	11,012
Int. Basins Catalonia	16,490	734	792	565	169	2,787	909	740	791
Balearic Islands	5,010	595	896	463	132	661	508	11	300
Canary Islands	7,440	302	1,057	247	55	409	681	101	417
Total Spain	**506,470**	**684**	**862**	**464**	**220**	**111,186**	**29,908**	**53,405**	**39,892**

Source: MIMAM (2000), MIMAM (Web Page). For groundwater own elaboration with data from MOPTMA-MINER (1994), ITGE (1995), López Geta (2000) and MIMAM (2000).

Note: Mediterranean Basin of Andalusia (B.A.) is the new name (from 2005) of the South Basin.
Data for the Balearic and Canary Islands correspond to their Basin Water Plans (from 1998).

As a result, Spain has traditionally adopted a supply-based water policy, focused mainly on the construction of large dams. As shown in other chapters of this book, some aspects of water resources management, such as demand management, water quality or groundwater resources have often been neglected by both the central and regional administrations.

3.1   *Surface water resources*

Spain's rivers can be classified according to the coast where they end. Rivers flowing into the Atlantic Ocean account for 58% of Spain's area (286,000 km²), whereas water courses ending in the Mediterranean and the Cantabrian[1] Sea occupy about 35% (182,000 km²) and about 5% (25,000 km²), respectively (see Figure 2).

Atlantic rivers are generally the longest, and slope gently towards the ocean. These rivers (Douro, Tagus, Guadiana, Guadalquivir) flow along the Earth's parallels, much like the mountain ranges do. On the contrary, most rivers that originate in coastal ranges, like the Cantabrian Mountains (strong and regular flows), and southern rivers (small and intermittent flows) follow the direction of the meridians, and are shorter. The rivers flowing into the Mediterranean, except for the Ebro (Spain's longest river at 928 km), come somewhere in-between. These rivers usually have their source in coastal mountain ranges, and most have small basins. Yet, some, like the Llobregat, Segura and Júcar, are of a remarkable size. The Llobregat, which rises in the Eastern Pyrenees, conveys quite a sizeable volume of water. The flows of the other two are not as great, and they experience seasonal flash floods and very dry summers.

Rivers in the Canary and Balearic Islands are characterized by intermittent regimes and steep slopes. Karst formations abound in the Balearic Islands. As a result, surface water often infiltrates and emerges downstream later on. Surface water courses are almost non-existent in the Canary Islands.

There are about 2,500 lagoons in peninsular Spain. Of extremely diverse origins—glaciar, endorreic, tectonic, volcanic, karstic and littoral—lagoons are unevenly distributed, and usually shallow and small in size. Most lakes are associated with enclosed endorreic or semi-endorreic areas. These form depressions in low permeability areas, which are flooded and where water infiltrates or evaporates only gradually (Arenillas & Sáenz, 1987).

3.2   *Surface water infrastructure regulation*

As already mentioned, artificial regulation of rivers by means of dams has traditionally been the key feature of Spain's water policy. The irregularity of the country's water inputs has encouraged water planners to opt for this water policy, often overlooking other kinds of management.

Spain is currently the world's fourth-ranking country in terms of number of dams (over 1,300, of which 450 were built before 1960 and over 100 before 1915), and is surpassed only by the USA, India and China. Spain can also be considered to be the country with the highest number of dams per million inhabitants. Strictly speaking, another four countries (Albania, Norway, Iceland and Cyprus) have a higher ratio, but these are not statistically representative as they have extremely small populations (Llamas *et al.*, 2001).

Dams have yielded significant benefits, and their proliferation can be justified by past social and economic contexts. It is usually argued that without dams Spain could only take advantage of about 9% of total runoff, while dams raise that figure to about 38% (MIMAM, 2000). However, as discussed in other chapters of this book, supply-based water infrastructures are increasingly being challenged by the new social, economic and environmental paradigms. As shown in Table 1 (last two columns), Spain's total water storage capacity amounts to about 53,500 Mm³, and the theoretical availability through this regulation is about 40,000 Mm³ (MIMAM, 2000). Nevertheless, this figure should be viewed circumspectly as there is a great deal of uncertainty surrounded the underlying hypothesis.

---

[1] The Cantabrian Sea is the name given to the Atlantic Ocean off the northern coast of Spain.

3.3   *Surface water monitoring*

Monitoring and maintenance of quantity and quality networks is currently the competence of river basin authorities (RBA). The Ministry of the Environment is responsible for data storage and reporting. In the case of basins falling within a single Autonomous Community, equivalent agencies are in charge of this mission.

An official gauging network (*Red Oficial de Estaciones de Aforo*, ROEA), made up of 730 stations (one every 700 km^2), 300 control points at dams with a capacity of over 10 Mm^3 and a further 180 control points in canals, quantifies and monitors flow and storage. Water information systems (*Sistemas Automáticos de Información Hidrológica*, SAIH) are also significant, and yield real-time information about the hydrological and meteorological situation of each basin (levels and flows of rivers and canals, reserves, dam outputs, precipitation and so on).

Water quality is monitored by an integrated water quality network (*Red Integrada de Calidad de las Aguas*, ICA), set up in 1993. ICA is made up of a series of conventional sampling stations, as well as a number of automated alert stations (*Estaciones Automáticas de Alerta*, EAA). The conventional sampling stations integrated and updated existing quality networks: COCA (*Control Oficial de la Calidad del Agua*, official water quality control), which gauges 40 different parameters; COAS (*Control Oficial de Abastecimientos*, official water supply control); and COPI, which caters for the needs of freshwater fisheries. These three networks account for over 1,000 control points. EAA comprises 200 stations, and is designed to yield real-time information on a series of parameters considered to be of special interest. This network is managed by an automated water quality information system (*Sistema Automático de Información de Calidad de las Aguas*, SAICA), and is designed to assure rapid evaluation of and decision making on water quality issues.

3.4   *Surface water quality*

It is common practice in Spain to use the General Quality Index (*Índice de Calidad General*, ICG) as a reference for evaluating surface water quality. ICG is a non-dimensional number calculated as a function of 23 quality parameters (9 *main* and 14 complementary parameters, the latter with a lower sampling density). ICG provides a 0 (contaminated) to 100 (clean) evaluation of water quality.

Table 2 provides information on surface water quality, based on annual ICG values, for nine inter-community basins. These are 2003 data (MIMAM, Web Page). The data for the Internal Basins of Catalonia refer to 1999 (these are the latest data made available by the Ministry of the

Table 2.   Number and percentage of surface water quality gauging stations, classified as per average annual ICG value.

River Basin	Classification according to the ICG									
	Excellent (85–100)		Good (75–85)		Average (65–75)		Poor (50–65)		Unacceptable (0–50)	
North	71	46%	52	34%	21	13%	9	6%	2	1%
Douro	40	54%	22	29%	10	13%	2	3%	1	1%
Tagus	20	13%	58	37%	46	30%	30	19%	1	1%
Guadiana	0	0%	20	36%	13	24%	11	20%	11	20%
Guadalquivir	3	9%	6	18%	5	15%	15	46%	4	12%
Mediterranean B.A.	1	25%	2	50%	0	0%	1	25%	0	0%
Segura	6	18%	10	30%	4	12%	11	34%	2	6%
Júcar	16	26%	14	23%	16	26%	14	23%	1	2%
Ebro	27	34%	33	41%	14	18%	4	5%	2	2%
Total	**184**	**28%**	**217**	**33%**	**129**	**20%**	**97**	**15%**	**24**	**4%**
Int. B. of Catalonia	5	8%	7	11%	26	40%	19	30%	7	11%

Source: MIMAM (Web Page). Data from 2003 (excepting Internal Basins of Catalonia, data from 1999).

Environment). Table 2 shows that the Guadiana and Guadalquivir basins, and to a lesser extent the Segura basin and Internal Basins of Catalonia, are experiencing the most significant surface water quality problems. Note that statistical data can be masked by factors such as weather conditions, and do not necessarily provide an accurate reflection of water quality issues.

Water quality has traditionally received less attention than water quantity. This has led to environmental changes, including dam building, interbasin transfers, intensive agriculture, urbanization, loss of riverine forestry, channeling or uncontrolled waste disposal. The Tinto (West Andalusia), Segura and Llobregat rivers have serious pollution problems, while the Guadiana, Ter (Catalonia) and Minho (Galicia) rivers are beset by problems caused by the number of dams and hydropower stations.

The oncoming obligations of the Water Framework Directive (WFD) imply a much-needed shift towards water quality protection (see Chapter 16 for a detailed review of the WFD). The WFD has at times been criticized for having an overly Central and Northern European outlook, emphasizing the importance of water quality issues over quantitative management. However, water quality degradation is probably the most important water management issue for the years to come.

4   GROUNDWATER RESOURCES

4.1   *Aquifer distribution in Spain*

Spain is a well-endowed country in terms of aquifer formations. Studies carried out for the WFD show that Spain's permeable surface area extends over 245,000 km^2, which is about half the country's total surface area. This figure does not include some low-permeability areas of great strategic importance, like Galicia or Madrid's mountainous areas.

The main detritic aquifers, spanning a surface area of over 100,000 km^2, cover the main depressions of the central plateau or *Meseta* (Douro and Tagus basins) and the Guadalquivir basin. The alluvial aquifers of these three basins, together with the Ebro river basin, have traditionally been subject to exploitation.

The main karst aquifers span 70,000 km^2. These are particularly important along the Mediterranean coast and in the southeastern area of the central plateau, as well as in some areas of Andalusia, the Pyrenees, the North and the Balearic Islands.

Volcanic aquifers are located almost exclusively in the Canary Islands (8,000 km^2). These are extremely important, as they constitute the archipelago's primary source of water.

4.2   *Groundwater reserves and renewable resources*

It is not easy to give a precise estimate of the total volume of water stored in Spain's aquifers. Depending on the study, estimates range from 150,000 Mm^3 to 300,000 Mm^3. However, actual reserves are probably much higher, since the existing calculations only take into account the volume stored to a depth of 100–200 m and do not consider *unofficial* hydrogeological units, whose reserves are at times abundant (Llamas *et al.*, 2001). In any case, the storage capacity of groundwater reserves is much higher than surface water infrastructures, whose full capacity is 53,000 Mm^3.

The reserves of many aquifers usually exceed the natural rate of recharge by one or two orders of magnitude. This feature has practical implications that are particularly important for a country like Spain, where evapotranspiration is high and droughts are frequent. However, groundwater planning and management has traditionally been neglected by the central administration, as reviewed in Chapters 14 and 15.

From a practical point of view, knowing how much water can be extracted from an aquifer is probably more important than the aquifer's total reserves. This is usually difficult to quantify, as it is conditioned by changing technical, social, economic and environmental factors.

An aquifer's renewable yield, sometimes termed *net recharge*, is the difference between aquifer recharge and evapotranspiration. We need to know this value to assess the effects of groundwater abstraction on total storage and on the aquifer's renewal capacity.

A mathematical model was developed to estimate Spain's renewable groundwater resources for the Water White Paper (MIMAM, 2000). As shown in Table 1, they amount to about 30,000 Mm^3/yr. Earlier studies had led to a 20,000 Mm^3/yr underestimate, though their authors acknowledged that they were subject to a great deal of uncertainty. Even the 30,000 Mm^3/yr figure is probably an underestimate, as the model does not take a realistic approach on some points: simulations are carried out under natural conditions, some low permeability areas are overlooked, and the model is unable to consider real groundwater behavior (Cruces, 1999).

4.3   *Groundwater monitoring*

Groundwater monitoring networks are divided into three groups: piezometric networks (measure the depth of water in aquifers), hydrometric networks (measure spring flows) and quality control networks (measure physical and chemical parameters that define the water quality).

Groundwater monitoring has traditionally been carried out by the Spanish Institute of Geology and Mining (IGME), which set up its network as of the 1970s. This means that there is a relatively long hydrogeological data series today. Its piezometric network is currently made up of about 3,000 points, while the quality network has about 1,800.

Recently, river basin authorities (RBA) have assumed powers over groundwater monitoring. Apart from the problems caused by the heterogeneity of surface and groundwater data sets, RBA do not usually have the technical expertise to deal with groundwater resources. The Ministry of the Environment is currently attempting to unify and redefine the existing monitoring networks. The aim of this initiative is to create an official groundwater network. In particular, the Groundwater Quality Monitoring Network (*Red de Observación de la Calidad de las Aguas Subterráneas*, ROCAS) is currently being redefined to comply with the requirements of the WFD and Eurowaternet (European Environmental Agency) (De la Hera *et al.*, 2004).

Aside from ROCAS, there is another permanent network (*Red de Observación de Intrusión*, ROI) that measures chlorine and conductivity to control seawater intrusion in coastal aquifers. This network has about 1,000 points.

4.4   *Groundwater quality*

Groundwater in Spain's karst aquifers generally contains calcium or magnesium bicarbonate, and has low to medium mineralization with ion concentrations that are usually fit for drinking. These waters are typical of the Northern basins, the Upper and Northern Douro, and the hydrogeological units of the Upper Guadiana, Guadalquivir, Andalusian Mediterranean Basin, the Júcar and Ebro basins, and the Internal Basins of Catalonia.

The chemistry of water in Spain's detritic aquifers varies significantly. In places like the Douro, Tagus, Guadiana and Guadalquivir basins, calcium or magnesium bicarbonated waters exist side by side with calcium or sodium sulphated or chlorated waters. These tend to be more mineralized than karst waters, and, very occasionally, may contain constituents in excess of drinking water standards.

Chemical groundwater quality in the Canary Islands is far from uniform. In coastal areas, water is highly mineralized, with a fairly high chlorine and sodium content. This is primarily a consequence of the arid climate. Salinity is lower in inland areas. CO_2 of volcanic origin sometimes results sodium or magnesium bicarbonated waters that are highly mineralized or have a lower pH (MOPTMA-MINER, 1994).

One of the traditional issues of concern with regard to groundwater is aquifer salinization. This process is usually due to seawater intrusion, dissolution of evaporitic materials or reuse of highly saline irrigation water. About 40% of the groundwater monitoring stations shows average chlorine concentrations of over 100 mg/L, the threshold used to indicate moderate to severe salinization. More than 55% of stations show sulphate concentrations greater than 150 mg/L.

Nitrogen pollution is usually due to diffuse pollution from agriculture, and is another recurring issue of concern with regard to groundwater quality. Nitrogen pollution is produced by the infiltration of water (rain or irrigation) carrying dissolved compost and pesticides. The WFD establishes a 50 mg/L limit for nitrates. About 20% of the control points yielded values in excess of this figure, and the Guadiana and Júcar basins were the most affected. According to the WFD, where readings are in excess of 25 mg/L (many of which are), authorities are obliged to carry out systematic monitoring every four or eight years. Thus, nitrate contamination is one of the main challenges facing Spain with regard to groundwater quality.

REFERENCES

Arenillas, M. & Sáenz, C. (1987). *Los ríos. Guía física de España* [Rivers. Guide to Spain's physical relief]. Alianza Publishing. Madrid, Spain.

Cruces, J. (1999). Evaluación de los recursos subterráneos en el Libro Blanco del Agua en España [Estimation of groundwater resources in the Spanish Water White Paper]. In: *Jornadas sobre las Aguas Subterráneas en el Libro Blanco del Agua en España* [Symposium on groundwater in the Spanish Water White Paper]. International Association of Hydrogeologists—Spanish Group. May 1999, Madrid, Spain: 11–22.

De la Hera, A., Alonso, A.M., Danés, C., Ruza, J., Espinosa, G., Sanz, E., Domínguez, A. & Diago, I. (2004). Proceso de diseño de la red de control de calidad de las aguas subterráneas en España según la Directiva Marco del Agua [Design of groundwater quality control network, according to the Water Framework Directive]. In: *VIII Symposium of Hydrogeology*, Zaragoza, Spain. Spanish Association of Hydrogeologists: 623–632.

INE (*Instituto Nacional de Estadística*, National Bureau of Statistics) (2004). Padrón municipal: cifras oficiales de población a 1 de enero de 2004 [Municipal Register: Official population figures as of January 1, 2004]. Available at www.ine.es and published in BOE (Spanish Official Gazette. December 30, 2004, Real Decreto 2348/2004).

ITGE (1995). *Identificación de Unidades Hidrogeológicas donde establecer reservas para abastecimiento urbano en el territorio peninsular* [Identification of hydrogeological units for urban supply in peninsular Spain]. Instituto Tecnológico Geominero de España (Spanish Geological Survey). Unpublished report.

Llamas, M.R., Fornés, J., Hernández-Mora, N. & Martínez Cortina, L. (2001). *Aguas subterráneas: retos y oportunidades* [Groundwater: challenges and opportunities]. Mundi-Prensa and Fundación Marcelino Botín, Madrid, Spain, 529 p.

López Geta, J.A. (2000). Estrategias de utilización de las aguas subterráneas en el abastecimiento de poblaciones [Groundwater use strategies for urban supply]. In: *Jornadas técnicas sobre aguas subterráneas y abastecimiento urbano* [Technical Symposium about groundwater and urban supply]. Instituto Tecnológico Geominero de España (Spanish Geological Survey). Eds: Fernández Rubio, R., Fernández Sánchez, J.A., López Camacho, B. & López Geta, J.A., 21–29.

MIMAM (*Ministerio de Medio Ambiente*, Spanish Ministry of Environment) (2000). *Libro Blanco del Agua en España* [Spanish Water White Paper]. Secretaría de Estado de Aguas y Costas, Dirección General de Obras Hidráulicas y Calidad de las Aguas. Madrid, Spain, 637 p.

MOPTMA-MINER (1994). *Libro Blanco de las Aguas Subterráneas* [Groundwater White Paper]. Monographs Collection. Ministerio de Obras Públicas, Transportes y Medio Ambiente, and Ministerio de Industria y Energía [Ministry of Public Works, Transport and Environment, and Ministry of Industry and Energy]. 135 p.

RELEVANT WEBSITES

MIMAM (Ministerio de Medio Ambiente) [Ministry of the Environment]:
    http://www.mma.es
Hispagua (Spain's water information system):
    http://hispagua.cedex.es
River Basin Authorities:
    Ebro → http://www.chebro.es
    Duero → http://www.chduero.es
    Tajo → http://www.chtajo.es
    Guadiana → http://www.chguadiana.es
    Júcar → http://www.chj.es
    Guadalquivir → http://www.chguadalquivir.es

Catalan Water Agency → http://www.gencat.net/aca

Andalusian Water Agency → htpp://www.agenciaandaluzadelagua.com

IGME (Instituto Geológico y Minero de España) [Spanish Institute of Geology and Mining]:
http://www.igme.es

CEH–CEDEX (Center for Water Studies of the Center for Studies and Experimentation on Public Works):
http://www.cedex.es/hidrograficos/presentacion.html

INM (Instituto Nacional de Meteorología) [National Meteorological Institute]:
http://www.inm.es

INE (Instituto Nacional de Estadística) [National Statistics Institute]:
http://www.ine.es

Eurostat (EU Bureau of Statistics):
http://epp.eurostat.cec.eu.int

CHAPTER 3

The state of water ecosystems

Francisco García Novo, Julia Toja Santillana & Carlos Granado-Lorencio
University of Seville, Spain

1 SPAIN'S GEOLOGICAL SETTING

Long mountain ranges running East-West across the Iberian Peninsula cut off large inner plateaus from one another, as well as from the Mediterranean Sea and the Atlantic Ocean. Inner plateaus formed watersheds where large shallow lakes (1000–10000 km^2) evolved in the Miocene-Pliocene periods. The erosion of tributaries finally carved out narrow gorges through which they flow to the Atlantic, creating rather long rivers running East-West, such as the Tagus (1120 km), Douro (913 km), and Guadiana (744 km). The Ebro river (928 km) flows West-East to the Mediterranean Sea (see Figure 2 in Chapter 2). Former lake beds gave way to large plains with floodable areas, temporary ponds and wetlands. On the edges of the Peninsula, the Guadalquivir river (560 km) depression opens on to the Atlantic Ocean. It evolved as a large gulf where repeated progradation/regression of coast lines took place following upon eustatic fluctuation and local tectonics during the Cenozoic era and, especially, the Pleistocene period. Large marshes were associated with the silting up of the Guadalquivir gulf. Doñana National Park, the most valuable wetland in the Iberian Peninsula, is protecting what is left of former Guadalquivir river marshes.

There were other large marshes and coastal lagoons in the Guadiana, Tinto and Odiel estuaries and inside the bay of Cadiz in the Southwest. On the northwest Atlantic coast, large marine marshes exist in *rías* (sea-flooded valleys or drowned valleys), like the Ría de Vigo (Arcade marshes), Arousa, Betanzos, Pontesdeume. A number of large marshes and coastal wetlands abound all along the Mediterranean and Atlantic coastlines.

The Pardo inventory (1948) listed 2474 continental water bodies and offered the first survey of limnological literature in Spain. Very few wetlands have retained their natural conditions, because many were drained for agricultural purposes and the Peninsula's shallow water bodies dry up during severe drought periods. The Lagunas de Ruidera (in the provinces of Ciudad Real and Albacete) are 22 karstic ponds fed by the upwelling of the local aquifer. The Tablas de Daimiel (1928 ha), the largest pond, preserves a valuable wetland and is now protected as a National Park.

2 HUMAN-INDUCED ALTERATIONS IN CONTINENTAL WATERS

In a protracted fight against malaria, drainage policies during the 20th century drastically reduced the number of ponds and wetlands, which were often condemned to oblivion. Gallego *et al.* (1999) documented the loss of 35 out of 40 ponds in Azuaga (Badajoz) from 1895 to 1999. The living population only identified 5 of the last ponds. Reclamation of inner and coastal wetlands was enhanced by the Spanish administrations granting the land title to those supporting the initiative. Land reclamation, dam construction and the development of irrigation areas were the principal motivations of water policies from late 19th up to the end of the 20th century.

The irregular regime of rivers with recurrent flooding made it difficult to cultivate the fertile river terraces where irrigation was feasible. Fluvial navigation was severely limited and urban settlements on river banks were at risk. Zaragoza on the Ebro river or Valencia on the Turia river were repeatedly flooded. The important port of Seville, a key connection to the American Colonies during 16th to 18th centuries, was severely damaged, as were goods and ships, and the town suffered huge casualties due to unpredictable Guadalquivir flooding. River dams in Spain were built at an early date: Proserpina and Cornalvo supplied water to the Roman city of Merida; the Tibi dam (Alicante) was built from 1580 to 1594. However, it was during the 19th century that the construction of dams and canals to harness water resources spread (see Chapters 9 and 19).

During the second half of the 20th century an intense dam building program created over 1300 reservoirs, storing large water masses located in all the major Spanish basins. Drainage networks were reduced to the small tributaries that flow into reservoirs or regulated rivers. The course of the river is sometimes obliterated by a chain of reservoirs leaving no free-flowing sections, as occurs across long stretches of the Sil, Minho, Douro or Tagus rivers. The flow of the Ebro, Júcar, Segura or Guadalquivir rivers is regulated to supply water to agriculture. Even though 80% of precipitation is concentrated in the October–March period, the regulated Guadalquivir's flow is lowest in winter, peaking in summer to meet watering demands.

Hydraulic policies during the 20th century have induced overriding changes in rivers and wetlands. The Janda and Antela lakes and many wetlands have been drained; most rivers have been impounded creating new lentic environments all over the country. River flow, sediment transport, aquatic habitat connectivity, and their biota have been thoroughly modified. The sediment transport disruption was of paramount importance. The silting up of reservoirs reduced storage capacity. The Barasona reservoir (92 Mm^3, Huesca) retained 18 Mm^3 of sediments. A regeneration procedure to remove sediments was applied with a thorough follow-up (*Limnetica* 1998). Small reservoirs became silted up or developed long alluvial plains at the tributary entrance, eventually turning into valuable wetlands (tamarisk woodland at Bornos reservoir, Cadiz). Reduced sedimentation failed to compensate for Ebro delta sinking, resulting in marine progradation. In fact, this was one of the major environmental concerns of the Ebro transfer proposed in 2001 and cancelled in 2004 (see Chapter 19). The suspended load of the Douro river fed the Portuguese coast south of Porto before dam building (see chapter 18, to learn more about transboundary issues between Spain and Portugal). The coast is now suffering from erosion and the Espinho village has lost several streets and a railway station to the sea.

The list of changed continental waters in Spain is much longer (see Chapter 4). From a social viewpoint, dam construction and river regulation have been of the highest value, storing water for urban, industrial and agricultural supply and reducing flooding, damages and the risk for riverine settlements. On top of this century-long program of massive basin reengineering, cities, farmers and industries have been spilling chemicals, pollutants and untreated wastewater for about as long. This did not start to subside until the early 1990s (see Chapter 4). Low water tariffs for all users prevented investments in treatment facilities and operational infrastructures to reduce spill impacts (see Chapter 8, focusing on urban water issues, and Chapter 13, reviewing water tariffs generally).

3   RESERVOIRS VS. LAKES

Reservoirs often share have both fluvial and lake characters. Reservoirs are markedly dissymmetrical with a shallow area similar to the incoming tributary and a deep area close to the dam where vertical transport dominates horizontal transport. Like rivers, they are transient water bodies with high turnover rates that under extreme precipitation may exceed the value of 5 in a year. The Gergal reservoir (35 Mm^3, Seville) exceeded turnover figures of 10 for April 1979. Under these conditions water keeps flowing horizontally in the reservoir. Vertical surface level fluctuations increase when stored water is diverted and tributary debit drops to a minimum during summer. Torrejón, a long (40 km) reservoir flooding a Tagus gorge, keeps a steady water level but flows alternately in either direction following the hydroelectric exploitation of neighboring reservoirs.

One thing that reservoirs have in common with lakes is the large mass of steady water with little suspended matter that often stratifies leaving very large volumes where oxygen depletion may occur. The lower turnover rate of deep layers in the reservoir induces a vertical differentiation of the water column that is again a lake characteristic.

During the early years after reservoir filling the decay of organic matter and nutrient release from submersed soils drives a large nutrient budget to impounded water. In a later phase bottom sediments act as nutrient sinks and the reservoir evolves to oligotrophy, slowly attaining a steady state if the water level remains stable and turnover rates were low. The Mediterranean climate prevailing in the Iberian Peninsula causes highly irregular river debit with long drought periods and short high-flow intervals usually associated with heavy sediment loads. The outcome is the silting up of reservoirs, again changing the environmental conditions.

The study of Spanish reservoirs began with a general survey of 100 large reservoirs (about 8% of existing reservoirs at the time) carried out from 1973 to 1975 by a team led by Professor Ramón Margalef (Margalef *et al.*, 1976). 15 years later a second survey was carried out by Armengol (Armengol *et al.*, 1991). Using physical and chemical variables and chlorophyll content, factor analysis identified two major groups of reservoirs: type I low in solutes and type II waters with high solute concentrations.

A more detailed analysis of data identified six groups with minor divisions related to water management (level fluctuation, turnover rates). Phytoplankton and zooplankton communities agree well to physicochemical groups (Margalef, 1983).

Results from the second survey (Armengol *et al.*, 1994) matched previous groups at higher divisions. However, it was noted that the mineralization of waters during drought periods in reservoirs in the south, such as Andalusia, can lead to changes in their composition and shifts from type I, low mineral content waters, to type II, high mineral content waters.

The 1973–75 and 1987–88 surveys of Spanish reservoirs quoted above uncovered the geographical and climatic patterns of reservoirs. Long-term studies of reservoirs, lakes, ponds and rivers have provided complementary information.

4   FLUVIAL ENVIRONMENTS AND FISH FAUNA

Rivers offer a range of hydrologic regimes depending on watershed climate and geomorphology. Local fish diversity depends on the regional pool and site factors through a series of ecological filters: environmental tolerance (variance) and biotic factors, impinging upon potential fish diversity.

Addressing rivers as ecosystems, temporal changes induced by environmental processes can cause pulses, such as fluctuations in precipitation, run-off and temperature. At high water the aquatic media will expand vertically and horizontally allowing fish to colonize the river margins and also exploit their production surplus. The effect of flooding on fish populations largely depends on the scale, recurrence and river sector where it occurs.

Fish play a significant role in fluvial ecological processes due to their dispersion (migration strategies). Fish act as biotic links between the main channel and the margins of the river bed. Flooded river bed habitats act as temporary refuges, spawning areas, and nurseries for larvae.

Winter high waters spread the aquatic system, allowing fish populations to expand and making some individuals to drift downstream. In summer periods, where precipitation is scarce and temperatures are high, the river may be discontinued, leaving the subsurface flow that feeds ponds in river bed where fish gather. Aquatic communities resettle in ponds suffering from a steady increase of water conductivity, temperature and eutrophia often coupled to rising ammonia levels and oxygen depletion at night due to bacterial activity. Shallow water makes ponds an easy site for fishing birds and opportunistic vertebrates, such as fox or wild boar, which exploit the marooned fish population.

Autumn rains restore the river's surface flow and the recolonization of the flooded river bed begins. This pattern of winter flow peaks and a large summer interval of drought have selected the fish communities of Spanish rivers. The most successful taxa with a wealth of lifestyles belong to *Cyprinidae* family, also present in Eurasia, Africa and North America.

The number of fish species in continental waters is remarkably low at 51 species. Recent studies on the genetics of morphological species are uncovering numerous crypto-species and their numbers are likely to increase. Fish introduced in natural habitats now add up to 28 species. The continental waters of the Mediterranean basin now have fish belonging to 17 families with 70 introduced species, 58 of which are now well established. About 80% of endemic taxa now survive in communities where there is at least one introduced species (Elvira & Almodóvar, 2001).

The lifestyles of native species in the riverine environments show a fine tuning to the patterns of the fluvial system. Fish populations exhibit a low number of age groups, rarely reaching the age of 15 and often not exceeding 3 to 4 years. Sexual maturity is reached at 2 to 3 years, and reproductive effort will peak at 2 to 3 years soon after maturity, thus providing the population with high fecundity rates. Males strongly dominate the younger cohorts, females being the prevailing gender in a later phase.

Growth patterns closely match aquatic system productivity. An early growth peak occurs in the spring, providing for gonad development previous to spawning. A higher peak occurs in the summer, coupled to body growth. Strategies of space use follow two distinct patterns. Most species have a limited vital domain of a few hundred meters of river channel, where they complete their life cycle. A few species (all cyprinids) are rheophyllous, migrating upriver for reproduction late in winter when debit is high. Fish associations show low diversities and a clear cut dominance by one or two species. When co-dominance occurs, one of the species usually exhibits rheophyllous behavior (potamodromous).

The fish taxa of the hydrographic network have co-evolved with fluvial ecosystems over a long time span. Recent changes induced by reservoirs made Spanish rivers deviate from the original patterns. Dam building has direct effects on fish such as fragmentation of the river continuum, loss of fluvial habitats, development of lentic, lacustrine environments, and disturbance of the original trophic chains. Downriver, the flow peaks when reproductive migration occurred were changed by large periods of regulated or minimum river flow and occasional debit peaks during intense rainfall periods when reservoirs were full and unable to further store water. The reduction of river discharges into the estuary makes it difficult for marine species to trace the river mouth and complete their migratory cycle. A drop in nutrients, biomass and sediment budget will be felt in the coastal sediment balance, biota and productivity (Granado-Lorencio, 2000).

Fish associations during the early stages of a reservoir are dependent on watershed species assemblage upriver. Active colonizers are opportunistic species taking advantage of the low water turbulence and the eutrophia associated with initial reservoir stages. The early phase of fish population development is often coupled to a collapse of *riverine species* or *obligate riverine species*, while *lacustrine species* increase rapidly. In Spain, the later group is dominated by introduced species. Habitat reduction and invasive species are the leading causes of fish biodiversity loss. Carp, mosquito fish and pumpkin seed sunfish are the most frequently introduced species. Dams acting as migration barriers deplete native migratory fish populations, as has happened with eels, salmon or sturgeon, which have disappeared from many rivers (Granado-Lorencio, 1998).

The low diversity of native fish fauna makes it easier for new species to become established. In the foreseeable future native stock species are likely to dwindle and introduced species will dominate the fish communities of reservoirs and large rivers.

New aquatic environments are created by technical developments. One example is the aqueducts supplying raw water to industries, energy plants or agriculture. The inner coating of pipes is colonized by an association of sessile filtering species building a web of filaments: *Cordilophora caspia, Mytilopsis leucophaeta* and *Corbicula fluminea*. Shellfish aquaculture and the discharge of ballast water from tankers have been pinpointed as sources of aquatic species introduction.

5   WETLANDS: THE PRODUCTIVE WATERS

Surviving wetlands are hot spots for diversity, aquatic biomass and productivity. Also, they act as sources of wild species, replenishing Spain's dwindling populations and endangered aquatic habitats and the bird populations of Africa and Europe (Dirección General de Obras Hidráulicas, 1991). Coastal wetlands serve as hatcheries for many coastal fish and marine invertebrates such as crustaceans.

Wetlands can be divided into hidden seepage or crypto-wetlands (González Bernáldez, 1992b) and floodable areas. In hidden seepage areas groundwater is lost through transpiration not reaching the soil surface. The uncontrolled growth of groundwater use, mostly for irrigation since the 1970s, has had an impact on the recharge regimes, significantly reducing the wetland area (see Chapter 14). Floodable areas show open surface waters including littoral wetlands. Water regime and composition depend on the geological setting and origin of the wetland: coastal formations (lagoons, marshes, estuaries), impeded drainage in erosive formations (glacial ponds, wind hollows), abandoned river beds, discharge areas of upwelling systems, karst formations and others (González Bernáldez, 1992a).

Shallow aquatic ecosystems are colonized by a wealth of species, some with large populations. High biodiversity apparently depends on a large primary productivity providing broad-based trophic chains. The accumulation of organic matter modifies the wetlands' nutrient balances, water regime and soil level. Biomass is dominated by photosynthetic organisms, like phytoplankton and vascular species rooted to the bottom and emerging at the surface (helophytes).

Alonso (1987) published a classification of 2000 Spanish ponds, divided into 12 groups based on the chemical characteristics of water that yield an approximate match with 5 broad types of zooplankton communities. Mineralization, alkalinity and water regime are predominant environmental factors. Alonso stressed the importance of small ponds in semiarid climates, such as the Ebro valley, where unusual communities with a wealth of endemic species have survived.

Several plant life forms may coexist in floodable areas often showing a ringed structure: an outer fringe of trees—alder, ash, elm, poplar, tamarisk—, with an inner circle of willow and shrub species with climbing vines, closer to the water's edge. Bracken and other fern species may be present. In the warmer climes of southern Spain, *Nerium oleander* may form strips on the banks, and rush and other species grow up around the wetland depending on the flooding interval and low to high water conductivity values. Other species increase with water acidity or alkalinity. Dense stands of reeds and sedges eventually cover wide areas of shallow waters. Irises stand temporary flooding, growing closer to the banks.

Primary productivity in wetlands was estimated at 2 kg dw m^2 yr^{-1} (dw, dried weight; maximum 4–5 kg dw $m^{-2}yr^{-1}$) as algal mats, fruits and seeds, tubers or leaves sustaining the net productivity of invertebrates, fishes and amphibians. Large concentrations of birds gather to exploit the peak productivity. Wetlands offer a breeding ground for many species that will later migrate, in particular to Northern Europe or Central Africa. Shorebirds, ducks, cranes, sterns and cormorants account for most species. An estimated 680,000 coots are the most abundant species. The flamingo population commutes between several wetlands in the Western Mediterranean, the main breeding colony being at Fuentepiedra (Málaga).

Some species, such as crested coot, bittern, little bittern, night heron, marbled teal or white headed duck, are dwindling. Others, such as glossy ibis and the osprey, are now recovering from a crisis. The censuses of 44 aquatic species gave an average figure of 1,500,000 wintering birds for 1980–2001, which is 44% of the waterfowl in the Western Palaearctic region (Martí & Del Moral, 2003).

6   THE ONSET OF CONSERVATION POLICIES

Significant progress was made in water technologies under the Roman Empire (from approximately the 3rd century BC to the 4th century: long channels, aqueducts, sewers (*cloacae*), and field drainages. In the 8th century, Islamic invaders introduced Middle Eastern irrigation technologies, including large water wheels, and boosted qanat construction. A total of 48 dams were built before 1900. Large irrigation canals (Canal Imperial de Aragón, Acequia de la Huerta), or the long Canal de Castilla barge canal were built in the 16th and 17th centuries. Dam construction increased continuously until the 1970s, and then it slowed down (see Chapter 19).

The historical transformation of wetlands was confined to some favorable sites, such as marshes to build salt evaporation ponds (Cádiz, Valencia) or to create paddy fields (Ebro river delta, Albufera de Valencia lagoon). It was early in the 20th century when steam machines were available for large scale drainage interventions that spread over the country. The largest works (Guadalquivir river marshes, Potreros or Vegas del Guadiana (Badajoz), Janda lake (Cádiz) and Antela lake (Orense) were undertaken by government.

Some coastal marshes have undergone a profound transformation since the 17th century to build industrial estates, yards, ports or urban settlements: the Corunna, Ferrol, and Bilbao marshes have largely disappeared. A naval base, shipyards, industrial estates and salt evaporation ponds spread over the marshes of Cadiz Bay. Murcia's Portman Bay housed a Roman port (Portus Magnus). Tailings from Cartagena mines filled up the bay, now turned into a mineral wasteland. Similarly, the Tinto river marshes have been filled with tailings from the Rio Tinto mines or with acid sludge from phosphate fertilizer plants, which is high in heavy metals, uranium and other radioactive elements. During the second half of the 20th century, small marshes and lagoons of the Mediterranean coasts were lost to tourist development. Casado and Montes (1995) have estimated a 60% loss of wetland surface (from 280,000 to 114,000 ha) for the 1948–1990 period. 45% of remaining marshes disappeared between 1987 and 2000 (Anuario, 2008).

Fortunately, the infant wave of conservationism of the 1970s prevented the completion of a Guadalquivir marshes drainage project and led to the creation of the Doñana National Park (1969) to safeguard the remaining 30,000 ha of the former wetland. Some wetlands, such as Laguna de Peñalara (1930, 768 ha, Madrid), Lagunas de Ruidera, (1933, 3772 ha, Ciudad Real) or Lago de Sanabria (1946, 22365 ha, Zamora), were protected at an early date and later declared Natural Parks. New protected areas declared since 1969 incorporated wetlands: the Tablas de Daimiel National Park (1973, 1928 ha, Ciudad Real); the Delta del Ebro (1983, 8445 ha, Tarragona), Aiguamolls de l'Empordà (1983, 4731 ha, Girona), Marismas del Odiel (1984, 7185 ha, Huelva), Albufera de Valencia (1986, 21039 ha), S'Albufera de Mallorca (1988, 1687 ha), Bahía de Cádiz (1989, 9662 ha) Natural Parks; the Marismas de Santoña y Loja Nature Reserves (1991, 3866 ha, Santander), or PEIN protection plans (Lago de Bañolas, 1992, 1030 ha, Girona).

The protected area of Spain encompasses 11.8% of land surface area. 258 out of 1115 protected areas correspond to wetlands or fluvial sectors. 14.3% of total river length is now protected (Anuario 2007). However, 87% of wetlands (estimated at 2559) are still unprotected. Out of 391 Ramsar-compliant wetlands, only 38 are now protected (SEO/BirdLife report). Reservoirs have been largely ignored and only 4 have warranted protection.

Dam construction is now in decline, wetlands or riverbanks have been protected, and there are few drainage works in progress. Water quality is becoming a national concern. Waste water plants operate in towns with populations of 2000 and are now incorporating the smaller settlements. As a matter of fact, the 2004 National Hydrological Plan, which amended the 2001 Plan regarding the Ebro transfer (see Chapter 19), maintained a broad and comprehensive program of water treatment around the country.

Pristine waters are now a scarce environment and their associated species are in retreat. Eutrophication, river fragmentation and a growing water demand impair the functioning of aquatic environments and their biota. Continental wetlands often compete with local agriculture for subterranean waters (see Chapter 14). Upwelling is reduced or discontinued due to a drop in the water table through pumping to the point of drying out the wetland. This has occurred at the Tablas de Daimiel National Park, now fed by an artificial surface water supply (Llamas, 1988). Introduced species are spreading in aquatic environments, and fish fauna is seriously endangered. The zebra mussel is becoming a very serious concern in the basins that flow into the Mediterranean Sea, including those discharging to the Atlantic by means of the inter-basin transfer connection from the Ebro.

Doñana National Park, Spain's foremost protected site, has been suffering from the overexploitation of Almonte-Marismas aquifer no. 27 (Suso & Llamas, 1990). Earlier ecological restoration of Natural Park wetlands was carried out on the Abalario Lagoon (300 ha, Sousa & García Murillo, 1999) and Algaida Marshes (52 ha, Gallego Fernández & García Novo, 2002). Doñana National

Park restoration was launched under the Doñana 2005 Project. Wetland restoration is gathering momentum in Spain (García Novo & Marín, 2006).

In view of the human-induced processes summarized above, perhaps the major political challenge that Spain is facing to implement the WFD is to define the target conditions for all water bodies. As Chapter 16 details, the WFD's key implementation is to define both the current and desirable status of water bodies, and develop action programs that help bridge the gap between the two (see the discussion in Chapter 4 about setting as-is and should-be conditions). While national and regional nature conservancy initiatives are becoming effective in protecting specific sites, parks and areas, the WFD must be enforced on all water bodies. As the most significant water bodies in Spain are considered 'heavily modified', programs to restore their ecological status must be submitted to the European Commission by 2009. Alternatively, Spain could file for permission for deviations from the schedule for achieving this objective or for less stringent quality targets. To do this, it must provide cost-benefit analyses that show that the costs of achieving a good quality status of critical water bodies would be disproportionate.

7   CONCLUDING REMARKS

Spanish fresh water ecosystems have undergone profound transformations over the last century. River flows have been altered, seriously modifying the geomorphological characteristics, wildlife and water parameters. Over the last 20 years, efforts at taking updated and detailed inventories of wetlands, lakes and reservoirs attest to the devastating effects of centuries of human-induced impacts. The history of water policy in Spain is characterized by the massive construction of dams and reservoirs. Scientists began studying the ecology of these water bodies, complementing the work of hydrologists and engineers who focused only on dam operation. Very recently, water managers have realized that a vast and rich panoply of water bodies actually provide habitats for biodiverse ecosystems of great value to a semi-arid country. While most of them are in some sense artificial, they do to some extent make up for the massive loss of wetlands and natural lakes to drainage and the development of irrigated agriculture.

Environmental policies have been reversing this trend over the last two decades, but restoring aquatic ecosystems is not a top water policy priority. Progress so far has been achieved by increasing the proportion of treated wastewater spills, and implementing programs for the most valuable and emblematic ecosystem conservation. Yet, meeting the objectives of the WFD will deliver important environmental benefits once the basin policy programs are approved and implemented (Chapter 16). Whereas water policy was implemented in the past through piecemeal initiatives targeting specific areas or habitats, the WFD must be applied to all water bodies. It remains to be seen how the water quality targets are set, as very few of Spain's water bodies will be restored to the natural conditions.

REFERENCES

Alonso, M. (1987). Clasificación de los complejos palustres españoles *Bases científicas para la protección de los humedales de España. Real Academia de Ciencias.* Madrid: 65–78.

Alonso, M. (1998). Las lagunas de la España Peninsular. [Lakes in Peninsular Spain] *Limnética* 15: 1–176.

Anuario 2007. (2008). Europarc-España. Fundación BBVA. Madrid.

Armengol, J., Riera, J.L. & Morgue, J.A. (1991). Major ionic composition in Spanish reservoirs *Verh. Internat. Verein. Limnol.* 24: 1363–1366.

Armengol, J., Toja, J. & Vidal, A. (1994). Successional rhythm and secular changes in Spanish Reservoirs In R. Margalef (Ed.) *Limnology now. A paradigm of planetary problems.* Ed. Elsevier. Amsterdam: 237–253.

Casado, S. & Montes, C. (1995). *Guía de los lagos y humedales de España.* [Guide of lakes and wetlands in Spain]. J.M. Reyero Editor. Madrid.

Dirección General de Obras Hidráulicas (1991). *Estudio de las zonas húmedas continentales de España* [Study of the continental wetlands in Spain] INITEC. MOPU. Madrid.

Elvira, B. & Almodóvar, A. (2001). Freshwater fish introductions on Spai: facts and figures at the beginning of the 21st cerntury *Journal if Fish Biology* 59: 323–331.

Gallego, J.B., García Mora, R. & García Novo, F. (1999). Small wetlands lost: a biological conservation hazard in Mediterranean landscapes. *Environmental Conservation* 26(3): 190–199.

Gallego, J.B. & Garcia Novo, F. (2007). High intensity *vs.* low intensity restoration alternatives of a tidal marsh in Guadalquivir Estuary, SW Spain *Ecological Engineering* 30: 112–121.

García Novo, F. & Marín Cabrera, C. (2006). *Doñana: water and biosphere.* UNESCO/Ministerio de Medio Ambiente. Madrid. 366 p.

González Bernáldez, F. (1992a). *Los paisajes del agua: terminología popular de los humedales.* [Water landscapes: popular terminology for wetlands] J.M. Reyero Editor. Madrid.

González Bernáldez, F. (1992b). Ecological aspects of wetland/groundwater relationships in Spain. *Limnetica* 8: 11–26.

Granado-Lorencio, C. (Ed.) (1998). *Conservación, recuperación y gestión de la Ictiofauna continental ibérica.* [Conservation, restoration and management of Iberian continental Ichthyofauna] Publicaciones de la Estación de Ecología Acuática. Seville.

Granado-Lorencio, C. (2000). *Ecología de comunidades.* [Ecology of communities]. Serv. Publicaciones, Universidad de Sevilla. Sevilla.

Llamas, M.R. (1988). Conflicts between wetland conservation and groundwater exploitation: two case histories in Spain: *Environmental Geology* 11: 241–51.

Limnetica 14. (1998). Special issue devoted to studies of Barasona reservoir.

Margalef, R. (1983). *Limnología* [Limnology] Editorial Omega. Barcelona. 1110 p.

Margalef, R., Planas, D., Armengol, J., Vidal, A., Prat, N., Guiset, A., Toja, J. & Estrada, M. (1976). *Limnología de los embalses españoles* [Limnology of the Spanish Reservoirs], Publicaciones de la D.G. Obras Hidráulicas MOP. Madrid.

Martí, R. & Del Moral, J.C. (Eds) (2003). *La invernada de Aves acuáticas en España.* [The wintering of aquatic birds in Spain]. Dirección General de la Biodiversidad-SEO/Birdlife Madrid.

Martín Duque, J.F. & Montalvo, J. (Eds) (1997). *Agua y Paisaje: Naturaleza, cultura y desarrollo.* [Water and landscape: Nature, Culture and Development] Multimedia Ambiental Madrid.

Pardo, L. (1948). *Catálogo de los lagos de España* [Catalogue of Lakes in Spain]. Ministerio de Agricultura. Madrid.

Schmidt, G. & L. De Stefano. (2009). Major processes degrading freshwater resources and ecosystems **(This volume)**

Sousa, A. & García Murillo, P. (1999). Historical evolution of the Abalario Lagoon Complexes (Doñana Natural Park, SW Spain) *Limnetica* 16: 85–98.

Suso, J.M. & Llamas, M.R. (1990). El impacto de la extracción de aguas subterráneas en el Parque Nacional de Doñana [The impact of groundwater pumping in the National Park of Doñana]. *Estudios Geológicos* 46(3–4): 321–323.

Toja, J., Basanta, A. & Fernández Ales, R. (1992). Factors controlling algal blooms in the complex of water supply reservoirs of Seville. In Prat, N. & Duarte, C. (Eds). Limnology in Spain. *Limnetica* 8: 261–277.

CHAPTER 4

Major processes degrading freshwater resources and ecosystems

Guido Schmidt
Environmental consultant, Tecnoma S.A., Madrid, Spain

Lucia De Stefano
Postdoctoral Researcher in the Program in Water Conflict Management & Transformation of Oregon State University, USA

1 INTRODUCTION

Spain has a broad variety of landscapes together with a relatively low population density compared to other Western European countries and has traditionally hosted a rich biodiversity. However, the socio-economic development that occurred over the past 50 years has meant that freshwater ecosystems are facing significant problems related to water quantity, pollution and habitat fragmentation. This ecosystem decline follows patterns similar to the global trends identified by the Millennium Ecosystem Assessment (MEA, 2005) and affects not only the collective natural heritage but also ecosystem services provided by freshwater, including those whose beneficiaries are neither strong economic actors nor are represented in the decision-making bodies (e.g. recreational users, fishermen) (MEA, 2005).

The current freshwater status and its declining trends clearly show that Spain is far from achieving the so-called "sustainable development", where water and freshwater habitats are conserved and managed to meet "the needs of the present without compromising the ability of future generations to meet their own needs" (WCED, 1987).

Habitat loss and degradation are considered to be the primary proximate cause for biodiversity world-wide (Stedman-Edwards, 2000). However, socioeconomic root causes are the set of factors that lay behind proximate causes and actually drive biodiversity loss (Table 1). These deeper causes are often difficult to identify and remedy because they are far removed in either space or time, from the actual incidence of loss (WWF, 1999). Reverting current trends can only be set off by taking into account these complex issues in a systematic and meaningful way, applying a coherent DPSIR-structured scheme (Drivers-Pressures-Status-Impact-Response; see for instance OSE, 2008) or other analysis frameworks (e.g. Wood *et al.*, 2000) that consider proximate and root causes at different temporal and territorial levels.

The following pages outline the existing causal relationships between the declining status of Spain's freshwater ecosystems and the different socioeconomic root causes that directly stem from water-related policies, regulations and uses.

2 TRENDS IN LAND USE AND CONSUMPTION PATTERNS

Over the last decades, Spain has based its economic development on significant changes in land-use and consumption patterns, particularly for water and energy (OSE, 2006; OSE, 2008; Prieto *et al.*, 2008). There has been a strong land-use-change trend from natural areas towards irrigated farmland, and, in parallel, from farmland towards artificial surfaces (infrastructures, housing, etc.). Between

Table 1.   Proximate and root causes of freshwater biodiversity and species decline in Spain.

Proximate causes	Root causes
Habitat loss and degradation:	**Social change and development biases:**
• Occupation or drainage of wetlands and floodplains	• Uncontrolled growth of urban development
• Alteration of instream river flows	and tourism
• River fragmentation	• Unsustainable development models
• Accelerated erosion-sedimentation processes	**Public policies, markets, and politics:**
Pollution and climate change:	• Policies that promote biodiversity loss
• Diffuse water pollution	(e.g., production-focused agricultural
• Point water pollution	subsidies; intensification of irrigation; urban
• Precipitation decrease	sprawling)
• Temperature increase	• Ineffective enforcement of legislation
• Decrease of dilution capacity	• Inadequate water pricing policies
	• Data opacity
Overharvesting:	• Unequal participation in decision-making
• Excessive water abstraction	• Vested political interests behind water supply
• Fishery in overstressed ecosystems	**Macroeconomic policies and structures:**
Species and disease introduction:	• Subsidized export of irrigated
• Introduction of invasive species	overproduction
	• Promotion of irrigated biofuels
	Demographic change:
	• Increase in quality of life standards
	• Lost of the physical, cultural and emotional
	link between people and freshwater
	ecosystems (low social appreciation of the
	value of water ecosystems)

Source: Based on Stedman-Edwards (2000).

1987 and 2000, irrigated farmland has increased by 10.3%, and, despite nature protection legislation, 2,537 hectares of wetlands have been transformed into other land uses[1]. Artificial surfaces have increased by an annual 1.9%, summing up to a 29.5% increase for the 1987–2000 period.

Some of the key data for water use trends between 1987 and 2006 are shown in Figure 1. While precipitation diminished, water resources drivers, population, tourists, artificial surfaces, irrigated farmland, as well as the abstraction of water resources and surface water storage capacity—increased during the same period of years.

Water consumption has increased significantly over the last decades[2] and the main drivers for this increase have been irrigated agriculture, urban development and energy production. Similarly, the growth of these economic sectors has brought about the degradation of the quality of water resources, both from direct pollution and for the declining capacity of water bodies to dilute contamination.

Agriculture is currently the main water user in Spain (≥80% of total water consumption) and the increased water demand has been driven by a number of causes: agricultural subsidies that support crop production, a pricing… system that does not recover the costs or encourage water savings, non-compliance with water-related legislation, as well as deficient coordination of law enforcement between the competent River Basin Authorities (RBAs) and regional agricultural agencies.

Despite its recent reforms and the strengthening of compliance regulations, the EU's Common Agricultural Policy (CAP) still encourages water consumption by paying higher subsidies to

---

[1] 69% towards irrigated farmland and 31% towards artificial land (OSE, 2006).
[2] Even in recent times the abstracted water increased annually by 2% on average (OSE, 2008).

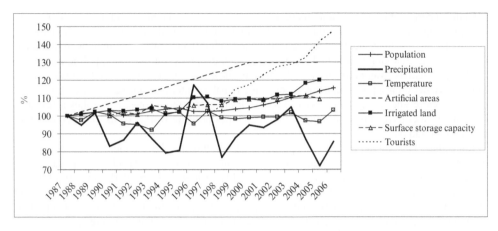

Figure 1. Evolution of significant water-related trends in Spain between 1987 and 2006. Modified from Prieto *et al*. (2008).

irrigated crops than to non-irrigated crops; by co-funding new irrigation systems; by not requiring farmers to be in full compliance with water legislation as a prerequisite for receiving CAP subsidies; and, finally, by subsidizing the export of overproduction, often due to irrigation (e.g. sugar beet), to non-European countries (OECD, 2008). Moreover, subsidies are still not fully decoupled from production, creating an incentive for farmers to produce irrigated crops even when not demanded by the market. Additionally, EU Rural Development funds are also employed to support highly-water-demanding crops like cotton, through, for example, agri-environmental schemes, or to promote water-intensive bioenergy crops. The impact of agriculture on the status of freshwater resources is exacerbated by point and non-point pollution, which is normally more intense and devastating in irrigated lands.

Uncontrolled urban and tourist development is another important driver of water demand: for instance, urban water consumption increased twice as much as the GDP from 1996 to 2001 (MIMAM, 2004). Even though this sector consumes, in absolute terms, far less water than agriculture, it does put great pressure on water and freshwater ecosystems, because demand is concentrated in confined areas (e.g. along the Mediterranean coast) and has sizeable seasonal peaks (EEA, 2003). Moreover, improper or absent treatment of urban wastewater rapidly degrades surface resources, especially in areas with reduced precipitation, where treated waters often makes up a substantial part of the instream flows.

Finally, pressure on water resources coming from energy production is twofold: river fragmentation and alteration for hydropower production, and consumption and modification of water physico-chemical characteristics by the refrigeration employed for thermal, solar and nuclear power plants.

The combination of these drivers has resulted in reduced instream flows, the drying-out and degradation of wetlands and the intensive use of groundwater resources. A large number of environmental effects are directly related to the water resource regulation and abstraction, causing increased vulnerability to pollution, modified erosion-sedimentation and other ecological processes, and species decline and invasion; without a natural flow regime, no river maintains its values and its ecosystem services (EEA, 2007).

3 WATER POLICY: INCREASED WATER-SUPPLY AS A POLITICAL, ECONOMICAL AND SOCIAL PRIORITY

When analyzing the dynamics of Spain's economic development of the last decades, it is easy to understand that a large proportion of politicians, public officials, economic stakeholders and even the general public, consider an increase of water supply as the main and primary solution to ever-growing water demands. Further water development is claimed without previously analyzing other political or management options, causal chains, side-effects or the economic or environmental viability of satisfying new heavily subsidized water demands. Furthermore, politicians define the key problem as 'water scarcity', 'lack of water' or a 'structural water deficit' and seldom as "unbalance between supply and demand or 'unsustainable water use' during election campaigns often use 'more' and 'cheap water' as a campaign play to garner more votes. This policy is based on expanding water development through three main types of water works: dams, transfer schemes, and desalination plants.

Since the 1940s, public investments pushed dam building as a key element to overcome poverty. With more than 1,300 large dams, Spain now ranks fourth in the world in terms number of large dams (WCD, 2000). Nonetheless, the National Hydrological Plan (NHP Act 2001) foresaw the construction of one hundred more dams, despite the negative potential effects on protected areas[3], their questionable economic viability (Arrojo, 2001), and the lack of any Strategic Environmental Assessment (SEA) of such a large collection of works of Public Interest.

This dam-based water development system has been complemented by the construction of a network of intra-basin and large inter-basin water transfers (Cantabria-Ebro-Cantabria, Tagus-Segura-Mediterranean, Guadiana-Odiel-Tinto, Guadalquivir-Mediterranean, Júcar-Vinalopó, Ebro-Tarragona, and Ter-Barcelona), that redistributes water over the Spanish territory to attend water demands from irrigation farming, and urban and touristic areas (WWF, 2007a). Although the cancellation of the Ebro-Segura transfer project in 2004 was meant to be a strong official recognition of economic, social and environmental concerns, over the last few years a new list of transfer projects (Tinto-Guadalquivir, Ebro-Barcelona, mid-Tagus-Segura) has popped up in the political arena and in the media, mainly due to the intense political pressure applied on the Central Government by increasingly more powerful regional governments.

Spain is the fourth producer of desalted water in the world, with a production capacity of 1.5 million m³ per day and between 700 and 900 mainly small and privately owned plants in the whole country (WWF, 2007b). In 2004 the central Government approved the called *Programa AGUA*, which represented the start of a 'new era' of public-funded water manufacturing. *Programa AGUA* foresees to provide the Mediterranean coast with 621 Mm³/y of additional desalted water, converting Spain in the western country with the largest desalination capacity. However, serious doubts exist about the willingness of users—especially farmers—to pay for desalted water unless it is heavily subsidized and while there are still other—legal or illegal—sources of water available (WWF, 2007b).

Over the last five years, water decision-makers have added to the traditional large-infrastructure approach some new and complementary water supply measures that are typical of a mature water development and allocation system (Carles, 2004; EEA, 2009). For example, water savings by improved irrigation technology have been claimed to be a centerpiece of today's drought and water scarcity management and significant investments have been made over the last few years. The law supporting these vast public spending programs (Royal Decree 287/2006 Plan for Irrigation Modernization[4]) presents irrigation modernization as a means to release water resources for urban and environmental uses. However, modernization projects deeply lack transparency in terms of how saved water volumes are calculated *ex ante* and *ex-post*, and of where that saved water actually ends up. This raises the question as to whether the real objective of those projects is about mitigating 'water scarcity' and shifting resources from the agricultural sector towards other uses, such as industrial and recreational purposes.

Urban water treatment is being improved due to legal obligations deriving from EU legislation and specific norms have been issued to legislate on the re-use of those 'regenerated waters'.

---

[3] It is estimated that the planned dams will affect 28 riverine forest, freshwater and moors habitats protected under the EU Habitats Directive, as well as several Spanish Natural Parks (WWF, 2003).

[4] Real Decreto 287/2006 de 'Plan de Choque de Modernización de Regadío'.

Additionally, water markets are being increasingly used to shift water between uses, particularly in situations of extreme drought.

From this description it is clear that Spain's water policy focuses mostly on meeting the ever-increasing economic water demands that cause freshwater degradation. However, some still marginal and often isolated initiatives have been developed in the context of the EU Water Framework Directive (WFD) implementation process, to try to reverse current ecosystems degradation trends.

Since 2008, Spain has had one of the world's strongest legislative packages for establishing environmental instream flows regimes (Guidelines for Water Planning, 2008[5]) and has kicked-off studies for a two-step process of identification and public discussion of these required regimes in key river-stretches of the main basins, with the Catalonian Regional Water Agency as a front runner. Nonetheless, the current complex technical, administrative, legal, economic and political circumstances and priorities do hamper the adequate implementation of the new legislative package. The results of the calculation and negotiation of the new flow regimes will appear possibly too late for their integration into the upcoming 2009–2015 River Basin Management Plans (RBMP).

In parallel, since 2007 the Spanish Ministry for the Environment, Rural and Marine Affairs is promoting a remarkable national strategy for river restoration with an annual 0.2 bn € budget. The strategy not only focuses on technical projects, but also strengthens the capacity of RBAs, involves volunteers activities and develops communications and awareness-raising campaigns that explain the WFD objectives of achieving good ecological and chemical status of all water bodies by 2015 (www.restauracionderios.org).

Furthermore, after some serious legal warnings from the EU, Spain is increasingly dealing with urban point-source water pollution and is progressively improving its urban wastewater treatment capacity (OSE, 2008) with EU co-funding. However, the functioning and maintenance of these purifying systems is not always properly managed and controlled. Non-point pollution, in particular by nitrates[6] and pesticides and more than 10,000 industrial spills per year (MIMAM, 2000) are nonetheless causing severe damage to the environment.

4   GOVERNANCE FAILURES IN WATER AND RIVER BASIN MANAGEMENT

Spain faces the challenge of balancing of water demands from the economic sectors escalating while grappling with the increasing degradation of freshwater resources. Several aspects of its water governance undoubtedly hinder this process. This section outlines what can be held as the most pressing issues: lack of information transparency, poor law enforcement, inadequate water pricing policies, outdated participatory structures and regional strains on river basins unity.

Although largely incomplete, the publication of White Book on Water in 2000 (MIMAM, 2000) implied a significant step forward in compiling and publishing key data on water resources, policy and management. Further valid information has been prepared in the frame of the WFD RBMP development, the Spanish Sustainability Observatory (OSE, 2006; and OSE, 2008) and the six pan-Iberian water congresses organized by the New Water Culture Foundation (www.unizar.es/fnca); this information is increasingly available online at the servers of the different administrations, Hispagua (hispagua.cedex.es) and iAgua (www.iAgua.es).

Nonetheless, many experts still report a lack of transparency of basic data, such as the water resources availability (Estevan & Naredo, 2004), the estimation of future water demands[7] or the real

---

[5]ORDEN ARM/2656/2008, de 10 de septiembre, por la que se aprueba la instrucción de planificación hidrológica.

[6]Trend data that show that nitrates are at least affecting negatively a 38% of Spain's groundwater bodies (OSE, 2008).

[7]For instance, the approval of the controversial Melonares dam in the Guadalquivir basin was justified with significant and unexplained overestimates of future urban water demands. The dam project foresaw a water demand of 195 Mm^3/y by 2012 (Moral Ituarte, 2005) while in 2008 the consumption was 125 Mm^3/y only.

water consumption, e.g. due to the absence of metering devices in most of the irrigated farmland, the swinging of official data about the irrigated farmland areas[8] and the scarce data available on CAP subsidies[9]. This data uncertainty in some cases supports vested manipulation of water problems and solutions, including irregular administrative procedures and corruption (Martínez & Brufao, 2006).

A lack of data lack often hides also administrative management failures; we want to illustrate this point for three sensitive issues: illegal water uses, theoretical water savings and water pricing.

Although the 'legislative arsenal' to tackle water problems is in many cases quite satisfactory, law enforcement comes up against serious difficulties. For example, the number of illegal bore-holes in Spain is estimated by non-official sources to be between 510,000 and 2,000,000[10] (WWF, 2006; Llamas, 2006a), and though the problem is openly acknowledged in official forums, no official nationwide numbers have been released. In the Doñana National Park, the Guadalquivir River Basin Authority has been for years aware of the existence of illegal groundwater abstraction but it started mapping illegal boreholes only in 2003. A pilot study in a particularly sensitive part of the Doñana wetlands catchment area found that all the inspected agricultural farms were abstracting groundwater completely or partially illegally (CHG, 2003). Similarly, in the area of the Daimiel Wetlands National Park, the RBA acknowledged that the number of illegal boreholes largely outweighed the legal ones (22,000 vs 16,000, CHGua, 2005) only after many years of blatant law breaching and once the wetlands had fallen into an 'ecological coma' difficult to revert (Llamas, 2006b).

As mentioned before, during the past few years the Spanish Government has boosted irrigation systems modernization projects as a means to release water resource for other priority uses. However, those measures were taken without kicking off the necessary administrative procedures to shift the saved volumes from irrigation towards the legal priorities of environmental streamflows or urban water supply. Presently, the only practical result of irrigation modernization is the intensification of irrigated farmlands, increasing the pressures (land occupation, pollution) on freshwater ecosystems (Cánovas, 2008; WWF, 2008).

In a similar way, and although considered by the WFD as a key measure for good water management, water pricing—in particular for farming—is poorly documented and implemented. Surveys indicate that only a very small proportion of irrigation farmers (ranging between 2.7% in the Tagus basin and 36.84% in the Júcar basin) pay tariffs directly related to their water consumption, and the majority pay per irrigated surface area or with a fixed tariff (MARM, 2008), undermining the basis of the 'polluter and user pays' principle. Although farming is by far the largest water user, it is expected that for the 2010–2015 period a general exemption for the WFD cost-recovery obligation will be applied to Spanish farmers.

Spain is ponderously adapting its participatory means to the requirements of the EU legislation and of its modern society. The governance bodies of the RBA still reflect the historical predominance of economic water users and exclude stakeholders without water rights, preserving the basic decision-making structures that reflected the power-sharing structure of the Spanish society in the first half of the 20th century. Non-governmental stakeholders are only consulted under the EU-obligatory procedures (RBMP, SEA, and Environmental Impact Assessments) and an increased participation fatigue (Reed, 2008) in the recently created regional or sub-basin civil platforms (Tagus, Ebro, Júcar, Genal, etc.) and environmental NGOs.

Traditionally the Spanish water governance system has held the river basin as the management unit, anticipating by many years the requirements of the WFD in terms of adequate scale of water management. Nonetheless, during the past few years this model has experienced an accelerated

---

[8] For instance, Carreño et al. (2008) compile different sources that list data for the Segura basin that vary from 248,069 to 315,646 hectares.
[9] Farmsubsidy (2009) claims that 1% of the major Spanish CAP benefactors get 22% of the overall agricultural subsidies and 10% perceive a 60%.

process of regionalization of water competence. The increased role of regional administrations in river basin management is perceived by many actors (irrigators, engineers, environmentalists) as a threat to a coherent water management inside the basin limits, in particular because their policies for new infrastructures, more water and more irrigation are putting increased and uncoordinated stress on the current water systems (Tecniberia, 2009).

5 OUTLOOK

As described above, the Spanish water policy is dominated by a water-supply focus, applied via a combination of water infrastructures and governance flaws. Such an approach acts as a driver for significant and manifold impacts on water quantity, water quality, freshwater and coastal ecosystems and their services for human well-being. The last years have shown the strong inertia of Spain to any water policy changes that the WFD, supported by the new water culture movement, should introduce in terms of sustainability, ecosystem conservation, law enforcement, economic rationality (cost-recovery of externalities) and major social occupation of river spaces. The initial expectations based on the 2015 WFD-milestone of good (ecological and chemical) status are held to be unrealistic for many Spanish water bodies, and in the best of the cases this first planning cycle sets the baseline for more ambitious 2021- and 2027-deadlines.

Although water technology will evolve and ensure a more efficient resource use and, in a prosperous economy, new sources of desalted water, this is by far too little to deal with increased water demands and the additional pressure that climate change will put on the freshwater systems. It is easy to perceive that the society will become more conscious about the values and benefits (security, provision, health, recreation, culture) from rivers and other freshwater ecosystems, and that Spain will face increased water conflicts between water-using sectors, territories and social groups (Prieto *et al.*, 2008). The WFD legal requirements alone cannot change in a decade an established mentality and a water policy that during more than 50 years has succeeded to meet the socio-economic needs of the Spanish society. However, the Directive has triggered new debates, restlessness and questions about where water policy should head for from now on. This questioning, even if not sufficient to make a difference in the ecosystems status today, is a necessary process for any future change. The level of maturity of the Spanish society will have the last say in that process. In the meanwhile, most possibly the ecosystems will be the weakest part of increasing conflicts on water, and the country will continue facing biodiversity loss and the decline of ecosystem services.

REFERENCES

Arrojo Agudo, P. (2001). *El Plan Hidrológico Nacional A Debate* [The National Hydrological Plan in Discussion] Bakeaz. Centro de Documentación y Estudios para la Paz) ISBN: 8488949448. ISBN-13: 9788488949448.

Cánovas, J. (2008). Modernización de los regadíos. Ahorro de Agua. [Modernisation of irrigation areas] *Foros de ciencia y tecnología. Jornada sobre ahorro, eficiencia en el uso del agua y gestión de la demanda.* Alcalá de Henares, 2008.

Carles, J. (2004). La Directiva Europea de Aguas y la reforma de la Administración Pública [Water Framework Directive and the Reform of the Public Administration]. In Arrojo, P. (Coord.). *El agua en España: Propuestas de futuro* [Water in Spain: Proposals for the Future]. Ediciones del oriente y del mediterráneo. Guadarrama (Madrid).

Carreño, F., Martínez, J., Miñano, J., Suárez, M.L., Robledano, F., Vidal-Abarca, R. & Esteve, M.A. (2008). Indicadores de sostenibilidad del agua: Caso cuenca del Segura. [Water sustainability indicators: the case of the Segura Basin] *VI Congreso Ibérico sobre Gestión y Planificación del Agua.* 4–7 December, 2008. Vitoria/Gasteiz, Spain.

Confederación Hidrográfica del Guadalquivir (CHG) (2003). *Análisis de las extracciones de agua subterránea en la cabecera de la cuenca del arroyo de la Rocina.* [Analysis of the groundwater pumping in the headwaters of the basin of the stream Rocina] Seville, Spain.

Confederación Hidrográfica del Guadiana (CHGua) (2005). *Plan del Alto Guadiana del Alto Guadiana* [Plan for the High Guadiana] (PEAG). Borrador Documento de Directrices 4.07.2005. p. 42.

Estevan, A. & Naredo, J.M. (2004). *Ideas y propuestas para una nueva política del agua en España.* [Ideas and proposals for a new water policy in Spain] Bakeaz, Colección Nueva Cultura del Agua. Bilbao, Spain. Summary at http://www.unizar.es/fnca/docu/docu41.pdf (accessed January 10, 2005).

European Environment Agency (EEA) (2003). *Europe's environment: the third assessment.* State of the environment report No. 3.

European Environment Agency (2007). *Europe's environment: The fourth assessment.* State of the environment report No. 1/2007.

European Environment Agency (2009). *Water resources across Europe: confronting water scarcity and drought.* EEA Report No. 2/2009.

Farmsubsidy (2009). Uncovering farm subsidies in Spain: A dossier of work in progress. www.farmsubsidy.org.

Llamas, M.R. (2006a). *La contribución de los avances científicos a la solución de los conflictos hídricos.* [The contribution of the scientific progress to the solution of water conflicts] University of Alicante, opening lecture 2005–2006, June 14, 2006 http://www.rac.es/ficheros/doc/00241.pdf (Accessed April 3, 2009).

Llamas, M.R. (2006b) Entrevista a Ecosostenible [Interview in Ecosostenible]. January 2006 http://www.rac.es/ficheros/doc/00238.pdf (Accessed April 3, 2009).

Llamas, M.R., Fornés, J.M., Hernández-Mora, N. & Martínez Cortina, L. (2001). *Aguas subterráneas: retos y oportunidades* [Groundwater resouces: challenges and opportunities]. Fundación Marcelino Botín. Ediciones Mundi-Prensa, Madrid.

MARM (2008). *Estudio de los costes asociados al uso del agua de riego. Cánones, tarifas y derramas pagados por los regantes.* [Survey of the associated costs of irrigated agriculture. Levies, Tariffs and Contributions paid by farmers], Madrid.

Martínez, J. & Brufao, P. (2006). *Aguas limpias, manos limpias. Corrupción e irregularidades en la gestión del agua en España* [Clean waters, clean hands. Corruption and irregularaties in water management in Spain] Fundación Nueva Cultura del Agua & Bakeaz. Zaragoza, Bilbao.

Millennium Ecosystem Assessment (MEA) (2005). *Ecosystems and Human Well-being: Synthesis.* Island Press, Washington DC (USA).

Ministerio de Medio Ambiente (MIMAM) (2000). *Libro Blanco del Agua en España* [White Book on Water in Spain] Dirección General de Obras Hidráulicas y Calidad de las Aguas, Secretaría de Estado de Aguas y Costas.

Ministerio de Medio Ambiente (MIMAM) (2004). *Perfil Ambiental de España 2004* [Environmental Profile of Spain 2004]. Ministerio de Medio Ambiente, Dirección General de Calidad y Evaluación Ambiental, Madrid.

Moral Ituarte, L. del. (2005). *La gestión del agua en Andalucía. Aspectos económicos, políticos y territoriales* [Water managemente: Economic, Political and Land-use Aspects]. Mergablum, Seville, 133 págs. ISBN 84-96378-18-7.

Observatorio de la Sostenibilidad en España (OSE) (2006). *Cambios en la ocupación del suelo en España. Implicaciones para la sostenibilidad* [Changes in land-use in Spain. Implications for sustainable development] http://www.sostenibilidad-es.org/Observatorio+Sostenibilidad/esp/prensa/noticias/CambiosUsos_Esp_CCAA.htm (accessed April 4, 2009).

Observatorio de la Sostenibilidad en España (OSE) (2008). *Agua y sostenibilidad. Funcionalidad de las cuencas.* [Water and sustainability. River Basin Functions]. www.sostenibilidad-es.org/NR/rdonlyres/BD7E1400-6378-4AC5-84B9-B199B5EFB614/2947/aguaespañol.pdf (accessed April 4, 2009).

OECD (2008). OECD Environmental Outlook to 2030. ISBN: 9789264040489 www.*oecd*.org/environment/outlookto2030 (Accessed April 4, 2009).

Prieto, F., Ruiz, P. & Martinez, J. (2008). Prospectiva 2030 en los cambios de ocupación del suelo en España y sus impactos en el ciclo hidrológico. [2030 Prospective for land-use changes in Spain and its impacts on the hydrological cycle] *VI Congreso Ibérico sobre Gestión y Planificación del Agua.* 4–7 December, 2008. Vitoria/Gasteiz, Spain.

Reed, M. (2008). Stakeholder participation for environmental management: A literature review. *Biological Conservation 141*: 2417–2431.

Stedman-Edwards, P. (2000). A Framework for Analyzing Biodiversity Loss. *The Root Causes of Biodiversity Loss.* Wood Alexander, Pamela Stedman-Edwards and Johanna Mang (Eds). Earthscan Publications Ltd, London and Sterling, VA.

Tecniberia. 2009. *Integrated Water Management in Spain.* Madrid.

World Commission on Dams (WCD) (2000). *Dams and Development. A new framework for decision-making.* Earthscan Publications. London, UK. http://www.dams.org/report/ (accessed March 31, 2009).

Wood, A., Stedman-Edwards, P. & Mang, J. (Eds) (2000). *The Root Causes of Biodiversity Loss*. Earthscan Publications Ltd, London and Sterling, VA.

World Commission on Environment and Development (WCED) (1987). *Our common future.* Oxford: Oxford University Press.

WWF (World Wide Fund for Nature, also known as World Wildlife Fund) (1999). *Root Causes of Biodiversity Loss: An analytical approach*. Washington DC. USA.

WWF (2003). *Damming Nature -Impacts of SNHP dams on Natura2000 sites*. Madrid, Spain. http://www.panda.org/downloads/europe/natura2000eng2.pdf. (accessed March 31, 2009).

WWF (2006). *Illegal water use in Spain. Causes, effects and solutions.* Madrid, Spain. http://assets.panda.org/downloads/illegal_water_use_in_spain_may06.pdf (accessed March 31, 2009).

WWF (2007a). *Pipedreams? Inter-basin water transfers and water shortages*. http://www.panda.org/who_we_are/wwf_offices/south_africa/?uNewsID=107300 (accessed March 31, 2009).

WWF (2007b). *Making water. Desalination: Option or distraction for a thirsty world?* http://wwf.org.au/publications/desalinationreportjune2007.pdf. (accessed March 31, 2009).

WWF (2008). *Regadíos y Directiva Marco del Agua*. Madrid, Spain. *www.unizar.es/fnca/docu/docu231.pdf*. (accessed March 31, 2009).

CHAPTER 5

Water uses in transition

Josefina Maestu
Directorate of Water, Ministry of the Environment, Spain

Carlos Mario Gómez
University of Alcala del Henares, Spain

1   GENERAL TRENDS OF WATER USE IN THE ECONOMY

Water abstractions to satisfy the requirements of economic activities put tremendous pressure on water ecosystems in Spain. Total abstractions represent 34.7% of available surface and groundwater resources. This rate is several times higher than in other OECD countries with greater water endowments or the potential for rainfed agricultural activities[1].

The quantity of water used by the economy and the resulting ecosystem quality reflects the types and importance of existing market and non-market economic activities. In this way, the composition and trends of water use in Spain can be interpreted as the result of the country's underlying economic transformations.

According to the Water Satellite Accounts published by the Spanish National Institute of Statistics (INE), the functioning of the national economy required the abstraction of 37,650 Mm^3 (million cubic meters) from the environment in 2001, and this volume had grown at an annual average rate of 1.6% over the previous five years. In spite of this increasing pressure on water ecosystems, the above figures need to be compared with the real Spanish GNP (gross national product) that grew at an average rate of 3.85% over the same period. This comparison reveals a positive trend in overall water productivity and a reduction in the quantity of water used per unit of final production. The steady trend of the Spanish economy towards higher water productivity is easier to appreciate when yearly water abstractions are compared with real GNP. Accordingly, there has been a clear dissociation between growth and overall water abstraction over the last two decades at least, meaning that increasing water demand and economic growth has somehow been made compatible with a constant or lower pressure on water resources (see also OECD, 2004). Over the last decade the water requirements per unit of GNP fell at an annual rate of 2.25%, reaching 24.7 m^3/€1000 in 2001 instead of the 27.6 m^2/€1000 required in 1997.

In other words, the effect of economic growth on water abstractions (or the scale effect that tends to increase water use at a rate proportional to production increases) has been counterbalanced by a number of factors (see Figure 1). First, the most water-intensive sectors and subsectors now have a relatively smaller share of GNP. Second, the use of water saving technologies by the faster growing production sectors (such as manufacturing and services) has increased, as has the diffusion of the best available techniques in more traditional activities (such as agriculture and livestock farming). Third, the conveyance efficiencies of all water supply services have improved. These services can now meet a substantially higher final water demand with a somewhat lower quantity of water abstracted from natural sources thanks mainly to technically sounder methods of water storage, transport and distribution. Total water abstraction in 2001 was 5% lower than in 1988. In 2001, the provision of one cubic meter required the abstraction of 1.75 m^3 from streams and underground sources. The combined national, and regional actions to enhance the technical

---

[1] See OECD (2004). The average intensity of water abstractions over available water resources is 14.2% in Europe and 19% in the United States.

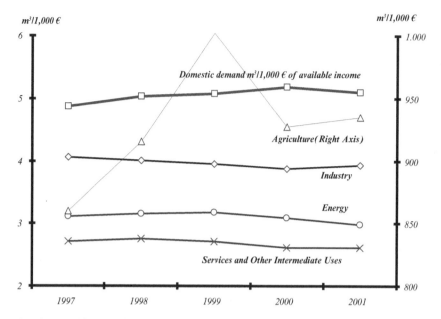

Figure 1.   Water requirements by economic sector.
Source: National Institute of Statistics: Water Satellite Accounts; Economic Accounts.

Table 1.   Water requirements and production growth 1997–2001 (Mm³).

	1997	2001	Average annual rate of growth
WATER ABSTRACTIONS[a]	**34,735.9**	**37,652.9**	**1.61%**
For Irrigation Operations	21,423.3	22,183.6	0.70%
Rest of Agriculture, Livestock and forestry	1,975.8	2,367.7	3.62%
Rest of the Primary Sector	147.8	173.2	3.17%
Manufacturing Industry	1,149.2	1,373.3	3.56%
For Drinking Water Production	4,392.9	5,383.1	4.07%
Power Generation[b]	5,530.4	6,029.9	1.73%
Others	116.4	142.1	4.00%
Distributed to Water Users[c]	**21,319.1**	**22,486.3**	**1.33%**
Agriculture, livestock and Forestry[d]	18,279.6	18,460.9	0.20%
Rest of Primary Sector	29.2	31.9	1.72%
Manufacturing Industry	343.9	373.8	2.34%
Domestic Supply	2,322.8	2,508.6	4.66%
Energy (Consumptive use)	41.0	48.0	3.95%
Construction, Services and Others	792.6	935.6	4.66%

Source: INE: Water Satelite Accounts 1997–2001. Tables 6 and 7.
[a] Includes water withdrawn as a primary input.
[b] Water used for refrigeration of thermal and nuclear power generation plants.
[c] Includes all water considered as an intermediate input or as a final consumption good.
[d] Includes water distributed to Irrigation Associations (Comunidades de Regantes) and other drinking and non drinking water used by the sector. The differences with water abstractions for irrigation is explained by water losses in water transport from the withdrawal to the distribution point.

efficiency of irrigation distribution networks affecting 0.8 million hectares could have already saved 1,543 Mm^3. In Figure 1, the 1999 peak for agriculture was due to low agricultural output in money terms and large water availability due to unusually high precipitation in the 1997–99 period.

The main water user in the Spanish economy is irrigated agriculture. It uses more than four of every five cubic meters of total water abstracted. Nevertheless, the share of agricultural output and income in the Spanish economy fell from more than 5% during the early 1980s to around 2.6% in 2007. Agriculture's share in water requirements decreased from 85% to close to 80% of total water consumption from 1997 to 2001. Contrary to this, urban water consumption, although representing only one eighth of total water consumption, is growing at a much faster annual rate of 4.7%. The third most important activity demanding water is the manufacturing industry, with less than 2.5% of the total and an average growth of 3.6%. The rest of water demand is distributed across other uses, mainly the tertiary sector and the building industry (with less than 4%), and consumptive uses for power generation (with less than 0.3% of the total) (see Table 1).

2   WATER IN AGRICULTURE

The comparative advantages of Spanish agriculture derive from its weather and location. In a context where sunlight, soils and market proximity are suitable for farming, the availability of water storage and irrigation facilities becomes a key factor for guaranteeing the financial viability of Spain's Mediterranean agriculture. The income from a typical irrigated hectare is six times greater than that of an average rainfed hectare. For this reason, incentives for developing new irrigation projects and modernizing the existing irrigation infrastructure have been a factor explaining the constant growth of the irrigated area (3.4 million hectares in 2008) for at least the last 100 years.

According to the National Institute of Statistics' Regional Accounts, agriculture accounted for 2.6% of Spain's total GNP and 4.5% of employment in 2007. Ten years earlier agriculture's share of GNP had been 4.7% and of total employment, 7.1%. There are two important points here. First, in spite of its declining importance, agriculture is still the main economic activity in many rural areas, producing the main input for other transformation activities, such as the agro-industry[2]. Second, the above trends do not only reflect agriculture's declining importance in the Spanish economy, but also the effects of an important underlying shift towards more productive agriculture (as shown by the fact that the fall in employment was higher than the decrease in production, meaning that labor productivity has so far increased). This trend is more marked in regions such as Andalusia, where farming is the economic activity with the greatest productivity gains in the regional economy.

While labor productivity increased in the Spanish economy, water productivity exhibited the opposite trend. Between 1997 and 2001, the intensity of water use in agriculture increased from nearly 800 m^3 to 875 m^3 for every 1,000 euros of agricultural production (see Figure 1).

These trends could be better interpreted in terms of the dual Spanish rural economy, where highly competitive and market-driven agriculture exists side by side with traditional and institutionally driven production. The significant differences in water productivity across crops and regions are a noteworthy indicator of the importance of this dualism. As shown in Table 2, crops with a high yield compared with the water applied can achieve from 2 to 43 euros of gross added value per cubic meter (mainly for irrigated flowers and greenhouse vegetables in Catalonia, along the East Coast, and in Spain's Southeast). By contrast, with a few rare exceptions, the gross value added per cubic meter for crops with a low yield per water applied is less than 0.04 euros or even goes into negative figures (mainly cereals and other EU Common Agricultural Policy subsidized products).

---

[2]There are, however, sizeable regional differences because agriculture's share of value added in most regions is higher than the average for the whole country. In regions such as Extremadura and Andalusia, for example, agriculture contributes an important part of total value added (10–13%).

Table 2.   A sample of lower and higher productive water uses in agriculture in southeast Spain in 2005.

Crop	Region	Surface (hectares)	Average irrigation efficiency[a] %	Water requirements[b] m³ per hectare	Gross value added € per m³
Wheat	Andalusia	44,733	78%	3,057	−0.15
	Murcia	3,434	83%	3,232	−0.26
	Valencian Community	3,126	66%	3,974	−0.02
Maize	Andalusia	54,969	75%	7,789	0.01
	Murcia	186	66%	7,156	−0.02
	Valencian Community	1,930	66%	6,010	0.00
Pepper	Andalusia	13,243	64%	5,379	0.45
	Murcia	1,736	87%	4,757	4.70
	Valencian Community	1,249	76%	4,994	2.75
Green Beans	Andalusia	5,148	80%	5,346	30.66
	Murcia	62	78%	5,346	2.97
	Valencian Community	737	82%	3,556	4.19
Tomatoes	Andalusia	17,293	80%	6,607	3.85
	Murcia	2,364	80%	7,347	2.71
	Valencian Community	1,373	90%	5,367	7.50
Pepper (Greenhouse)	Andalusia	3,504	90%	5,367	6.08
	Murcia	297	90%	5,367	9.78
	Valencian Community	684	90%	5,367	6.34
Flowers (Carnation)	Andalusia	582	90%	7,132	21.00
	Murcia	170	90%	7,661	25.61
	Valencian Community	136	90%	6,084	43.05
Flowers (Roses)	Murcia	25	90%	7,187	31.45

Source: MAPA, Recopilación de Datos de Superficies Ocupadas por los Cultivos Agrícolas (Hojas 1T). Group of Economic Analysis, Ministry of Environment, and MMAMRM (2008).
[a]The Average Irrigation Efficiency measures the water that reach the crop as a percentage of the quantity of water introduced in the irrigation system.
[b]Water requirements is a measure of the volume of water that need to be applied at a farm level. This measure does not take into account water lost in transport from the withdrawal point to the irrigation association and in the distribution from the irrigation association to its members.

Even when irrigation's efficiency problems are recognized, government and society widely perceive the construction of new and the modernization of the old irrigation facilities is the main policy instrument for rural development. This can be put down not only to the above important rural income gains, but also to its role in the conservation of Spain's rural heritage and landscape, maintaining social capital, and as means of preventing the depopulation of marginal areas.

The increase in agriculture's water use intensity has been counterbalanced by at least two positive trends. On the one hand, agronomic vocation and the scarcity of water in a number of areas have driven farmers' responses to different types of institutional policies, incentives and market conditions. The recent rapid development of a highly intensive agriculture specialized in high value-added products, mainly in the South and the Mediterranean regions, has also helped to offset the overall average trends in productivity. On the other hand, improvements in on-farm watering techniques have also been a positive factor for improving productivity. Traditional gravity irrigation (flood irrigation) is still present across 60% of the agricultural area and is the prevailing technique in marginal agricultural areas. New irrigation areas have generally opted for sprinkle or localized (drip) irrigation. Sprinkle irrigation is the main form of irrigation on the inland plains,

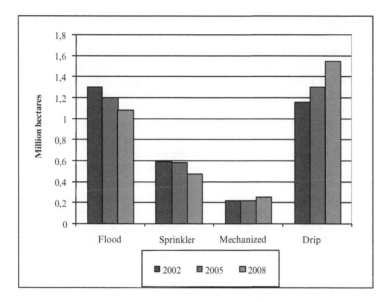

Figure 2.   Trends in irrigation techniques Spain.
Source: Ministerio de Medio Ambiente y Medio Rural y Marino (MERMA, 2008).

whereas the more efficient localized irrigation has a greater presence along the Mediterranean coast and in the South. The trend away from gravity and flood irrigation and towards sprinkler and localized irrigation techniques is characteristic of market-driven agriculture, where yields and market prices rather than subsidies and institutional support are the key factor in the farmers' decision making (see Figure 2). The move from gravity to drip irrigation means extra investment and maintenance costs for farmers. On average sprinklers increase production costs by 130 euros per hectare, and the most efficient localized technique by 142 euros per hectare (Camacho, 2003). Improvements in watering techniques are only a sound option when the water price effectively paid by farmers is high enough for water cost savings to compensate for the extra investment needed to improve farm operations. This is more likely to happen in areas where water comes from groundwater sources at a cost greater than 0.08 €/m^3 to the farmer for an average use of 2,000 cubic meters per hectare.

There is some evidence of water prices (groundwater abstraction costs and prices charged to farmers by collective irrigation water suppliers) being correlated with the enhanced technical and economic efficiency of water use. Irrigation water prices effectively paid by farmers vary with the sources of water used (groundwater or surface water) and the state of conservation of supply infrastructures. There are important local differences in the costs of irrigation water services that are reflected in prices. In Murcia, Almeria and Valencia, where agriculture is specialized in high value added crops and aquifers, more than two thirds of water services cost farmers on average 0.08 euros per m^3 (and up to 0.39 €/m^3 in some aquifers). On average farmers pay from 0.01 to 0.02 euros per m^3 for surface water diverted from streams and large-scale infrastructures (dominant in areas with low value crops).

In any case, for traditional and modern agriculture, farmers' decisions may be constrained in both cases by institutional arrangements regulating water property rights. Organized farmers using surface or groundwater for irrigation are required to have "concession rights" granted by the river basin authorities (RBAs). Apart from that, the effective volume of water available for any crop season depends on both legally prescribed maximum-per-hectare quotas defined by the basin authority and the availability of resources, which are subject to the rainfall. Maximum-per-hectare water allowances are the main instruments for preventing overexploitation of water resources, but they also place constraints on farmers' possibilities for moving towards crops with greater water requirements. Additionally, they may also act as an incentive for improving technical efficiency, but do not necessarily imply reduced pressure on the water environment. In any case, the availability of water for crops is not only uncertain from season to season. Because the law gives priority to domestic and urban water uses over rural uses, the impact of drought on water availability is higher in agriculture than in other uses connected to urban water supplies.

The evolution of Spanish irrigated agriculture is primarily linked to trade barriers, production subsidies and the many public actions that determine the sector's financial incentives. The changes envisaged in these factors will determine the prospects of the rural economy. This way the size and composition of the agricultural product and the associated water requirements will depend on how the new CAP subsidies, at least partially decoupled from production, affect the different crops and how eco-conditionality is applied. Decoupling could have significant effects, at least on the production of cereals, olives and grapes. Indirectly, changes in relative prices could lead to a reallocation of land, water and other factors towards other more competitive products.

It is still uncertain what impact the progressive liberalization of agricultural markets, including the implementation of the World Trade Organization (WTO) and the Euro-Mediterranean convention agreements, as well as the enlargement of the European Union (some interim results are available from the European Union (EC, 2009)), will have. However, fewer incentives for traditional and water-intensive crops might not necessarily lead to lower pressures on water ecosystems, as excess water may be reallocated to profitable crops in the new context. Nevertheless, changes in structural policies and in the international environment are expected to result in increases in water use productivity.

An important aspect of agriculture as a water user is related to the use of pesticides, fertilizers and other chemical inputs that reach water bodies through surface runoff and leakage. The impact of diffuse pollution on the ecological quality of waters depends on many factors, such as the application of the right doses according to best practices, precipitation, soil slope, water irrigation techniques, and the availability of wastewater collection and treatment. Surveys carried out by the Ministry of Agriculture show important regional differences in fertilizer doses applied to different crops on farms. Taking into account the cultivated land registered in the 1999 census, the average dose of fertilizers was estimated at 67 kg, 37 kg and 28 kg per hectare of nitrates, phosphorus and potassium-based fertilizers, respectively (MAPA, 2004).

Nitrate pollution, from both current and past practices, has attracted much of the European water policy's attention. In 2000, 27 groundwater bodies had an anomalous nitrate concentration (MIMAM, 2000), and many areas of Spain are currently declared vulnerable to nitrate pollution from agrarian sources. Nitrate control programs need to be applied by the regional governments not only to agriculture but also mainly to animal farming.

3   URBAN WATER USE AND HOUSEHOLD CONSUMPTION

The different kinds of urban water consumption account for a small but growing share of total water demand that came close to 10% of the total in 2002 and is expected to reach 15% in 2015. Household drinking water consumption accounts for nearly 70% per cent of urban uses. Manufacturing firms, shops, restaurants, hotels and other services are responsible for 20% of urban water demand. The remaining 10% of water demanders are public organizations providing public goods (such as gardening and cleaning of public spaces). All these uses have been increasing at an average rate of 4.6% per year, and are, in one way or another, the consequence of the growth of the Spanish economy. Over the last two decades, per capita income increases were positively related to higher average household water consumption[3] but also to an increased demand for water

---

[3] Estimates of the drinking water demand function in the Júcar River Basin show that water responds negatively to water prices and positively to income increases. Price elasticity has been estimated at –0.6 and income elasticity at 0.4 in the Júcar River Basin (see Ministry of the Environment (2004) *Economic Characterization of Water Use in the Júcar River Basin: Pilot Study*).

Table 3.   Trends in urban water consumption 1997–2001.

Autonomous communities (reported only the most populated ones)	Population 2001	Per-capita water consumption lit/person/day 2001	Average rates of growth 1997–2001		
			Total urban water consumption	Population	Per-capita water consumption
Andalusia	7,403,968	171.15	11.21%	0.46%	10.63%
Aragon	1,199,753	121.13	7.49%	0.20%	7.23%
Castile and León	2,479,425	142.03	5.13%	–0.23%	5.42%
Castille-La Mancha	1,755,053	152.71	14.41%	0.49%	13.80%
Catalonia	6,361,365	235.70	6.89%	0.87%	5.80%
Valencian Community	4,202,608	150.46	5.42%	0.94%	4.24%
Galicia	2,732,926	151.90	10.25%	–0.07%	10.34%
Madrid (Community of)	5,372,433	173.00	4.78%	1.35%	3.10%
Basque Country	2,101,478	204.20	10.45%	0.03%	10.40%
TOTAL SPAIN	**41,116,842**	**184.61**	**8.02%**	**0.72%**	**7.12%**

Source: National Institute of Statistics and own elaboration.

to provide public goods, like water for cleaning public spaces, for public gardens and parks and for other urban amenities. On top of this, there is the growth of water demand from tourism, and a higher than average population growth over the last few years explained not so much by vegetative growth than by the influx of migration.

The impact on urban consumption is highly influenced by the urbanization process. This tends to concentrate water pressure along the Mediterranean coastline and in the metropolitan areas of the centre and the north of the country. Per-capita water consumption has increased since 1996 at an average annual rate of 2% and was estimated at 171 liters per day in 2004, with sizeable regional differences ranging from a maximum 200l/person/day in Castille-La Mancha to a minimum 142 liters/person/day in the Balearic Islands and the Basque Country (INE, 2008).

Recent and significant increases in water prices have slowed down the rapidly growing final water demand. Their effects on water abstractions have been partially offset by the efficiency gains in the domestic water supply system. Regional distribution of water consumption is positively correlated with water prices. Over the last decade water prices increased every year by more than 4% and averaged 0.98 €/m³, again with sizeable differences between the higher prices paid in the Canary Islands (1.44 €/m³), the Balearic Islands (1.44 €/m³) and Murcia (1.53 €/m³) and the lower prices in Castilla-León (0.67 €/m³) and Castilla-La Mancha (0.9 €/m³). Price differences are correlated with geographic differences in urban water provision costs, as reflected by the latest available estimates from 1996. In 1996, the cost was above 2.5 €/m³ in the Canary Islands and under 1 €/m³ in Galicia and Castilla-León[4]. The difference between the total water distributed and effectively received by final users fell from 32% in 1990 to 19.4% in 2000. Reductions in this unregistered water are explained partially by improvements in consumption measurements and fraud detection, but mostly by the technical improvements in water distribution networks.

4   SERVICES AND TOURISM

Service activities in the Spanish economy, growing at a yearly rate of 4.1%, account, as a whole, for 60% of Spanish production and only 3.5% of total water demand. Being the most dynamic

---

[4] Cost recovery for sewage collection and treatment services is lower than for drinking water distribution services. Its costs are 34% of the water provision cost but only 25% of the average price.

sector of the economy, its higher than average growth is one of the reasons why the overall water intensity of the Spanish economy has been declining over the last few decades. The most significant water use in this category is related to tourism and recreational activities. The opportunity for recreational activities explains the recent growth of houses for non-permanent residents (one of every three new dwellings built in the last ten years in a country where the number of new dwellings built in 2006 and 2007 was higher than in France, Germany and Italy together).

Although its share of overall water demand is lower than one per cent, recreational water uses are one of the most rapidly growing activities. This category includes swimming pools, theme parks, and mainly golf courses. Old and new projects to develop such facilities are viewed as a way to increase local revenue, as an alternative to the decline of traditional manufacturing activities. They are also looked upon as a means to increase returns on the tourism infrastructure by reducing the high demand seasonality, particularly in the Mediterranean areas, improving the quality of tourist packages by attracting high purchasing power tourists and increasing the demand for complementary services, like transport and restaurants. This explains the reasoning behind plans to develop 45 new golf courses in Murcia and Valencia, in spite of the extremely limited water availability (see Sanz-Magallón, 2005). The average water requirements of a Mediterranean golf course are similar to some cereal crops (between 6,500 and 10,000 cubic meters per hectare depending on rain and local conditions). Nevertheless, the financial return of a typical golf course is substantially higher and has been estimated at from €12,000 to €48,000 per hectare in Murcia and at 15 €/m^3 in the Valencia Region as compared to some eurocents for cereals (between 26 and 10 eurocents in the Júcar River basin and Extremadura, respectively) (see Sanz-Magallon *et al.* 2004).

5   INDUSTRIAL AND ENERGY WATER USE

The manufacturing industry, which accounts for nearly 17% of total Spanish production, consumes only 1.5% of total water. In spite of a growth rate similar to Spanish production as a whole, this sector exhibits a positive trend in water efficiency, as revealed by the steady reduction of water requirements from 4 to 3.75 cubic meters for every 1000 euros of production from 1997 to 2001. This efficiency improvement is explained by both the faster growth of relatively low water user industrial activities and the enhancement of water productivity in many of the industrial water uses (see Figure 1).

The energy sector is also a significant water user, accounting for a mere 0.2% of water consumption, but with sizeable non-consumption water requirements. In the energy sector, water is used to cool nuclear and thermal plants and for hydropower production. Impacts relate to hydromorphological indicators and to temperature. There are 1,136 river stretches or sections housing hydropower facilities in Spain. There are also 1,200 dams used for hydroelectricity generation, which, depending on the hydrological year, store an average of 56,500 Mm^3. Of these, 30% are multi-purpose dams. It is estimated that dams for hydropower production have flooded 2,825 square kilometers (MERMA, 2008). Total hydropower capacity is estimated at 35,743 GWh/year. Apart from power generation, an average of 4,915 Mm^3 of water is abstracted every year from natural streams and dams to cool fossil fuel and nuclear-based power generation plants. This is an amount equivalent to 0.63 Mm^3 per MW of installed generation capacity (MIMAM, 2005).

At market prices, the production of hydropower was estimated (MIMAM, 2005) at €21,000 million in 2003 (valued at average market prices of 3.58 cents/kWh for flow energy and 4.31 for regulated energy). Hydropower is considered to play an important role in the stability of the Spanish electricity system given its flexibility for meeting peak electricity demands. The ratio of hydroelectricity production to existing water resources has been estimated at a pressure index of 0.32 GWh/year per Mm^3.

The economic analysis of water use for electricity generation should consider not only the benefits associated with the activity, but also the economic value of the forgone benefits. It is hard to put a price on forgone benefits because there is often no market value for natural areas and landscapes; their recreational value or for the impacts of biodiversity changes due to alterations of natural flow regimes, etc. The private opportunity costs of the measures taken to compensate for or reduce impacts according to environmental impact assessments, and especially those related to the maintenance of minimum flows, have been valued by the power industry association (UNESA) to be equivalent to 1,200 GWh per year in Spain for installations of more than 5 MW. This is 3.3% of the production potential. At market prices, this amounts to €49 million per year. The substitution costs would be approximately €9,840 billion to build alternative installations and €1,430 billion per year in fuel costs. The sector estimates that hydroelectricity production saves about 2.8 million equivalent barrels of oil and helps avoid other forms of pollution that have negative effects on health and other effects via air emissions.

6   FINAL REMARKS

Overall sector water use reflects the trends of economic growth and has led to greater pressure on water ecosystems mainly through greater water abstractions or pollution emissions, as well as because of the changes in the water flow regime and morphology of rivers. Water use (and pollution emissions) has increased in all sectors. However, the share of water use by different sectors has changed because of the high growth rate of urban uses in relation to other users.

Water productivity has increased as a whole, as economic growth has decoupled from the use of water resources. This is explained by some positive trends, such as increased technical efficiency in water distribution services in cities and agriculture, and improved technical efficiency in water application on farms. This is often supported and promoted by the European Union and public funding in urban areas and for agricultural irrigation networks.

Improvements in water productivity have been clear in urban water use and industry, where income increases leading to higher per capita consumption have offset price increases (reflecting the higher costs of providing wastewater collection and treatment services). Technological innovation in industry and the decline of traditional and larger water-consuming sectors also explain greater productivity gains in industry.

Productivity of water use in agriculture has been declining mainly as a result of institutional factors and most likely because of the downward trend of agricultural product prices. Yet market-driven agriculture has contributed substantially to increases in productivity and is expected to continue to do so where new competitive markets can lead to crop substitution.

REFERENCES

Barbero Martín, A. (2004). El seguimiento del Plan Nacional de Regadíos [The follow-up of the National Irrigation Plan]. *Agricultura* (June): 554–558.

Camacho, E. (2003). *Máquinas de Riego* [Irrigation Machines]. Córdoba, Spain: Universidad de Córdoba.

European Commission (2003). Prospects for agricultural markets in the European Union 2003–2010. Brussels, Belgium: Directorate General for Agriculture.

INE (2003). *Encuesta sobre el suministro y tratamiento del agua del año 2001* [Survey on urban water supply and wastewater treatment], Madrid, Spain: Instituto Nacional de Estadística.

INE (2008). Indicadores Sobre el Uso del Agua. [Water use indicators] Madrid. Spain. Instituto Nacional de Estadística.

MAPA (2004). *Pilot survey on the use of fertilizers in agriculture*. Madrid, Spain: Ministerio de Agricultura, Pesca y Alimentación.

MERMA (2008). *El agua en la economía española: situación y perspectivas. Informe integrado del análisis económico de los usos del agua.* [Water in the Spanish economy: balance and perspectives]. Ministry of Environment and Rural and Marine Affairs. Madrid. Spain.

MIMAM (2000). *Caracterización de las fuentes agrarias de contaminación del agua por nitratos* [Characterization of the agricultural sources of nitrate pollution]. Madrid, Spain: Ministerio de Medio Ambiente.

MIMAM (2005). Informe Integrado de los Usos del Agua en España. Grupo de Análisis Económico de la Directiva Marco del Agua. Primer Borrador [Integrated report on Water uses in Spain. Economic Analyses Group of the Water Framework Directive]. Madrid, Spain: Ministerio de Medio Ambiente, November.

MIMAM (2006). *Política del Agua: Balance. Documento para la Conferencia de Presidentes*. [Water Policy: A Balance. Document for the Presidents' Conference]. Madrid, Spain: Ministerio de Medio Ambiente, December. http://www.mma.es/portal/secciones/novedades_web/pdf/politica_de_agua_balance_28_ 12.pdf (looked on Jan. 31, 2007).

OECD. (2004). *Environmental Performance Reviews. Spain*. Paris, France: Organization for Economic Co-operation and Development, Paris.

Sanz-Magallón, G. & Serrano, J. (2004). Agua-*Ocio-Deporte. Una Valoración Socioeconómica y Medioambiental* [Water-Leisure-Sports. A Socio-economic and environmental valuation]. Alicante, Spain: Fundación COEPA.

Sanz-Magallón, G. (2005). Una aproximación al valor del agua utilizada en los campos de golf de las comarcas de Levante y Sureste [An approximated calculation of the water value used in golf courses in the comarcas of Levant and Southeast]. *Revista Española de Estudios Agrosociales y Pesqueros*, 2005, 205: 99–124.

CHAPTER 6

Water footprint and virtual water trade in Spain

Maite M. Aldaya
Twente Water Centre, University of Twente, The Netherlands

Alberto Garrido
Department of Agricultural Economics and Social Sciences, Universidad Politécnica de Madrid, Spain

M. Ramón Llamas
Department of Geodynamics, Complutense University, Madrid, Spain

Consuelo Varela-Ortega, Paula Novo & Roberto Rodríguez Casado
Department of Agricultural Economics and Social Sciences, Universidad Politécnica de Madrid, Spain

1 INTRODUCTION

As the most arid country in the European Union, water resources management in Spain is an issue as important as controversial. In this country, even if water resources are unevenly distributed and, in some regions drought conditions are increasing, the crisis is one of water governance rather of than physical scarcity. The estimation and analysis of the water footprint of Spain, from a hydrological, economic and ecological perspective, is very useful to facilitate an efficient allocation of water and economic resources. This analysis can provide a transparent and multidisciplinary framework for informing and optimising water policy decisions, contributing at the same time to the implementation of the EU Water Framework Directive (2000/60/EC) (WFD). This is particularly relevant since Spain is the first country that has included the water footprint analysis into governmental policy making in the context of the WFD (Official State Gazette, 2008).

The water footprint (WF) is a consumption-based indicator of water use (Hoekstra & Chapagain, 2008). The WF of an individual or community is defined as the total volume of freshwater that is used to produce the goods and services consumed by the individual or community (ibid.). Closely linked to the concept of water footprint is the virtual water. The virtual water content of a product (a commodity, good or service) refers to the volume of water used in its production (Allan, 1997). Building on this concept, virtual water 'trade' represents the amount of water embedded in traded products. International trade can save water globally if a water-intensive commodity is traded from an area where it is produced with high water productivity (resulting in products with low virtual-water content) to an area with lower water productivity (Hoekstra and Chapagain, 2008). Nevertheless, just a small amount of international virtual water trade is due to water scarcity (Yang & Zehnder, 2008). International trade in agricultural commodities mainly depends on factors such as availability of land, labour, technology, the costs of engaging in trade, national food policies and international trade agreements (Hoekstra & Chapagain, 2008). At national or regional level, a nation can preserve its domestic water resources by importing products instead of producing them domestically. This is particularly relevant to arid or semi-arid countries with scarce water resources such as the case of Spain, which imports water-intensive low-economic value crops (mainly wheat, maize and soybeans) while it exports water-extensive high-economic value commodities adapted to the Mediterranean climate, essentially olive oil, fruits and vegetables. Apart from stressing its potential contribution to water savings, it is also important to establish whether the water used in the production of a given crop proceeds from rainwater stored in the soil as soil moisture evaporated during the production process (green water) or from surface water and/or groundwater evaporated as a result of the production of the crop (blue water) (Falkenmark & Rockstrom, 2004). Compared

49

to blue water, the opportunity cost of green water use is lower since it cannot be easily reallocated to other uses besides natural vegetation or alternative rain-fed crops (Hoekstra & Chapagain, 2008).

The present chapter analyses the water footprint and virtual water trade in Spain assessing both green and blue water of the different socioeconomic sectors from a hydrological and economic perspective. The analysis aims to contribute to achieve a more efficient allocation of water resources. First of all, it provides a general overview of the water footprint and economic value of the different sectors in Spain, focusing afterwards on the agricultural sector, which is the main water user. Second, the virtual water trade and policy implications are analysed. Finally, it concludes that the current idea of water scarcity in Spain is mainly due to mismanagement in the agricultural sector providing interesting lessons for arid and semiarid countries. This mismanagement is due to several reasons such as the persistence of the former idea of food self-sufficiency, the still imperfect World Trade Organization (WTO) regulations, the absence of appropriate economic instruments for water management, national policies that promote irrigated agriculture to contribute to regional stability and agricultural commodity prices.

2   AN OVERVIEW OF THE WATER FOOTPRINT AND ECONOMIC VALUE OF THE DIFFERENT SECTORS IN SPAIN

Spain is the most arid country in Europe and the one that consumes one of the largest volumes of water per capita after the US and Italy, amounting to about 2300 m³/capita/year (Chapagain & Hoekstra, 2004). According to Chapagain & Hoekstra (ibid.), total water requirements (green and blue) by the different economic sectors in Spain are about 100 km³/year, that are distributed as shown in Table 1.

These figures, based on national averages, are taken as a first approximation. More detailed studies provide more accurate data as shown in the next section.

According to Chapagain & Hoekstra (ibid.), urban water supply amounts to 5% of the total water used with a value of 4.2 billion euros (MIMAM, 2007). The industrial sector represents 15% of the total water use (from which more than half corresponds to virtual water 'imports'), 14% of the Gross Domestic Product (GDP) (123 billion euros; INE, 2008) and 16% of the economically active population (3.1 million jobs; INE, 2008). Urban water supply and industrial sector figures refer to blue water uses—not necessarily consumptive—and are in line with the values given by official statistics (MIMAM, 2000; 2007).

The agricultural sector, considering green and blue crop consumption and livestock water use, represents about 80% of the total water use in line with Chapagain & Hoekstra (2004) (2/3 with national water and 1/3 with 'imported' virtual water) (Table 1) and Rodríguez Casado et al. (2009). According to this author, Spain is a net virtual water 'importer' concerning agricultural products, whereas a net virtual water 'exporter' when considering livestock products (fish water footprint has not been included as there are no estimates available yet). The agricultural sector, however, just contributes with about 3% of the GDP (about 26 billion euros, including livestock and fisheries, according to INE, 2008) and employs 5% of the economically active population (1 million jobs, following INE, 2008). Since agriculture is by far the main (green and blue) water user in Spain, this sector is at the centre of the present study. Thanks to factors such as globalization, availability of cheap and fast transport, guaranty of groundwater irrigation against climate variability and environmental regulation among others, the Spanish farmers are becoming increasingly competitive.

3   AGRICULTURAL WATER USE

Concerning the crop water consumptive use (or evapotranspiration) of agriculture in Spain, there are remarkable differences between the results of the first and more recent estimations (Table 2). On the one hand, crop water consumptive use estimated by Chapagain & Hoekstra (2004) is higher than that of Rodríguez Casado et al. (2009) probably because of the greater detail of the more

Table 1.   Virtual water flows and water footprint of Spain, Italy, US and India (period 1997–2001).

	Spain	Italy	US	India
Population (10^6)	40.5	57.7	280.3	1007.4
Urban water supply				
km³/year	4.2	8.0	60.8	38.6
m³/cap/year	105.0	136.0	217.0	38.0
Crop evapotranspiration				
National consumption (km³/year)	50.6	47.8	334.2	913.7
Idem (m³/cap/year)	1251.0	829.0	1192.0	907.0
For export (km³/year)	17.4	12.4	139.0	35.3
Idem (m³/cap/year)	430.0	214.0	495.0	35.0
Industrial uses				
National use (km³/year)	5.6	10.1	170.8	19.1
Idem (m³/cap/year)	138.0	176.0	609.0	14.0
For export (km³/year)	1.7	5.6	44.7	19.1
Idem (m³/cap/year)	42.0	97.0	159.0	6.0
Virtual water 'import'				
Agricultural products (km³/year)	27.1	60.0	74.9	13.8
Idem (m³/cap/year)	671.0	1039.0	267.0	14.0
Industrial products (km³/year)	6.5	8.7	56.3	2.2
Idem (m³/cap/year)	1605.0	150.8	208.9	21.8
Re-export of imported products	11.4	20.3	45.6	1.2
Idem (m³/cap/year)	281.0	351.0	163.0	1.0
TOTAL WATER FOOTPRINT				
km³/year	94.0	134.6	896.0	987.4
m³/cap/year	2325.0	2332.0	2483.0	980.0

Source: Modified from Chapagain & Hoekstra (2004) in Llamas (2005).

recent study. The latter uses regional level climate data and differentiates between rainfed and irrigated farming for the water consumption estimations, whereas the former assumes that every crop water requirements are satisfied. This assumption, however, is not always fulfilled in Spain where rainfed farming covers an area of more than 80% of the total utilised agricultural area. On the other hand, official numbers from the Spanish Ministry of the Environment are the lowest (MIMAM, 2007) (Table 2), probably due to the fact that official figures focus on the blue water consumption, that is, the total amount of irrigation water that is lost to crops' evapotranspiration, without taking into account green water evapotranspiration. Incorporating the concept of green water into the bigger picture makes it possible to understand water implications of land cover change and water scarcity problems of rainfed agriculture (Falkenmark & Rockstrom, 2004). In order to achieve an effective land use planning, green water analysis should be considered within an integrated land and water resource approach.

Within the agricultural sector, irrigated agriculture uses about 80% of blue water resources (MIMAM, 2007). Concerning the economic aspects, however, irrigated agriculture is a vital component of the agricultural sector. Even if it just occupies about 20% of total crop area, it produces 60% of the total Gross Value Added (GVA) of this sector (MIMAM, 2007) (Figure 1). This fraction is higher than the global average. Worldwide, the Gross Value of irrigated agricultural production is 46%, which makes up 28% of the harvested cropland (Comprehensive Assessment of Water Management in Agriculture, 2007). The economic productivity (€/ha) in irrigated agriculture in Spain is about five times higher than that of rainfed agriculture (Plan Nacional de Regadios, 2009).

Table 2.   Estimated values of internal or domestic water consumptive use in Spain's agricultural crop production after different sources.

Source	Agricultural water consumption[1] (Mm³)	Blue water consumption[2] (Mm³)	Green water consumption[3] (Mm³)
MIMAM (2007)[4]	–	11,897	–
Chapagain & Hoesktra (2004)[5]	50,570	–	–
Rodríguez Casado *et al.* (2009)[6]	26,824	15,645	11,177

[1] Agricultural water consumption refers to the total crop water evapotranspiration.
[2] Blue water consumption is the total amount of irrigation water evapotranspirated by the crops.
[3] Green water consumption represents the total amount of soil water evapotranspirated by crops.
[4] Average figures for the year 2001 (average rainfall year).
[5] Average figures for the period 1997–2001.
[6] Average figures for the years 1998, 2001 and 2003.

By comparing blue water requirements (consumptive use) and supply, water use average efficiency turned out to be about 65% in 2001 (MIMAM, 2007). This figure has recently diminished since the implementation of the National Irrigation Plan, which is undertaking the modernisation of irrigation systems and improving water use technical efficiency (Plan Nacional de Regadios, 2009). These water savings, however, are possibly relative savings as the irrigated area has also increased, water continues to be priced by area and not by volume consumed in most systems; and the previously water 'lost' due to inefficient irrigation might be used downstream.

4   TOWARDS AN EFFICIENT ALLOCATION OF WATER RESOURCES

Spanish agriculture has comparative advantages as a result of its soil availability, sunshine hours, lower labour costs and its strategic location for the access to the European Union markets. Spain has no barriers to trade with other EU Member States. On the whole, Spain benefits from this advantage producing high value crops adapted to the Mediterranean climate, such as vegetables, citrus trees, vineyards and olive trees (Figure 1).

First of all, it has to be highlighted that rainfed grain cereals in Spain occupy more than 5 million hectares as shown in Figure 1. In the year 2001, grain cereals were the main land and water users in Spain, utilizing 47% of total arable land and 32% of blue water resources (Figures 1 and 2) (MIMAM, 2007). In economic terms, however, they generated the lowest Gross Value Added (GVA) value, which was about 6% GVA of irrigated agriculture according to MIMAM (2007) data. Nevertheless, the analysis should not just focus on economic aspects but also address social and environmental factors. On the other hand, vegetables, citrus trees and fruit trees are very productive in economic terms and require a relatively small amount of land and water. These are, however, mainly grown with blue water resources. The best opportunities and economic yields are obtained when these are grown in areas where blue water resources are less abundant.

Similar trends are obtained when analysing the water apparent productivity (Figure 2). When looking at the productivity per crop type, vegetables (including greenhouse crops such as horticultural, flowers and ornamental plants) present the highest values per water unit (with about 3.5 €/m³). With lower values, other profitable crops are vineyards and temperate climate trees. It has to be highlighted that vineyards, as well as being one of the most profitable crops, are very well adapted to the Mediterranean ecosystem. Finally, with remarkably lower values, grain cereals, industrial crops and pulses display an average productivity of less than 0.3 €/m³. Accordingly, the apparent productivity of vegetables is more than six times higher than that of cereals.

A mere 4% of all blue water used in irrigated agriculture accounts for 66% of total Gross Value Added. Conversely, close to 60% of the water used in this sector produces a slight 5% of total GVA in agriculture. This means that Spain is mainly producing blue water-intensive, low-value crops.

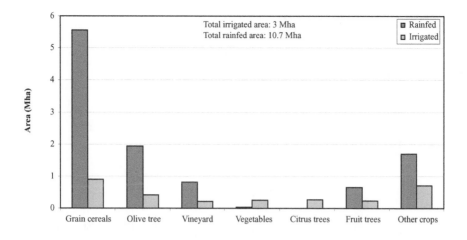

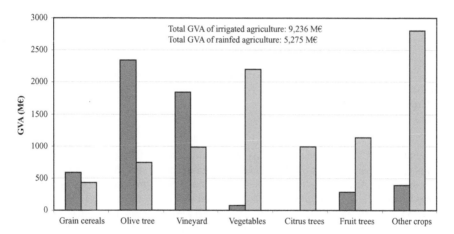

Figure 1.   Total area (Mha) and Gross Value Added (GVA) (M€) comparing rainfed and irrigated agriculture per crop in Spain for the year 2001.

Source: Based on date from the Spanish Ministry of the Environment (based on 78% of total irrigation in Spain) (MIMAM, 2007).

In this sense, in order to achieve a win-win solution for increasing productivity, enhance rural employment opportunities and improve the livelihoods of the rural population while protecting the environment, a more efficient allocation of water resources is desirable. Even if Spain has already achieved a fairly high level of accomplishment of the policy of 'more crops and jobs per drop', it still struggles to attain 'more cash and nature per drop'.

Even if not considered in the analysis of the Spanish Ministry of the Environment (MIMAM, 2007), most probably high value crops are watered with groundwater resources or combining ground and surface water (Hernández-Mora *et al.*, 2001). This fact of forgetting or ignoring the relevance of groundwater irrigation is a frequent attitude in many countries that only recently is beginning to change (Llamas & Martinez-Santos, 2005; Shah, 2008). In line with existing data, groundwater irrigated agriculture has a higher productivity when compared to irrigation using surface water (Hernández-Mora *et al.*, 2001). This difference can be attributed to the greater control and supply guarantee that groundwater provides, which in turn allows farmers to invest in modern and efficient irrigation techniques and cash-crops farming practices without the risk of water shortages during dry periods. Generally farmers who are groundwater users bear all financial costs,

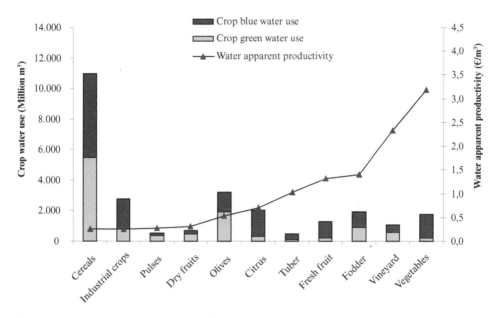

Figure 2.   Water apparent productivity (€/m³) and blue and green crop water use in Spanish agriculture for the year 2006.
Source: Garrido *et al.* (2008).

both, operation and maintenance as well as investment costs. In fact groundwater users usually pay a higher price per volume of water used than irrigators using surface water that is, in general, largely subsidized. At the same time, the higher financial costs groundwater-using farmers bear motivates them to look for more profitable crops that will allow them to maximize their return on irrigation investments (Hernández-Mora *et al.*, 2001). Along these lines, the groundwater role is significantly different from the surface water role. However, most water footprint studies have not hitherto differentiated between surface and groundwater (e.g. Chapagain & Hoekstra, 2004; Rodríguez Casado *et al.*, 2009). This distinction is crucial to inform water policy decisions, and to follow the environmental requirements of the Water Framework Directive (Aldaya & Llamas, 2009; Hernández-Mora *et al.*, 2007).

5   VIRTUAL WATER TRADE IN SPAIN: SOLVING WATER SCARCITY PROBLEMS

Agricultural commodity trade in relation to water is an issue that has rarely been dealt with. Overall, Spain is a net virtual water 'importer' concerning agricultural commodities. According to Chapagain & Hoekstra (2004) Spain 'imports' about 27 km³/year and 'exports' 17 km³/year, resulting in a negative balance of 10 km³/year. Spain exports high economic value and low virtual water content crops, such as citrus fruits, vegetables or olive oil, while it imports virtual water intensive and low-economic value crops, such as cereals (Novo *et al.*, 2009; Rodríguez Casado *et al.*, 2009). This not only has a huge potential for relieving local hydrologic, economic and political stress in Spain (Allan, 2006) but it is also very relevant for the national economy and water balance. Cereal grains can thus be crucial commodities for food security to water scarce importing countries (Yang *et al.*, 2006). Spanish cereal production is just 5% of total European production. In this sense Spanish demand would always be supplied by other EU producers or security stocks. This, however, does not imply that importing food is the only response the water scarce countries and regions should and can take. Furthermore, in the real world, even if the potential of trade to 'save' water at national level is substantial, most international food trade occurs for reasons not related to water resources (Comprehensive Assessment of Water Management in Agriculture, 2007). In line with Yang & Zehnder (2008), globally, less than 20% of the total virtual water trade is due to water scarcity. In this sense, 80% of the virtual water trade is mainly due to pure commercial factors. International trade in agricultural commodities mainly depends on factors such as availability of land, labour, technology, the costs of engaging in trade, freight costs, national food policies and international trade agreements (Aldaya *et al.*, 2008; Hoekstra & Chapagain, 2008).

Figure 3 shows that in Spain the composition of virtual water 'imports' and 'exports' are fairly stable in the period studied (1997–2005). Spain's cereal imports make up about 70% of all water agricultural imports, whereas livestock exports represent 55% (Rodríguez Casado *et al.*, 2009). Both are obviously linked and respond to Spanish natural endowments, land and climate, and its close integration in the EU economy. Water scarcity as such does not explain why Spain 'exports' virtual water through livestock products. This is explained to a greater extent by lesser enforcement of environmental legislation related to livestock production, more empty territory and a great deal of economic integration. But clearly without the option to import cereals and feedstock, the livestock sector would not have grown to the extent it did in the last 10 years.

6   INCORPORATING THE WATER FOOTPRINT INTO POLICY MAKING

In the last twenty years Spanish water policy has changed dramatically (Garrido & Llamas, 2009). Over the last decades, the priorities of the Spanish society have been changing and, moving away from the traditional supply-enhancing water policy, environmental factors have become increasingly important. There is a growing need to integrate nature conservation, social equity and economic growth into the process of decision making. For the time being and almost in the entire world, water footprint analysis has focused on hydrological aspects. A significant innovation of this work is to emphasize the imperative challenge of considering economic and ecological aspects, with the aim of going towards a policy of 'more cash and nature per drop'.

The water footprint analysis, thus, from a hydrological, economic and ecological perspective, differentiating green and blue ground and surface water, provides a transparent and multidisciplinary framework for informing and optimising water policy decisions, contributing at the same time to the implementation of the EU Water Framework Directive (2000/60/EC) (see Chapter 16 for a more detailed analysis).

In this context, Spain was the first country in the EU to adopt the water footprint evaluation in governmental policy making. In September 2008, the Spanish Water Directorate General, under the competence of the Ministry of the Environment and Rural and Marine Affairs, approved a regulation that includes the analysis of the water footprint of the different socio-economic sectors as a technical criterion for the development of the River Basin Management Plans, that all EU Member States will have to accomplish by 2009 (and every six years thereafter) as part of the requirements of the Water Framework Directive (Official State Gazette, 2008).

The Water Framework Directive sets the clear objective of achieving the 'good ecological status' of all water bodies in the EU (surface as well as groundwater) by 2015 and the strong recommendation of full cost recovery for water services including environmental and resource costs. This, theoretically, is going to have a direct effect on irrigation agriculture and agricultural systems. According to Garrido & Varela-Ortega (2008), the implementation of the WFD might result in a reduction of irrigated area and, thus, blue water consumption, and a better use of soil and water resources, with important impacts on land planning and management. These expected results may vary considerably across regions and irrigation systems.

Along with the WFD, the Common Agricultural Policy (CAP), has had over the years a clear impact on irrigation agriculture, on cropping patterns and hence on water use. Along the decades of the 80's and 90's the CAP programs encouraged irrigation expansion and intensification as larger production-coupled subsidies were granted to the farmers for their intensively irrigated crops. This coupled aid scheme induced water consumption most acutely in arid and semi-arid regions

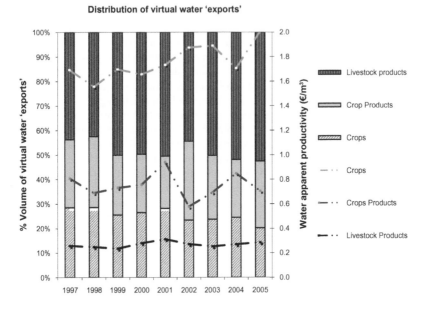

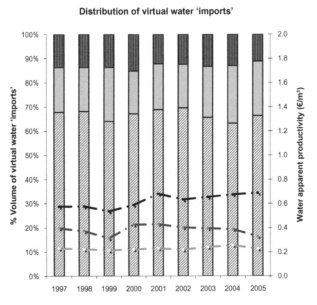

Figure 3.   Volume of virtual water 'export' and 'imports' (%) and water apparent productivity (€/m³) for the years 1997–2005.

across the EU, predominantly along the Mediterranean coastline and its hinterland (Varela-Ortega, 2008; Garrido & Varela-Ortega, 2008). This situation produced clear socio-economic benefits to the rural population but, on the other hand, it engendered negative consequences to aquatic ecosystems (Baldok *et al.*, 2000; Martinez-Santos *et al.*, 2008; Varela-Ortega *et al.*, 2008). Responding to the WTO agreements, the CAP evolved from the Mc Sharry reform of 1992, to the reforms of Agenda 2000 and lastly to the Luxemburg reform of 2003. These two last reforms, have included progressively environmental and nature protection regulations with the aim of achieving a more compatible agricultural production with the protection of ecosystems. The 2003 reform, in force since 2005, makes a step further by establishing a system of subsidies or direct payments, that are decoupled from production (to cereals, oilseeds, protein crops and olives) and substituted by a single farm payment (that also includes livestock support aids) (Spain is still under a 25% coupled scheme except for land set-aside that is fully decoupled). These payments are currently tied to the requirement by all farmers in the Member States to comply with specific environmental regulation as well as specific nature-protection farming and tillage operations under a 'cross compliance' scheme (EC, 2003). Lastly, the CAP has made a step further in the newly proposed reform 'the CAP Health Check' that strengthens environmental requirements and specifically includes water management, climate change and biofuels as the main challenges to be addressed by all member states (EC, 2008). Therefore, this new policy context implies the need for achieving a well-balanced and sustainable integration of agricultural, environmental and water sectors (Varela-Ortega, 2008).

Following the 2003 reform, irrigated acreage of intensive water—consuming crops such as maize and legumes is being reduced in favour of winter cereals and oilseeds of lower water requirements and to olive (under CAP subsidies) and vine (with no subsidies) that are well adapted to specific region-based farming conditions and water-saving modern irrigation technologies as well as market opportunities (Garrido & Varela-Ortega, 2008; Varela-Ortega, 2008). Alongside, other water intensive crops such as tobacco and sugar-beet have also diminished in surface due to their subsequent reduction in their price support schemes and market reforms. Vineyard and olive tree irrigated production is increasing significantly (using more than 800,000 irrigated hectares in 2006) (MAPA, 2008). It is expected that significant changes in crop distribution will continue to occur in the near future. These significant and gradual changes of cropping patterns in irrigated acreage result from several factors, including CAP production-decoupled payments, an increased obligation to comply with environmental requirements, investment in irrigation and water transportation technologies and more market-driven farming decisions.

Significant changes in water demand can occur not only by changing the amount of irrigated area but also by modifying the cropping patterns. In the EU, cropping patterns have been profoundly influenced by farm and trade policies (Varela-Ortega, 2008). This is in line with recent water footprint studies, which recognize that water resources management is intimately linked to the structure of the global economy (Hoekstra & Chapagain, 2008). Currently, due to more decoupled modes of farm income support, EU farmers are responding more to market signals. Most of these originate in the global markets, offering a broader opportunity to enhance the connections and synergies between food markets and farm trade and water policies.

One of the most relevant consequences of the water footprint and virtual water 'trade' knowledge in arid and semiarid countries such as Spain is the change in the water security and food security concepts, paradigms that have hitherto prevailed in the minds of most policy makers. This comes from a food self-sufficiency tradition that will probably change in the near future. Previous works support that water crisis is a problem of water management in relation to various aspects, such as obsolete irrigation systems or excessive blue water use for growing low economic value crops (Llamas, 2005; UNDP, 2006; Comprehensive Assessment of Water Management in Agriculture, 2007). Along these lines, the water footprint analysis is providing new data and perspectives that are enabling to form a more optimistic outlook of the frequently spread looming 'water scarcity crisis'.

7 CONCLUDING REMARKS

Water scarcity in Spain is mainly due to the inefficient allocation of water resources and mismanagement in the agricultural sector, such as the use of large amounts of blue water in virtual water intensive but low economic value crops. Nevertheless, the Spanish water footprint should be analysed in its time and spatial dimension as well as considering the sectorial and geographical standpoints. Furthermore, we cannot forget about the multifunctionality of agriculture.

On the whole, there seems to be enough water to satisfy the Spanish agricultural sector needs, but a necessary condition is to achieve an efficient allocation and management of water resources. This will take some time since crop distribution in Spain is determined by several factors such as the Common Agricultural Policy or the WTO regulations. The mentioned transition will require the action of the Spanish Government by embracing transparency and encouraging an active and effective public participation. This is already happening in Spain in lieu of the application of the WFD.

The water footprint analysis, hydrological, economic and ecological, at a river basin level provides most valuable information to facilitate an efficient allocation of water resources to the different economic and environmental demands. There is no blueprint. The Spanish context is characterized by regional differences on green and blue water resource availability. Along these lines, virtual water studies, taking into account not only green and blue (ground and surface) water systems but also trade policies, can contribute to a better integrated management of water resources.

Finally, this analysis, performed in industrialized countries such as Spain, can help to move forward from a policy of 'more crops and jobs per drop' towards 'more cash and nature per drop'. Achieving this new paradigm would mean a win-win solution to the conflict between farmers and conservationists, allowing the preservation of the environment without damaging the economy of the agricultural sector.

REFERENCES

Aldaya, M.M., Hoekstra, A.Y. & Allan, J.A. (2008). Strategic importance of green water in international crop trade. *Value of Water Research Report Series* No. 25, UNESCO- IHE Delft, The Netherlands.
Aldaya, M.M. & Llamas, M.R. (2009). Water Footprint analysis (hydrologic and economic) of the Guadiana river basin within the NeWater project. *Third Edition of the United Nations World Water Development Report (WWDR-3)*.
Allan, J.A. (1997). 'Virtual water': a long term solution for water short Middle Eastern economies?' *Water Issues Group, School of Oriental and African Studies*. University of London. London. Available from: www.soas.ac.uk/research/our_research/projects/waterissues/papers/38347.pdf [Accessed 15 March 2009].
Allan, J.A. (2006). Virtual Water, Part of an invisible synergy that ameliorates water scarcity. In *Water Crisis: Myth or Reality?* (Rogers, Llamas y Martinez, edts.) Balkema Publishers, pp. 131–150.
Baldock, D., Dwyer, J., Sumpsi, J., Varela-Ortega, C., Caraveli, H., Einschütz, S. & Petersen, J.-E. (2000). *The Environmental Impacts of Irrigation in the European Union*. Report to the European Commission. European Commission., Brussels.
Comprehensive Assessment of Water Management in Agriculture (2007). *Water for Food, Water for Life: A Comprehensive Assessment of Water Management in Agriculture*. London: Earthscan, and Colombo: International Water Management Institute.
Chapagain, A.K. & Hoekstra, A.Y. (2004). Water Footprints of Nations. Value of Water Research Report Series No. 16, UNESCO- IHE. Delft, The Netherlands.
EC (European Commission) (2003). *Council Regulation (EC) No 1782/2003 of 29 September 2003 establishing common rules for direct support schemes under the common agricultural policy and establishing certain support schemes for farmers*. Office for Official Publications of the European Union, Luxemburg.
EC (European Commission) (2008). *Health Check of the CAP (Guide)*. Commission of the European Communities, COM, 2008. 306 Final. Brussels.
Falkenmark, M. & Rockström, J. (2004). *Balancing water for humans and nature: The new approach in ecohydrology*. Earthscan, London, UK.
Garrido, A. & Llamas, M.R. (2009). Water Management in Spain: An Example of Changing Paradigms. In Dinar, A. & Albiac, J. (Eds.) *Policy and Strategic Behaviour in Water Resource Management*. Earthscan, London, pp. 125–146.
Garrido, A. & Varela-Ortega, C. (2008). 'Economía del agua en la agricultura e integración de políticas sectoriales' Panel de Estudios. Universidad de Sevilla—Ministerio de Medio Ambiente.
Garrido, A., Llamas, M.R., Varela, C., Novo, P., Rodríguez Casado, R. & Aldaya,. M.M. (2008). *Water footprint and virtual water trade: policy implications*. Observatorio del Agua. Fundación Marcelino Botín. Santander.
Hernández-Mora, N., Llamas, M.R. & Martínez, L. (2001). Misconceptions in Aquifer Over-Exploitation. Implications for Water Policy in Southern Europe. In: *Agricultural Use of Groundwater. Towards Integration*

between *Agricultural Policy and Water Resources Management*. Ed. C. Dosi. Kluwer Academic Publishers. Dordrecht, The Netherlands, pp. 107–125.

Hernández-Mora, N., Martinez-Cortina, L., Llamas, M.R. & Custodio, E. (2007). Groundwater Issues in Southern EU Member States. Spain Country Report. [online] Available from: http://rac.es/2/2_ficha.asp?id=119&idN3=6&idN4=40 [Accessed 15 March 2009].

Hoekstra, A.Y. & Chapagain, A.K. (2008). *Globalization of water: Sharing the planet's freshwater resources*. Blackwell Publishing. Oxford, UK.

INE (2008). National Statistics Institute. [online] Available from: http://www.ine.es/ [Accessed 15 March 2009].

Llamas, M.R. (2005). Los colores del agua, el agua virtual y los conflictos hídricos Discurso inaugural del año 2005–06 [Water colours, virtual water and water conflicts. Opening address of the 2005–06 Course]. *Revista de la Real Academia de Ciencias Exactas, Físicas y Naturales*. Madrid. Vol. 99, No. 2, pp. 369–389.

Llamas, M.R. & Martínez-Santos, P. (2005). Intensive Groundwater Use: Silent Revolution and Potential Source of Social Conflicts. *Journal of Water Resources Planning and Management, American Society of Civil Engineers*, September–October 2005, pp. 337–341.

MAPA (2008). Anuario de Estadistica Agroalimentaria, Ministerio de Agricultura, Pesca y Alimentacion [online] Available from: http://www.mapa.es/es/estadistica/pags/anuario/introduccion.htm [Accessed 15 March 2009].

Martinez-Santos, P., Llamas, M.R. & Martinez-Alfaro, P.E. (2008). Vulnerability assessment of groundwater resources: A modelling-based approach to the Mancha Occidental aquifer, Spain. *Environmental Modelling and Software* 23: 1145–1162. DOI 10.1016/j.envsoft.2007.12.003.

MIMAM (*Ministerio de Medio Ambiente*, Spanish Ministry of Environment) (2000). *Libro Blanco del Agua en España* [Spanish Water White Paper]. Madrid, Spain.

MIMAM (*Ministerio de Medio Ambiente*, Spanish Ministry of Environment) (2007). El agua en la economía española: Situación y perspectivas. [Water in the Spanish Economy: current affairs and perspectives] Spanish Ministry of the Environment [online] Available from: http://circa.europa.eu/Public/irc/env/wfd/library?l=/framework_directive/implementation_documents_1/wfd_reports/member_states/spain/article_5/completo_nivel1pdf/_EN_1.0_&a=d [Accessed 15 March 2009].

Novo, P., Garrido, A. & Varela-Ortega, C. (2009). "Are virtual water "flows" in Spanish grain trade consistent with relative water scarcity?" *Ecological Economics*, in press.

Official State Gazette (2008). Approval of the water planning instruction. Ministry of the Environment and Rural and Marine Affairs. Official State Gazette 229. 22nd September 2008. [online] Available from: http://www.boe.es/boe/dias/2008/09/22/pdfs/A38472-38582.pdf [Accessed 15 March 2009].

Plan Nacional de Regadios (2009). Ministerio de Medio Ambiente y Medio Rural y Marino, [online] Available from: http://www.mapa.es/es/desarrollo/pags/pnr/principal.htm [Accessed 15 March 2009].

Rodríguez Casado, R., Garrido, A. & Varela-Ortega, C. (2009). La huella hidrológica de la agricultura española. [Water footprint of Spanish Agriculture] *Ingeniería del Agua*, in press.

Shah, T. (2008). *Taming the Anarchy: Groundwater Governance in South Asia*. Resources for the Future, Washington, D.C.

UNDP (2006). Human Development Report 2006: Beyond scarcity: Power, poverty and the global water crisis. United Nations Development Programme. [online] Available from: http://78.136.31.142/en/reports/global/hdr2006/ [Accessed 15 March 2009].

Varela-Ortega, C. (2008). "The Water Policies in Spain: Balancing water for food and water for nature". Paper presented at the 6th Rosenberg International Forum on Water Policy, Zaragoza, Spain. In: Ingram, H. and Garrido, A. (eds) *Water for Food: Quantity and Quality in a Changing World*. Rosenberg International Forum on Water Policy. Routledge Publisher, Taylor and Francis Group; Abingdon, UK (in press).

Varela-Ortega, C., Swartz, C., Blanco, I. & Downing, T. (2008). Water policies and agricultural policies: An integration challenge for agricultural development and nature conservation. *Proceedings of the 13th IWRA World Water Congress. Global Changes and Water Resources: confronting the expanding and diversifying pressures*. Montpellier, France.

Yang, H., Wang, L., Abbaspour, K.C. & Zehnder, A.J.B. (2006). Virtual water trade: an assessment of water use efficiency in the international food trade. *Hydrology and Earth System Sciences* 10: 443–454.

Yang, H. & Zehnder, A.J.B. (2008). Globalization of Water Resources through Virtual Water Trade. *Proceedings of the Sixth Biennial Rosenberg International Forum on Water Policy*, Zaragoza, Spain.

III
Constraints and social perceptions

CHAPTER 7

Drought and climate risks[1]

Ana Iglesias
Department of Agricultural Economics and Social Sciences, Universidad Politécnica de Madrid, Spain

Marta Moneo
Potsdam Institute for Climate Impact Research, Potsdam, Germany

Luis Garrote
Department of Hydraulics, Universidad Politécnica de Madrid, Spain

Francisco Flores
Ministry of the Environment and Rural and Marine Affairs, Madrid, Spain

1 CHALLENGES TO WATER MANAGEMENT

This chapter explores the challenges of drought and climate risks to water management in Spain. In this section a summary of data is presented to describe some of the main pressures. In the following sections the chapter evaluates the risk of drought and the options for drought management, describing risk management protocols and policy options, and the additional pressure of climate change. Finally, in the conclusion section the chapter draws on the results presented in the previous sections to suggest some management implications.

Throughout the last century water management has proved unsustainable due to a combination of factors: (1) Population: Over the 20th century, population has tripled while water withdrawals have increased by a factor of about seven. (2) Pollution: The effects of industry and agriculture intensification have resulted in major pollution problems in many regions of the world; this is linked (together with scarcity) to degradation of aquatic ecosystems. (3) Governance: Poor governance as result of fragmented and uncoordinated management, top-down institutions and increased competition for the finite resource. (4) Climate variability and change: The impacts of drought and climate change in freshwater resources affect all sectors of society. Thinking ahead, water management needs to incorporate the principles of sustainable development and the changing values that society place into the use of the resource.

Water resources in Spain are limited, scarce, and difficult to predict from year to year. The average annual potential water availability per capita considering the total freshwater resources is 2,700 m^3 compared to 3,807 m^3 in the EU-15 and 7,000 m^3 worldwide (Iglesias *et al.*, 2007), but some Spanish regions have less than 1000 m^3 per capita and year, such as the Southeast regions and the Archipelagos (Iglesias *et al.*, 2009a). In addition, real available water resources in Spain are less than half of the total freshwater resources (Table 1). The country is diverse from the point of view of climate and water infrastructure (Table 1.). This results in important differences in the criteria needed to characterize drought across the territory. For example, in the basins where demand is above the available resources, drought usually results in crisis.

Regulated water resources account for 40% of the total natural resources, compared with 8% worldwide. The potential use of surface water under natural regime in Spain is only 7% (Garrote *et al.*, 1999). Groundwater use is intensive in many areas of the country contributing to

---

[1] Funding was provided by the EU MEDA Water Project MEDROPLAN. We acknowledge the support of the Tagus River Basin Authority.

Table 1.   Total freshwater resources, available resources, demands, and water reliability in the hydrological basins of Spain (see Figure 1 in Chapter 2 for locations).

	Total freshwater resources (km³)	Available resources (km³)(a)	Reservoir capacity (km³)	Regulated water (%)(b)	Demand (% of available resources)	Irrigation demand (% of total demand)	Population (millions)	Total resources per capita (m³/hab)
North (1)	44.2	6.8	4.4	15	37	42	6.7	6,542
Douro (2)	13.7	8.1	7.7	60	47	93	2.2	6,071
Tagus (3)	10.9	7.1	11.1	65	57	46	6.1	1,784
Guadiana (4)	5.5	3.0	9.6	54	85	90	1.7	3,298
Guadalquivir (5)	8.6	3.6	8.9	42	104	84	4.9	1,755
South (6)	2.4	0.54	1.3	21	268	79	2.1	1,135
Segura (7)	0.8	0.7	1.2	90	253	89	1.4	590
Júcar (8)	3.4	2.0	3.3	58	149	77	4.2	819
Ebro (9)	18.0	13.0	7.7	72	80	61	2.8	6,509
Inner Basins of Catalonia (10)	2.8	1.1	0.8	40	122	27	6.2	451
Baleares (11)	0.7	0.3		45	96	66	0.8	785
Canary Is. (12)	0.4	0.4		102	102	62	1.7	241
SPAIN	111.2	46.6	56.1	42	76	68	40.1	2,728

(a) Surface and groundwater. Overall groundwater contribution is under 20 percent of total.
(b) Regulated water: rate of available resources from total natural resources.
Source: Own elaboration with data from Vallarino and Garrote (2000).

an additional 10% of the total available resources. With limited and scarce water resources and demand rising due to demographic shifts, economic development and lifestyle changes, water management problems are significant even without drought events, due to the imbalance between availability and demand.

In Spain, groundwater resources play a vital role in meeting water demands, not only as regards quality and quantity, but also in space and time, and are of vital importance for alleviating the effects of drought (Garrido *et al.*, 2000; Llamas, 2000). However, groundwater pumping should be controlled because excessive use of the aquifers can cause overexploitation problems with the consequent negative environmental, social and economic impact. Direct use of groundwater in Spain is currently estimated at 5 km³/year, mainly for irrigation use (80%), but the water quality is easily deteriorated due to point-source pollution or diffuse pollution caused by agricultural and livestock activities (see Chapter 14; Estrela *et al.*, 1995).

Wetland area in Spain has decreased from over 1200 km² in the 1970s to less than 800 km² (see Chapter 3) in the present time (excluding the Guadalquivir marshlands). This decrease maybe in part related to recurrent drought episodes and surface water scarcity, and amplified by the excessive groundwater pumping to compensate for these problems.

2   DROUGHTS' CHARACTERISATION

Drought is a normal, recurrent feature of climate, consequence of a precipitation reduction over an extended period of time. Drought occurs in most climatic regimes, and is often described as a natural hazard. It is a temporary anomaly, unlike aridity, which is a permanent feature of the climate. Defining drought is therefore difficult; it depends on differences in regions, needs, and disciplinary perspectives (Wilhite, 2005; Iglesias *et al.*, 2009a; Iglesias *et al.*, 2009b). Meteorological drought is caused by a deficit in precipitation and hydrological drought is caused by the decrease or deficiency in ground water and reservoir levels when the meteorological drought is very intense or persistent. Whatever the definition, it is clear that drought cannot be viewed solely as a physical phenomenon, as its severity depends on the impact on people or ecosystems and their ability to cope and recover. Although drought may cause water scarcity—the extent to which demand exceeds available resources—human actions such as population growth or water mismanagement may also be the cause.

Droughts differ from other natural hazards in several important ways: no universal definition exists; its spatial extent is usually very large; slow-development that makes it difficult to determine the onset and end of the event; duration may range from months to years. Drought impacts are generally non-structural and also difficult to quantify (UNISDR, 2002). This is especially significant in regions where economic resources and technology can buffer the effect of negative environmental changes (insurance, public support). Therefore, characterization of drought episodes is complex, and includes both physical aspects and social consequences.

No single indicator or index can identify drought. Many efforts have been made to characterize drought by using a range of indices (Rossi *et al.*, 2003; Wilhite *et al.*, 2000; Vogt & Somma, 2000). Classical drought indices, such as the Standardized Precipitation Index (SPI) (Hayes *et al.*, 1999) or the Palmer Drought Index (Palmer, 1965), are widely used to characterize meteorological drought. These indices do not correlate well with hydrological drought periods or historical drought impacts, due to the effect of storage (Garrote *et al.*, 2003). Many of the more complex indices that take storage and management into account are not easily interpreted across the regions and cannot be validated with the data available over wide geographical areas. Therefore, managers of water resources tend to rely on precipitation and streamflow variables to determine the onset of alert, alarm, and emergency situations (see below: Current strategies for drought management).

Drought can have serious effects on the economy and the environment of Spain, on the population's well being. For example, the major drought of the mid 1990s affected over 6 million people, almost ten times more than the number of people affected by floods in Spain during the last fifty years. The economic damage caused by droughts in Spain during the last twenty years is about five times more than in the entire United States (CRED, 2007). The most recent drought period of 2004–2005 contributed to social unrest and triggered disputes over projected water infrastructure. Drought events affect water supplies for irrigation, urban, and industrial use, ecosystem's health, and give rise to conflicts among users that limit coherent integrated water resource management.

Drought events in Spain have been frequent since 1970 (Iglesias *et al.*, 2007) and climate experts begin to determine the realities of future water availability in the region (IPCC, 2007). The implementation of the new European Water Framework Directive (WFD) gives Spain the opportunity to develop integrated drought management plans that incorporate the extensive national experience in hydrological management with the new environmental challenges (see Chapter 16 for a detailed description of the WFD).

The time series of precipitation in many locations of Spain and the SPI calculated at different time intervals are extremely variable and correlated (correlation coefficient = 0.75 in the 1900 to 2004 time series). The SPI can establish thresholds for drought episodes that trigger management actions (Iglesias *et al.*, 2009a; De Luis *et al.*, 2000; Estrela *et al.*, 2000). Historical data show at least two periods with different precipitation trends, highlighting the importance of choosing the adequate reference period for developing indicators for management (see Figure 1). Precipitation in the latest period, from the 1960s has clearly decreased. The very low precipitation of the 1940s defined the historical drought during that period, with severe consequences for the economy. The structural water deficit of many areas in the country has been aggravated during three severe drought episodes (1975–76, 1981–82, and 1992–95), each more severe than the previous one (Figure 1). During these droughts, besides the collapse of irrigation water supply, urban water supply series were affected significantly.

Drought characterization in highly regulated systems is complex and calls for multiple indicators. The slope of accumulated deviation of precipitation from the mean (Figure 1) is an indicator that highlights drought patterns relevant to hydrological management. The Figure highlights the recurrent multi-year characteristic of drought periods in Spain and the value of continuous monitoring for preparedness. Also, this indicator is related to other variables that are more difficult to monitor, such as groundwater levels.

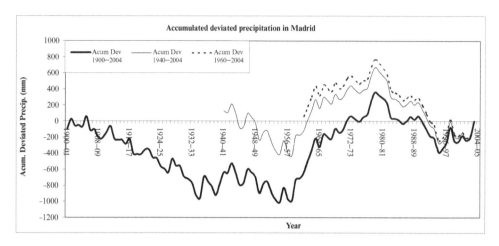

Figure 1.   Accumulated deviated precipitation in Madrid. (Data source: Tajo River Basin Authority).

3   LINKING DROUGHT INDICATORS TO WATER MANAGEMENT

Water managers are ultimately concerned by drought if it affects water supply. A robust indicator of hydrological drought is reservoir storage. Figure 2 shows the time series of inflow in Bolarque a reservoir near Madrid with a contributing basin of 7,420 km^2. The behavior of this series appears to be non-stationary, or, at least, highly variable with a period between wet and dry spells clearly beyond the possibilities of regulation for water supply (Garrote *et al.*, 2003). Data show a possible intensification of drought conditions in recent years, during the decades of 1980s and 1990s, also shown in Figure 1. If a linear trend is fitted to the data, the slope is clearly negative, with a decrease of more than 8 Mm^3/yr every year.

However, since droughts were also important during the 1940s and 1950s, the question arises as to whether recent droughts are a consequence of man-induced climate change or there is a multi-annual cycle of wet and dry conditions with a period of about 40 years over the time period analyzed. Classical drought indices, such as the SPI, have limitations for analyzing regulated systems. Although the correlation between streamflow and SPI over the time series is 0.54, the SPI threshold of −1 does not capture severe water shortages such as the one of the mid 1990s that led to emergency actions (see below: Supporting legislation).

The variability of streamflow at Bolarque reservoir in the high course of the Tagus Basin was not completely understood when a water-transfer facility was planned. The Inter-basin Tagus-Segura aqueduct diverts water from the Tagus river at Bolarque and transports it to the Segura, and Guadiana basins (see Chapter 19 for a more detailed description of this important infrastructure). The aqueduct was designed during the wet period using data that, at that time, seemed to be reliable, but the subsequent evolution of stream flow showed a significant reduction of mean flows and forced a change in the planned exploitation of the infrastructure. The Bolarque example also represents the complexity of the indicators that need to be included in a drought characterization system.

In some cases, the indicators do not reflect the real impact of drought. Figure 3 shows results from the simulation model used in the Tagus Hydrological Plan for the Henares basin (sub-basin of the Tagus Basin, east of Madrid), compared to SPI index in the basin. Shortages of water supplied to the demands of the Henares basin are poorly correlated with SPI due to the effect of reservoir storage. The fact that the system has enough regulation capacity to supply the demand with high reliability and relatively scarce failures is probably one of the reasons for this low correlation. When the demand exceeds the available water resources, as in Southern Spain (Table 1), drought episodes may result in failures in demand supplied (del Moral & Giansante, 2000).

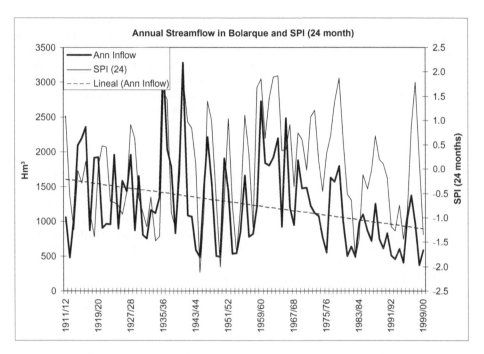

Figure 2.   Time series of inflow in Bolarque and SPI calculated for a 24 month time sale in Madrid (Data source: Tajo River Basin Authority).

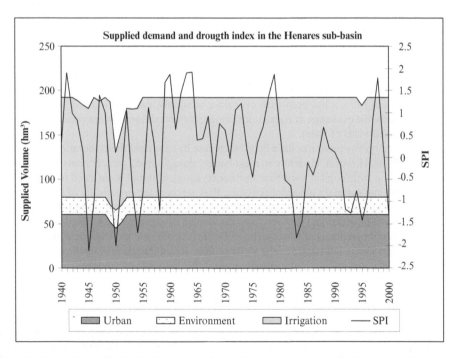

Figure 3.   Demand supplied in the Henares sub-basin (Tajo Basin) and historical droughts calculated with the Standardized Precipitation Index (SPI).

Source: Own elaboration with data from the Tajo River Basin Authority.

Storage and regulation by reservoirs do not always solve the problem of water scarcity in areas where dry periods are particularly damaging to the natural and human wellbeing. Eutrophication is a major problem in southern areas of Spain, where 40% of the reservoirs show biological oxygen demand, conductivity, and nitrogen and phosphorus concentrations well outside the adequate range (Estrela *et al.*, 1995). These water quality parameters usually get worse during dry periods due to the depletion of reservoir storage. This factor may play a significant role during crises since water from certain reservoirs may not be acceptable for human consumption.

4   MAJOR DROUGHTS IN THE LAST TWENTY YEARS

Since the early 1990s drought has affected more people in Spain than any other natural hazard. From the point of view social damage, the drought of 1991–1996 affected over 10 million people in Spain, causing more than 10 billion US$ (1995 value) of economic damage (CRED, 2007). During this drought outreach of the impacts was identified too late, amplifying the damages since the systems were not prepared to cope with the event. Southern Spain was most affected and water managers had to adopt unprecedented emergency measures, such as the urban water restrictions in the metropolitan areas of Seville (1,300,000 inhabitants), Granada, Jaén, Málaga and Toledo, in spite of the increment of the groundwater abstractions (up to 100% in Granada). The impacts in the agricultural sector were also very serious from the point of view of crop production, resulting in almost complete absence of irrigation water for cereal crops in half of Spain. Nevertheless, due to the extensive subsidy system provided by the European Union and the Spanish National Agricultural insurance system, farms' income was not affected to the same extent (Llamas, 2000). The most recent drought of 2004–2005, although similar in magnitude as characterised by drought indicators, has been less dramatic for the general population. This proves the value of the effort in communicating and managing risk over the last decade.

An example of effective drought planning during these two major episodes is the set of complex emergency actions undertaken in the Madrid metropolitan area (over 5 million people). The actions include: (1) Legal change of water rights and priorities. (2) Construction of an emergency work to link the reservoir to the urban distribution system in Madrid in 1994. (3) Exchange for the loss of water rights, providing additional sources of water to irrigators of an economic compensation. (4) Construction of water treatment plants to ensure lower water volume was of drinking quality. It is interesting to note that, although some infrastructure was built under pressure due to the emergency conditions in 1994, it has been used ever since to supply water for Madrid. In contrast, the case illustrates how the exchange of rights under emergency conditions always results in detriment of the ecological services of water.

In the past, emergency measures have been taken case by case until the elaboration of the Basin Hydrological Plans. However, currently in most basins, drought management plans are developed and are focused on hydrological variables. Further revisions of initial plans need to address both hydrological risk and societal vulnerability issues. Risk refers to the probability of drought occurrence and vulnerability refers to the characteristics of a system or social group in terms of its capacity to cope with, resist and recover from the impact of drought. Risk-based approaches to preparing for drought are focused on acquiring accurate probabilistic information about the events themselves. When this is not possible, the strategy fails. In contrast, understanding and reducing vulnerability does not demand accurate predictions of the incidence of drought.

5   CURRENT STRATEGIES FOR DROUGHT MANAGEMENT

In most cases analysed in Spain and Mediterranean countries, drought management strategies are grouped according to three levels of severity: Pre-alarm, alarm, and emergency (Iglesias *et al.*, 2009a; Garrote *et al.*, 2003; Iglesias *et al.*, 2009b). These severity levels are determined by established thresholds of indicators, and trigger different groups of measures. There are many examples of actions taken in response to drought over the past decades in Spain that fit into the general

Table 2.   Drought management at the basin level: Indicators and measures for different levels of drought intensity.

	Pre-Alarm	Alarm	Emergency
Drought monitoring and trigger indicators	Monitoring shows the initial stage of danger	Monitoring shows that drought is occurring and will have impacts if measures are not taken	Monitoring shows that impacts have occurred and supply is not guaranteed if drought persists
Objective	To ensure acceptance of measures to be taken in case of alarm or emergency by raising awareness of the danger of drought	To overcome the drought situation and to guarantee water supply while emergency measures can be put in place	To minimize damage, the priority is drinking water
Groups of operational actions	1. Low cost, indirect, voluntary 2. Non structural directed to influence water demand and avoid worse situations 3. Focus on communication and awareness 4. Intensification of monitoring and evaluation of worse case scenarios	1. Low cost, direct, coercive, direct impact on consumption costs 2. Non structural directed to specific water use groups 3. Water restrictions for uses that do not affect drinking water 4. Changes in management 5. Revision of tariffs 6. Rights Exchanging Centres	1. High cost, direct, restrictive, approved as general interest actions 2. Structural, new infrastructure, intra-basin, inter-basin and transboundary transfers 3. Non structural, such as permission for new groundwater abstraction points 4. Water restrictions for all users, including urban demand

Source: Own elaboration based on Iglesias *et al.*, 2009a; Garrote *et al.*, 2006.

framework summarised in Table 2. The operational management of urban supply companies, such as the CYII and most river basin authorities support the model presented in Table 2 (see Iglesias *et al.*, 2009a; Iglesias *et al.*, 2009b, for further details). In all cases, the main priority during the emergency level is to maintain drinking water supply and all structural and non-structural measures of high economic, social, or environmental cost are designed and taken in order to avoid water restrictions for urban demand.

The monitoring system includes hydrological (precipitation, reservoir level, runoff, water table in ground water, snowpack, etc.), socio-economic (macroeconomic and demography), and environmental (water quality and biodiversity) indicators. The indicators are evaluated in view of the demand and supply analysis and projections, scenario analysis for drought levels, and contingency and preparedness plans. In the case of River Basin Authorities, the basin plan is the instrument to implement the operational actions.

Specific drought management plans have been developed at different administrative levels. An example of detailed and operational plan is provided by the urban water supplier CYII that manages water for the metropolitan area of Madrid (Cubillo & Ibañez, 2003). The plan is based on the extensive understanding of the demands of the system and the linkages between the evolution of the stored volume and the level of guarantee established for water supply. The supply company establishes each year threshold levels that reflect the specific supply and demand characteristics of the system. In contrast to private or semi-private companies, the drought planning process is exposed to public scrutiny. Basin Authorities take decisions in collective bodies, although the executive responsibility is held by the Chairman of the Basin's Authority (see Chapter 12). Reservoir Release Commissions are responsible for the continuous management of reservoirs.

Under severe scarcity conditions, a Permanent Committee is appointed to manage the situation. In the basin case, concerned individuals and social or political groups can make allegations that affect the planners' decisions.

6   SUPPORTING LEGISLATION

The United Nations Convention to Combat Desertification (UNCCD, 1994), provides the framework for implementing drought mitigation strategies. The European Parliament and Council Water Framework Directive 2000/60/EC of December 20 (WFD) contains a series of principles that will affect drought policies in all EU Member States and the demarcation and description of basins' territories (Articles 3 and 5). In Spain, water management is established at the basin level since the late 1900s and the instrumental drought legislation reflects this institutional organization.

In general, decisions related to drought are taken in the context of formal legal system. The 2001 Consolidated Water Act is a modern and comprehensive water code that provides Basin Authorities the capacity and obligation to establish drought plans. All Spanish Hydrological Basin Plans were approved by the Royal Decree 1664/1998 (Real Decreto 1664/1998, of July 24th). There are legal provisions for emergency actions in case of crisis situations, such as extreme drought. Informal customs may evolve into formal decisions, for example, historical users of groundwater without formal rights may be legalized. The legislation does not provide explicit regulations about how to calculate the ecological discharge during in drought situations, this important question is being left to the discretion and responsibility of the various River Authorities.

The National Hydrological Plan Act 2001 lays down the basic principles of water planning at the national level enclosing all Water Basin Plans. The Law establishes the indicators of drought alert, and enforces the Basin Authorities to develop special drought management plans, including the emergency plans in the urban areas of the basin, that have to be finalized before the WFD is fully implemented. The indicator system includes the following categories: precipitation, streamflow, reservoir inflow, reservoir stock, piezometric level, and reservoir outflow. The Jucar River

Table 3.   Summary of the major categories of legislation related to drought in Spain.

Category	Date	Contents
Emergency measures. Most of them undertaken after the most severe drought periods as mitigation measures.	1995–2000	Acts, Royal decrees and orders created to mitigate the impacts of drought. Emergency supply measures Transfers of water between different river basins Measures for sub sectors of agriculture (apiculture, livestock, tree crops)
Scope definition. Definition of the areas where the emergency measures are applied.	1993, 2000, 2001	Definition of the criteria used to delimit areas affected by drought. Establishment of criteria for aid supply Final criteria used: Amount of rainfall and Stocking rate
Transboundary	1998	Albufeira Convention between Spain and Portugal for transboundary basins under the framework of sustainable water resources management and common environmental protection.
Insurance	2001, 2002	Definition of the conditions, application areas and other characteristics of drought insurance.

Source: Own elaboration.

Basin (Table 1) is the pilot basin for defining specific indicators for each category, but the final plan has not been formulated.

Table 3 summarizes the major categories of legislation related to drought. The legislation has evolved as consequence of severe historical drought episodes, such as 1993–95, and 2001.

A main advantage of the explicit linkage of legislation and management to the basin level is the opportunity to address directly the needs and problems of the natural hydrological system and the stakeholders represented in the Assembly of Users. For example, Basin Authorities can establish priority of users or right holders according to each situation, can approve works and projects needed to solve emergent scarcity problems, and can create Water Exchanging Centres, through which right holders can offer or demand use rights in periods of droughts or severe water scarcity situations (Article 71). This initiative must be proposed by the Environment Ministry and be approved by the Ministerial Cabinet.

7   CLIMATE CHANGE

Climate change is already happening. The Fourth Assessment Report of the Intergovernmental Panel on Climate Change (IPCC, 2007) clearly shows that the climatic variations over recent decades have had noticeable direct consequences in natural ecosystems, glaciers and agricultural systems in many regions. Many areas of the world are already struggling today with the adverse impacts of an increase in global average temperature. The scientific literature also suggests that observed changes in climate have affected the frequency and intensity of extremes (drought, floods, and heat waves). The alarming number of extreme weather events that have occurred during the last five years may be the consequence of climate change and suggest that climate change is resulting in the increase in natural climate disasters, at least in some regions. The IPCC defines climate change as a statistically significant variation in the state variables that define the climate of a region (such as temperature or precipitation) or in its variability persistent over an extended period of time (typically decades or longer periods). Although scientific projections of future climate evolution are highly uncertain and subject to numerous hypotheses, there is a growing concern among the scientific community about the impacts of climate change on drought magnitude and frequency. Global increases in temperature and changes in the hydrological cycle are expected to have a major impact on drought in the Mediterranean region (IPCC, 2007). Future risk assessments require the development of environmental change scenarios at different time-scales. These scenarios should include estimates not only of changes in the climatic baseline, but also estimates of possible future changes in socio-economics.

In Spain climate change projections indicate a decrease of precipitation in the southern regions, in some cases up to –40%, by the 2050s compared to 1961–1990 levels, or a small increase in precipitation in the northern regions, with changes in the annual precipitation patterns. In all cases, temperature increases of about 1.5°C are expected, and thereby increased evaporation and reduced soil moisture, resulting in more adverse regional climate conditions than presently experienced. Climate and hydrological experts begin to be aware of the implication of future water availability in the region (Iglesias, 2002; Iglesias *et al.*, 2002; Hisdal *et al.*, 2001; Lloyd-Hughes & Saunders, 2002).

The effects of climate changes on major water management determinants and expected social and ecological consequences are summarised in Table 4. Most studies agree that climate change will likely have the following common consequences in across the Mediterranean areas of Europe (EEA, 2007): (1) Increase demand for agricultural water in all regions due to expected increases in crop evapotranspiration in response to increased temperatures in all regions; (2) Increased water shortages, particularly in the spring and summer months, therefore increasing the water requirement for irrigation; and (3) Increased water quality deterioration due to higher water temperatures and lower levels of runoff in some regions, particularly in summer, imposing further stress in agricultural irrigated areas.

Under climate change, reservoir water inflow and water resources availability decreases –7 and –5% in average in all Spanish basins considering a range of climate change scenarios (Garrote *et al.*, 1999; Figure 4). In contrast, irrigation demand increases in all locations under the Hadley

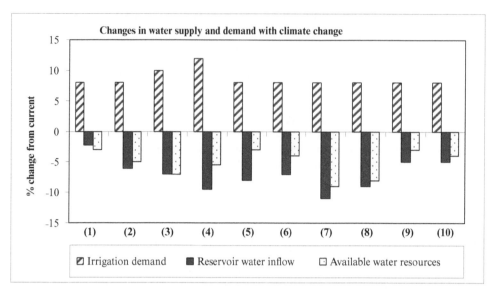

Figure 4.   Changes in available water resources, reservoir inflow, and irrigation water demand in the hydrological basins in Spain (see Table 1 for names and Figure 1 in Chapter 2 for locations).
Source: Garrote *et al.*, 1999; Iglesias & Minguez, 1997.

Table 4.   Effects of climate changes on main water management determinants and expected social and ecological consequences.

	Expected intensity of negative effects	Potential consequences for agro-ecosystems and rural areas	Confidence level of the potential agricultural impact
Water resources	Changes in hydrological regime. Differences in water needs. Increased water shortage.	Variations in hydrological regime. Decreased availability of water. Risks of water quality loss. Increased risk of soil salinisation. Conflicts among users. Groundwater abstraction, depletion and decrease in water quality.	High
Irrigation requirements	High in areas already vulnerable to water scarcity	Increased demand for irrigation Decreased yield of crops	High
Changes in water and soil salinity and erosion	High for southern countries.	Decrease in water quality from nutrient leaching. Decreased crop yields. Land abandonment. Increased risk of desertification. Loss of rural income.	High
Increased expenditure in emergency and remediation actions	High for regions with low adaptation capacity.	Loss of rural income. Economic imbalances.	Medium

Source: Adapted from Iglesias, 2009a.

Center Climate Scenario (Iglesias & Minguez, 1997; Figure 3). Irrigation changes were calculated in five agricultural sites and the results were extrapolated to represent the conditions of the water basins. The results indicate increases of water demand and reductions of water supplies that surely will affect ecosystem sustainability, implying substantial future changes in water management. Water resources systems will have to adapt to the slow evolution of climate. If projections become a reality, water scarcity is expected to rise in the next decades posing additional problems to water managers and users.

8   FINAL REMARKS

The structural water deficit of many areas in the country has been aggravated during the drought episodes in the last fifty years. Since the 1990s, Spain has improved drought preparedness strategies but has also experienced severe drought impacts. Drought indicators, although imperfect, contribute to understand the temporal characteristics of drought and to define alert situations. Past efforts to manage drought have built capacity to deal with similar situations, but have failed to solve the conflict among users, especially with the environment.

Water managers tend to view drought as the temporal deficit between water demand and water resources, and respond by managing the extensive hydraulic infrastructure of the country developed to increase the natural water availability. The structural response partially overlooks the importance of managing demand, increasing efficiency, or developing alternative sources of water.

Drought management needs to be integrated into the long-term strategies for water management. In Spain, water resources are managed at the basin level, giving the opportunity to respond directly to the needs and problems of the natural hydrological system with policy decisions. Basin authorities can establish priority of users or right holders, or can approve emergency works and projects according to each level of drought risk. Drought management plans tend to focus on the probability of occurrence of hydrological events, with very limited analysis of the capacity to anticipate, cope with, resist, and recover from drought of different systems, especially natural habitats, and social groups.

Climate change projections indicate an increased likelihood of droughts. Variability of precipitation—in time, space, and intensity—can directly influence water resources availability. The combination of long-term change (e.g., warmer average temperatures) and greater extremes (e.g., droughts) can have decisive impacts on water demand, limiting further ecosystem services. If climate change intensifies drought impacts, Spanish water delivery systems and control may become increasingly unstable and vulnerable. Water managers may find planning more difficult. Current water management strategies based on changes in mean climate variables should be revised to account for the potential increase in anomalous events.

There is a high degree of social and scientific awareness about the potential impacts of climate change and the need to adapt water management to hotter and more extreme conditions. It is certain that the need for increased spending as a result of intensified damage caused by extreme weather events will lead to a loss of rural income and economic imbalances between the more and less prosperous parts of Europe and also to environmental damage. Nevertheless, adaptation policies, strategies and concrete measures are fragmented and uncoordinated in most cases. This is in part due to the diverse perception and value that different society groups place on the issue of climate change and in part due to the difficulty in evaluating potential cost of inaction.

Societies have shown, throughout history, a great ability to adapt to changing conditions, with or without a conscious response by citizens and government (Mendelsohn *et al.*, 2004). However, it is likely that the changes imposed by climate change in the future will exceed the limits of autonomous-endogenous adaptation, and that policies will be required to support and enable different sectors of society to cope with similar changes. The White Paper on Climate Change Adaptation of the European Commission highlights water management as one of main priorities for developing national adaptation plans.

REFERENCES

Aquastat (2005). Database of the Food and Agriculture Organization of the United Nations. http://www.fao. org/. Last access: October 2005.

CRED (2007). Center for Research on the Epidemiology of Disasters of the University of Louvain and the United Nations Department of Humanitarian Affairs. International Disasters Data Base (EM-DAT), http://www.cred.be/. Last access: October 2007.

Cubillo, F. & Ibanez, J.C. (2003). *Manual of Water Supply*. Canal de Isabel II, Madrid, Spain.

De Luis, M., Raventós, J., González-Hidalgo, J.C., Sánchez, J.R. & Cortina, J. (2000). Spatial analysis of rainfall trends in the region of Valencia (East Spain). *International Journal of Climatology* 20(12): 1451–1469.

del Moral, L. & Giansante, C. (2000). Constraints to Drought Contingency Planning in Spain: The Hydraulic Paradigm and the Case of Seville. *Journal of Contingencies and Crisis Management* 8(2): 93–102.

EEA (2007). Climate change and water adaptation issues. EEA Technical Report No. 2/2007, 110 p.

Estrela, M.J., Peñarrocha, D. & Millán, M. (2000). A spatio-temporal analysis. *International Journal of Climatology* 20(13): 1599–1618.

Estrela, T., Marcuello, C. & Iglesias, A. (1995). *Water resources problems in Southern Europe: An overview report*. European Topic Centre on Inland Waters, ETC/IW-EEA.

Garote, L., Flores, F. & Iglesias, A. (2006). Linking drought indicators to policy. The case of the Tagus basin drought plan. *Water Resources Management* 21: 873–882.

Garrido, A., Iglesias, E. & Gomez-Ramos, A. (2000). El agua subterránea en la gestión de las sequías: El impacto económico de las sequías en la agricultura. [Groundwater in drought management: The economic impact of droughts in agriculture], *Revista de la Real Academia Español de Ciencias* [Review of the Spanish Science Academy] 94(2): 233–244.

Garrote, L., Flores, F. & Carrasco, J.F. (2003). The hydrologic regime of the Tagus basin in the last 60 years. *International World Resources Association Congress*. October 2003, Madrid.

Garrote, L., Rodríguez, I. & Estrada, F. (1999). Una evaluación de la capacidad de regulación de las cuencas de la España peninsular. VI Jornadas Españolas de Presas. Libro de Actas, Madrid, [An evaluation of the regulation capacity of continental Spain. Sixth Conference Spanish of Dams], 645–656.

Hayes, M., Svoboda, M., Wilhite, D. & Vanyarkho, O. (1999). Monitoring the 1996 drought using the SPI. *Bulletin of the American Meteorological Society* 80: 429–438.

Hisdal, H., Stahl, K., Tallaksen, L.M. & Demuth, S. (2001). Have streamflow droughts in Europe become more severe or frequent? *International Journal of Climatology* 21(3): 317–333.

Iglesias, A. (2002). Climate Changes in the Mediterranean: Physical aspects and effects on agriculture. In *Mediterranean Climate*, edited by H.J. Bolle. Springer, New York. 372 p.

Iglesias, A. (2009). Policy issues related to climate change in Spain. In Dinar A, Albiac J, (Eds) *Policy and strategic behaviour in water resource management*. Earthscan, Washington, DC, USA.

Iglesias, A., Cancelliere, A., Cubillo, F., Garrote, L. & Wilhite, D.A. (2009a). *Coping with drought risk in agriculture and water supply systems: Drought management and policy development in the Mediterranean. Springer*, The Netherlands.

Iglesias, A., Garrote, L., Flores, F. & Moneo, M. (2007). Challenges to manage the risk of water scarcity and climate change in the Mediterranean. *Water Resources Management*, 21, 775–788.

Iglesias, A., Garrote, L. & Martín-Carrasco, F. (2009b). Drought Risk Management in Mediterranean River Basins. *Integrated Environmental Assessment and Management*, 5(1): 11–16.

Iglesias, A. & Minguez, M.I. (1997). Modelling crop-climate interactions in Spain: Vulnerability and adaptation of different agricultural systems to climate change. *Mitigation and Adaptation Strategies for Global Change* 1(3): 273–288.

Iglesias, A., Ward, M.N., Menendez, M. & Rosenzweig, C. (2002). Water Availability for Agriculture Under Climate Change: Understanding Adaptation Strategies in the Mediterranean. In *Climate Change and the Mediterranean: Socioeconomic Perspectives of Impacts, Vulnerability and Adaptation*, Giupponi, C. & Shechter, M. (Eds). Edward Elgar Publishers, Cheltenham, United Kingdom, 318 p.

IPCC (2007). *Climate Change 2007: Fourth Assessment Report of the Intergovernmental Panel on Climate Change*. Cambridge University Press, Cambridge.

Llamas, M.R. (2000). Some lessons learned during the drought of 1991–1995 in Spain. In: Vogt, J.V. and Somma, F. (Eds). 2000. *Drought and Drought Mitigation in Europe*. Kluwer Academic Publishers, Dordrecht, pp. 253–264.

Lloyd-Hughes, B. & Saunders, M.A. (2002). Seasonal prediction of European spring precipitation from ENSO and local sea surface temperatures. *International Journal of Climatology*, 22: 1–14.

Mitchell, T.D., Carter, T.R., Jones, P.D., Hulme, M. & New, M. (2003). A comprehensive set of high-resolution grids of monthly climate for Europe and the globe: the observed record (1901–2000) and 16 scenarios (2001–2100).

Mendelsohn, Dinar, R.A., Basist, A., Kurukulasuriya, P., Ajwad, M.I., Kogan, F. & Williams, C. (2004). *Cross-sectional Analyses of Climate Change Impacts*, World Bank Policy Research Working Paper 3350, Washington, DC

Palmer, W.C. (1965). Meteorological drought. Research Paper No. 45. U.S. Weather Bureau, Washington, DC.

Rosenzweig, C., Strzepek, K., Major, D., Iglesias, A., Yates, D., Holt, A. & Hillel, D. (2004). Water availability for agriculture under climate change: Five international studies. *Global Environmental Change*, 14, 345–360.

Rossi, G., Cacelliere, A., Pereira, L.S., Oweis, T., Shataniawi, M. & Zairi, A. (eds) (2003). *Tools for drought mitigation in Mediterranean Regions*. Kluwer Academic Publishers, Dordrecht, The Netherlands.

UNCCD (1994). *United Nations Convention to Combat Desertification in Countries Experiencing Serious Drought and/or Desertification*, adopted in June 1994, entered into force on 26 December 1996, and ratified by 191 countries in 2004, United Nations.

UNISDR (2002). *Drought Living With Risk: An Integrated Approach to Reducing Societal Vulnerability to Drought*. UN International Strategy for Disaster Reduction (UNISDR), Ad Hoc Discussion Group on Drought. Geneva.

Vallarino, E. & Garrote, L. (2000). Posibilidades de aprovechamiento. Límites de la regulación. [Beneficial uses. Regulation limits] *Revista Obras Públicas* [Review of Public Works] Vol. 10(50): 11–24.

Vogt, J.V. & Somma, F. (eds) (2000). *Drought and Drought Mitigation in Europe*. Kluwer Academic Publishers, Dordrecht.

Wilhite, D.A. (ed) (2005). *Drought and water crises*. CRC Press, London, UK.

Wilhite, D.A., Hayes, M.J., Knutson, C. & Smith, K.H. (2000). Planning for Drought: Moving from Crisis to Risk Management. *Journal of the American Water Resources Association* 36: 697–710.

CHAPTER 8

Water supply in urban areas

Enrique Cabrera, Enrique Cabrera Rochera & Ricardo Cobacho
Institute for Water Technology, Polytechnical University of Valencia, Spain

1 SOUND URBAN SUPPLY MANAGEMENT: A LONG WAY OFF

Generally speaking, Spain's urban distribution systems are quite deficient. As a result, there is a long and costly road to go before modern and efficient systems are implemented, because agencies are reluctant to support infrastructure improvements. The starting point would be the establishment of a regulatory agency to lay down clear rules and enable a rigorous application of the cost recovery principle. However, these actions could turn out to be very unpopular if they are not adequately conveyed to the general public. To do this, it is necessary to raise awareness and educate society. A society must be aware of the need for water management rather than the provision of more and more water resources. There is a need for a new philosophy in which water supply to citizens, by far the most important concern, will play the leading role.

Water supply systems as we know them today appeared in the world over a century and a half ago. In Spain, the oldest system is probably Madrid's. The canal that was to bring water from the Lozoya river to the kingdom's capital was finished on June 24th, 1858, and was financed by popular subscription headed by Queen Isabel II. The water transfer had been approved in 1851 (Canal de Isabel II, 2005). Almost ten years later, the Compagnie des Eaux de Barcelone was established at Liège in 1867 to supply Spain's second city through the Dos Rius aqueduct.

Only a few years later, in 1882, the city of Seville appointed the Seville Water Company to retail water to the city under a 99-year concession (Del Moral, 1998), whereas the Sociedad de Aguas Potables y Mejoras de Valencia (Aguas de Valencia, 2005) was established in 1890 to supply the city from the Turia river. By the end of the 19th century, all of Spain's major cities were connected to a network. Soon the whole country came to appreciate the benefits of such a service, and most towns of any size had regular water supplies by the beginning of the 20th century.

Consequently, most water distribution networks in Spain are nearly 100 years old, as are many kilometers of their pipes. And since no administration has purposely ever taken upon itself to modernize and update these systems, the oldest pipes are often in poor condition. The poor tradition of system maintenance only adds to this problem, making for a foreseeable ending.

The modernization of these systems has never been a priority in Spain, where water policy has focused almost exclusively on the management and supply of water resources (Cabrera *et al.*, 2002). This policy is much more concerned about the quantity than the quality of water, and is more interested in promoting new than in maintaining existing works. This stands in stark contrast to Spain's level of economic development and per capita income.

Although urban and industrial use still accounts for only 18% of total demand, which amounts to 35,000 Mm^3/year, the economy, and the country itself, is highly dependent on this small percentage. Urban and industrial demand then totals 6,300 Mm^3, equivalent to 410 liters per capita per day. Nonetheless, there is still room for savings through the development of demand management policies. Considering the condition of our networks, the consumption of urban water could be halved in view of the ample margin of savings estimated for California using the latest technology (Gleick *et al.*, 2003). This applies also to pollution (originating mostly in urban and industrial environments).

1.1   *Distribution networks*

For water utilities, the distribution network is the component that requires most attention and investment. Even so, water agencies are poorly informed about the real state of pipes. The utilities' technical staff obviously knows what condition the pipes are in, but, in the absence of a national institution to monitor and control this service, there are no official data. Even though data are published in several sources, e.g. by the National Institute of Statistics (INE, 2005) or the Spanish Association for Water Supply and Wastewater (AEAS, 2005), they are neither consistent nor sufficient. In the above two cases, data are taken from utilities' responses to surveys designed according to their own criteria and often in pursuit of their own interests. One possible reason is that openly admitting mediocre network operation would do serious harm to the utilities' public image (in a country where the resource's value is so often stressed). A good solution would be to establish a specialized body (for instance, a regulatory agency) to audit these data.

Roughly speaking, the networks in larger cities are in better shape than in medium-sized and small towns, even though the volumetric efficiency (the quotient between the metered water and the injected water) is usually less than 70%. As there is no management culture in Spain, this is considered to be an acceptable value. Considering the typical features of Spanish networks—low pressure levels, high density of connections and high unitary consumptions—however, the Infra-structure Leakage Index (a commonly used indicator) would be far above that of correctly managed systems (ITA, 2004). The Situation is, as mentioned, even worse in medium-sized and small towns (population of under 40,000). Some of these networks often have losses of over 50% of the injected volume.

It is during drought periods that these low efficiencies become more evident. As soon as the drought leads to water rationing, supply is provided intermittently, a situation typical of developing countries (Lund & Reed, 1995). This happens for two reasons. The most important one is that a supply interruption is meant to stop further network leaks during the interruption. Given the poor state of the pipes, the savings are much greater than can be achieved by a user reducing consumption. Because these interruptions force users to store water (using tubs and buckets), when the service is restored users release this water into the wastewater system, and the whole thing backfires against interruption strategies. The second reason for service interruptions is psychological and promoted by the administration. The user, however, has to put up with the inconveniences, the loss of quality and the certainty of future interruptions (for the network will deteriorate quicker).

1.2   *Domestic tanks*

Network limitations are not confined to their shape and state. Their capacity to transport the water is also insufficient, and many networks have not been able to keep up with the growth rate of cities. Nowadays, pipe diameters are often incapable of satisfying user demand at peak hours. The rather clumsy solution to this problem is well known. New households are equipped with domestic tanks, which gradually fill up and can level out peak user demand. These tanks are often promoted by the utilities' management because they reduce complaints, and are seen by users as an improvement in supply reliability. Users that have experienced service interruptions in the past are well aware that these tanks can drastically reduce any inconvenience.

However, these domestic tanks can pose serious problems. The first and foremost concern is that the use of tanks seriously compromises water quality reliability as the transit time from the treatment plant to the consumption point increases notably (Cobacho *et al.*, 2008; Cabrera & García-Serra, 1997). Other problems associated with domestic tanks are:

a. Energy loss. Water is pressurized, taken to the tank at atmospheric pressure, and then pressurized again to be delivered to the user at the consumption point.
b. Metering inaccuracies. Meters placed upstream from the tank will be subject to low flow rates, which are usually responsible for high errors. If the meter is placed downstream from the tank, users may collect the water directly from the tank through an illegal connection.

c. Increased leakage. The tank's entry valve is often liable to failure and hardly ever receives any attention. When this valve does not close correctly, the water keeps entering the tank and spilling to the wastewater system.

d. Loss of information. In the presence of domestic tanks, for instance, it is impossible to apply night flow techniques to detect leakage. This technique compares the user demand pattern to night flows, when demand is almost non-existent. Tanks greatly distort these demand patterns and prevent the application of this technique. A similar problem occurs when applying methods to analytically discriminate real and apparent losses (Almandoz *et al.*, 2005).

Although this overview has not taken into account supply source problems (supply quality and reliability), it is important to note that the quality of surface and groundwater has deteriorated over the last decades (MIMAM, 2000). Urban and industrial wastes are responsible for the degradation of surface water, while nitrate leaching is responsible for most groundwater pollution.

1.3 *Other problems in urban water supply systems*

The problem of deficient and insufficient networks is a consequence of incorrect system management, and the solution is an exclusively technical one. There are further problems that have nothing to do with system management. These are quantity problems and are mainly located in southeastern Spain, although northern regions have also experienced serious difficulties (Silverio, 1999).

In southeastern Spain, the provinces of Alicante and Murcia are probably the most problematic areas. They are highly dependent on the Tajo-Segura transfer, which is being increasingly called into question by the area of origin in the region of Castille—La Mancha (see Chapter 19). Consequently, and now that the Ebro transfer, launched in 2001 (BOE, 2001), has been called off (BOE, 2004), the government is primarily relying on desalination, but water saving policies are not yet on the agenda. In light of the current situation, Spain's southeast will have to make bigger efforts at introducing new demand management policies. Note, in any case, the two capital cities of these provinces, Alicante and Murcia, are an example to follow. Their network efficiency levels are amongst the highest in Spain.

However, the most serious drought ever experienced in Spain (which began in 1991, lasted four years and affected 10 million users) occurred in southern Spain and centered on the city of Seville in the Lower Guadalquivir Basin (Palancar, 1998; see also chapter 7). Given the area's increasing population (with immigration and Northern European citizens retiring to the area) and the incontrovertible effects of climate change, these scarcity episodes are bound to happen more and more often (Chapter 7 focuses on drought and climate change). As water demand grows in areas where resources are scarcer, there is an evident need to manage these systems better. Southeastern Spain, where the highest urban demand is concentrated, is in fact the driest region of all (MIMAM, 2000). The same applies to agricultural demand.

Briefly, let us recall the causes that, we think, are responsible for the current state of affairs. First, politically motivated pricing does not lead to efficient management or encourage users to save water. Second, administrations seem to be concerned with the construction of waterworks, but have little interest in efficient management. Lastly, user awareness and perception of urban water supply problems are wrongly associated with the insufficiency of water resources at the wholesale level. We will refer to these three causes from the urban supply point of view. Another three chapters of this book have dealt with these three major issues separately.

2 THE NEED TO CONTROL AND STRUCTURE URBAN SUPPLY

The final objective of urban water supply is to deliver water of the highest standard at the lowest possible cost. Obviously, the first step should be to define such standards. In the case of Spain, it is quite striking that while most economic sectors are one subject to some regulatory framework or another, the problem of defining water supply standards has not yet been addressed. It is highly likely that this problem will be eventually tackled by the regional water agencies that are being set up. The first agency was established in Catalonia in 1998, and was recently followed by the Andalusian agency. It is still surprising, however, that the Agencia Catalana del Agua (2005) only deals with the regulation of the bulk water supply and wastewater services and makes no mention at all of urban supply. Although the municipalities have authority over water services, a higher institution should take responsibility for their regulation and control. Such is the case in the most developed countries (JWWA, 2005).

The non-existence of regulatory standards is negatively affecting the privatization process of urban water services. Where the rules of the game are unclear, the processes are not transparent and, worse still, the municipalities often use them as a source of funding to cover old budget deficits, asking the company to advance a sizeable sum (called a *canon*). This further complicates the future of an already deficient system. The company advancing the *canon* will recover the costs sooner or later through tariffs, which will not be used to make the necessary improvements. As a result, part of the tariffs will go to finance other projects, even though water is subsidized and not all costs are covered.

As a matter of fact tariffs are the second biggest problem. Spanish tariffs are more often related to political criteria than to the needs of investment in systems. Typically, the mayor of a city will propose an annual tariff revision with an increasing in line with the consumer price index so as not to risk unpopularity. Table 1 shows the tariffs in the different regions of Spain. There is no technical explanation for the price differences shown. Even though raw water is significantly more expensive in some utilities (for instance, water produced through desalination), the price differences are still too high or not well enough explained by any of the relevant variables.

With water tariffs set on political grounds, it is difficult to renovate the networks, and efficiencies are mediocre (with pipes having a life expectancy of 50 years, 2% of the network should be renovated every year). Additionally, it is also impossible to apply demand management policies. The leaked water is so cheap that the company prefers to let it leak than invest in the pipe renovation or refurbishment. In other words, the current prices favor a low economic level of leakage (ITA, 2009). Users also face a similar problem, since there are no real incentives to use water

Table 1.   Average total price (water supply and wastewater) of water in Spain (€/m³).

Year	2000	2001
Andalusia	0.59	0.64
Aragón	0.59	0.59
Asturias	0.51	0.54
Baleares	1.32	1.45
Canarias	1.58	1.66
Cantabria	0.53	0.53
Castille and León	0.42	0.46
Castille-La Mancha	0.44	0.48
Catalonia	0.91	0.94
Valencian Community	0.66	0.71
Extremadura	0.72	0.74
Galicia	0.54	0.60
Madrid	0.69	0.76
Murcia	1.12	1.12
Navarra	0.60	0.59
Basque Country	1.12	1.09
Rioja	0.41	0.42
Ceuta y Melilla	0.58	0.63
Spain (average)	0.73	0.77

Source: INE 2005.

rationally. In Spain it would be unimaginable to install a double circuit to reuse grey water from sinks and showers or to implement water harvesting, as happens in Northern European countries (Shirley-Smith & Butler, 2008).

Consequently, service improvement is in need of closer control and an adequate pricing policy. Users need to be informed and educated before these unpopular decisions can be implemented. Customers are happy with the current political prices—which also neglect environmental costs (Deoux, 2003)—, and politicians prefer not to raise tariffs. This is a vicious circle, which only a severe drought could break, demonstrating that system performance must be improved. Meanwhile, and without a proper environmental education, the problems will only get worse. As a matter of fact, the EU Water Framework Directive's cost recovery principle is probably more an inconvenience for politicians than the true road to efficiency.

There are two different sides to water sector crises, which are often the factor of change. The quantity problem (discussed above) and the quality problem, which led to changes in Germany (FMENCNS, 2001), a country that is decades ahead of Spain. The necessary environmental education should then explain the reasons for implementing the cost recovery principle, and also increase user participation in decision-making processes, while inviting them to raise their expectations. It is striking that, in the 21st century, Spanish users still believe that service interruptions are a consequence of droughts and not of improper management of the network. Although water must necessarily be rationed sometimes, the service should not be interrupted.

3 CONSIDERATIONS ON THE ROAD TO BE TAKEN

Of the many problems related to water policy in Spain, urban supply is probably the easiest one to solve. With Spain's average income most users could easily pay the prices set by the WFD. Sooner or later, tariffs will be raised, but not for the moment. In fact, during the former socialist period (2004–2008) this issue was deeply analyzed and discussed (MIMAM, 2007). However no real action has been implemented. And with the present economical crisis, most probably, prices will be maintained. With its technological level and the level of training of technical staff, the country is prepared for this measure. It all depends on the political decisions being made to implement the three actions mentioned in the last section. We will further elaborate on the first two.

The need to set up regulatory agencies (regional or national) with defined competencies is evident. An agency should establish the conditions of the water supply service, monitor its performance, check for full cost recovery and promote the environmental education of users. The central administration now relies on river basin authorities to control resources, especially surface water, and promote major hydraulic works. Another agency of equal rank is needed to monitor water use and promote efficiency. For example, Figure 1 shows the administrative structure in Western Australia (Martin, 2004). Resource management and use control are on the same level. As the institutional framework depicted in Figure 1 shows, the resource and basin level management and competencies tier is closely linked to the water industry's regulatory layer. In Spain, the wholesale and retail urban supply sectors are weakly linked to the basin authority, and their operations and programs are subject to different laws (see Chapters 10 and 11).

Regarding cost recovery, Table 2 shows a rough correlation between cost recoverability through tariffs and water use efficiency. From the available data, annual per capita domestic consumption and small business consumption can be assimilated. To do this, the data from Figure 2 (Merkel, 2003) are related to water demand values published by IWA (IWA, 2004). Although the IWA data are not homogeneous, the correlation shown in Table 2 is clear (it would be greater if domestic and commercial demand was disaggregated).

Furthermore, as Table 2 shows, consumption in countries with larger cost recovery rates is lower. Note that, according to IWA statistics, the water demand is much lower than in the Water White Paper (MIMAM, 2000), which is, in principle, a much more reliable source. The reference years (1997 vs. 2002) cannot account for this difference.

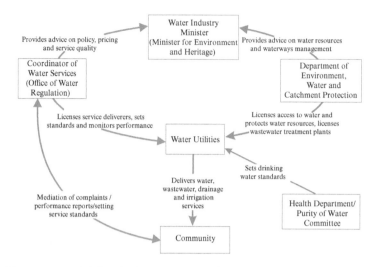

Figure 1.   The new water administration in Western Australia (Martin, 2004).

Table 2.   Degree of compliance with the full cost recovery principle (CWT/FCR) related to the domestic and small business demand (m³/inhabitant and year).

Country	Demand (m³ × 10⁶/ year)	Inhabitants (× 10⁶)	Consumption per capita	CWT/FCR (Figure 2)
Denmark	247	5.37	46.00	$0.8/0.9 = 0.89$
United Kingdom	3,300	59.23	59.72	$1.2/1.4 = 0.86$
Germany	3,790	82.54	45.92	$1/1.2 = 0.83$
France	3,210	60	53.50	$1.1/1.5 = 0.73$
Spain	2,237	40.85	54.76	$0.4/1.6 = 0.25$
Greece	784	10.96	71.53	$0.4/2.1 = 0.20$

Source: Compilation by authors.
Note: Greece and Portugal are not included in this table because the IWA statistics do not include these countries. Additionally, the consumption figures for France correspond to 1997, while the rest of the figures are from 2002.

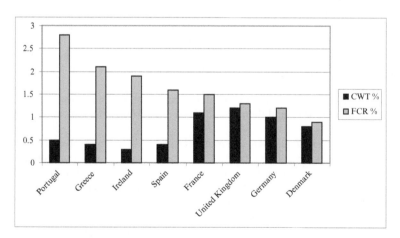

Figure 2.   Current Water Tariffs (CWT) versus tariffs that include Full Cost Recovery (FCR), as a percentage of the per capita income.
Source: Merkel, 2003.

The full cost recovery principle is not only necessary to increase efficiency. Transferring the economic resources from the state towards a managing municipality is an essential tool for decentralizing water management. These responsibilities should be handled by authorities closer to the users, a key principle that should drive water policy in the 21st century. This whole process should be supervised by a regulatory agency to ensure that the revenue from the water services is invested in improving the service and not to cover earlier debts generated in other areas of municipal responsibility.

Finally, note that the full cost recovery principle can and must be compatible with social initiatives. The important point is to balance income and expenses, regardless of how such income is produced. For instance, the first 10 m³ of consumption could be supplied at low prices, and tariffs could increase progressively with demand. Fortunately, as Chapter 13 shows, block-rate pricing is widespread around Spanish cities.

4 CONCLUDING REMARKS

Spain's water supply systems are still behind on many technical and management aspects. Yet there are notable exceptions. Generally speaking, they are municipalities with a better cost recovery tariff. The same could be said about the other services that close the city's water cycle: wastewater collection and treatment. Politically-based water charges are undoubtedly one of the main reasons for the current state of affairs. Urban water tariffs have been kept artificially low for the past 15 years thanks to subsidies from regional, national and even European funds. The enforcement of the European Water Framework Directive will certainly lead to a rise in urban tariffs. The application of the Directive will finally put an end to some inconsistent situations (where the worst managed systems receive the largest subsidies or where treatment plants that were given away "for free" are not fully operational because of high variable costs).

However, there are reasons for optimism. Spain should have no problems whatsoever in modernizing all these services within a reasonable time frame. The major obstacles to enforcing the Directive and tackling environmental problems are political unwillingness, vested interests and poor users environmental education. Unless a major crisis occurs—a drought, the end of European funding or a major environmental accident—the opposing forces will yield to the changing factors that are helping urban supply systems to become efficient, environmentally responsible and competitive.

REFERENCES

AEAS (Asociación Española de Agua y Saneamiento), [Spanish Water and Wastewater Association] (2005). http://www.aeas.es (accessed January 24th, 2005).

Agencia Catalana del Agua [Catalan Water Agency] (2005). http://mediambient.gencat.net (accessed February 28th, 2005).

Aguas de Valencia [Valencia Water Company] (2005). http://www.aguasdevalencia.es (accessed March 19th, 2005).

Almandoz, J., Cabrera, E., Arregui, F., Cabrera, E. Jr. & Cobacho, R. (2005). Leakage assessment through water networks simulation. *Journal of Water Resources Planning and Management* 131(6): 458–466.

BOE (Boletín Oficial del Estado) [Spanish Official Gazette] (2001). Ley 10/2001, de 5 de Julio, del Plan Hidrológico Nacional. *Boletín Oficial del Estado* no. 161. July 6th, 2001: 24228–24250.

BOE (Boletín Oficial del Estado) [Spanish Official Gazette] (2004). Real Decreto Ley 2/2004, de 18 de junio, por el que se modifica la ley 10/2001, de 5 de julio, del Plan Hidrológico Nacional. *Boletín Oficial del Estado* no. 148. June 19th, 2004: 22453–22457.

Cabrera, E. & García-Serra, J. (1997). *Problemática de los abastecimientos urbanos. Necesidad de su modernización.* [Problems of water supply systems. Modernization needs]. Valencia, Spain: Grupo Mecánica de Fluidos—Universidad Politécnica de Valencia.

Cabrera, E., García-Serra, J., Cabrera, E. Jr., Cobacho, R. & Arregui, F. (2002). Water Management Paradox in Southern Europe. The case of Spain. Paper presented at the 5th International Conference on Water Resources Management in the era of Transition. September, 2002. Athens, Greece. European Water Resources Association: 375–386.

Canal de Isabel II. (2005). http://www.fundacioncanal.com (accessed May 2nd, 2005).

Cobacho, R., Arregui, F., Cabrera, E. & Cabrera, E. Jr. (2008). Private water storage tanks: evaluating their inneficiencies. *Water Practice & Technology* 3(1), doi: 10.2166/WPT.2008025.

Del Moral, L. (1998). *El sistema de abastecimiento de agua de Sevilla: análisis de la situación y alternativas al embalse del Melonares*. [Seville's water supply system: analysis of the present situation and alternatives to the Melonares Dam]. Bilbao, Spain: Editorial Bakeaz.

Deoux, F. (2003). Definition and estimation of resource costs of an aquifer. Paper presented at the International Conference on Computing and Control for the Water Industry. *Advances in Water Supply Management*. Lisse, The Netherlands: Balkema Publishers: 563: 572.

FMENCNS (Federal Ministry for the Environment, Nature Conservation and Nuclear Safety) 2001. *Water Resources Management in Germany. Part I. Fundamentals*. Bonn, Germany. FMENCNS, Division WA I 1(B).

Gleick, P.H., Haasz, D., Henges-Jeck, C., Srinivasan, V., Wolff, G., Cushing, K.K. & Mann, A. (2003). *Waste Not, Want Not. The Potential for Urban Water Conservation in California*. Oakland, CA: Pacific Institute.

INE (Instituto Nacional de Estadística) [National Institute for Statistics] 2005. http://www.ine.es (accessed March 9th, 2005).

ITA (Instituto Tecnológico del Agua) [Institute for Water Technology] (2009). *Rehabilitación y Renovación de Redes de Agua*. Valencia, Spain: Instituto Tecnológico del Agua—Universidad Politécnica de Valencia.

IWA (International Water Association) (2004). *International Statistics for Water Services*. London, UK: International Water Association.

JWWA (Japan Water Works Association) (2005). *Guidelines for the management and assessment of drinking water supply service (JWWA Q 100)*. Japan: Japan Water Works Association.

Lund, J.R. & Reed, R.U. (1995). Drought Water Rationing and Transferable Rations. *Journal of Water Resources Planning and Management*. 121(6): 429–437.

Martin, B. (2004). A structure for water administration in 21st century. The case of Australia. In Challenges of the New Water Policies for the XXI century. Editors: Cabrera, E. & Cobacho, R. Lisse, The Netherlands: Balkema Publishers, 99–114.

Merkel, W. (2003). El Futuro de la Industria de Agua en el mundo [The future of the world water industry]. *Ingeniería del Agua*. 10(3): 337–353.

MIMAM (*Ministerio de Medio Ambiente*, Spanish Ministry of Environment) (2000). *Libro Blanco del Agua en España* [Spanish Water White Paper]. Secretaría de Estado de Aguas y Costas, Dirección General de Obras Hidráulicas y Calidad de las Aguas. Madrid, Spain, 637 p.

MIMAM (Ministerio de Medio Ambiente) (2007). *Precios y costes de los servicios del agua en España*. [Prices and costs of water services in Spain] Madrid, Spain: Ministerio de Medio Ambiente.

Palancar, M. (1998). Experiencias y conclusiones tras una larga sequía. Sevilla 1992–1995. [Experience and conclusions after the long drought. Seville 1992–1995]. In *Gestión de sequías en abastecimientos urbanos* [Drought Management in Urban Water Supply Systems]. Editors: Cabrera, E. & García-Serra, J. Valencia, Spain: Universidad Politécnica de Valencia, 521–533.

Shirley-Smith, C. & Butler, D. (2008). Water management at BedZED. Some lessons. *Engineering Sustainability* vol. 161, issue 2, pp. 113–122.

Silverio, A.L. (1999). Experiences and conclusions after a long drought in the metropolitan area of Bilbao. In *Drought management planning in water supply systems*. Editors: Cabrera, E. & García-Serra, J. Dordrecht, The Netherlands: Kluwer A.P., 388–433.

CHAPTER 9

Changing water discourses in a modern society

Leandro del Moral

Department of Human Geography, University of Seville, Spain

1 INTRODUCTION

Economic, social and cultural changes, the acute deterioration of the aquatic environment and a rise in interregional conflicts can be associated with the decline of the prevalent water policy in Spain. Criticism of this situation has widely spread across society, undermining the logic of the old paradigm. It has not, however, come up with an alternative model to date. Underlying this apparent stalemate, though, is a powerful dynamism of ideas and socio-political processes.

Fundamental topics for discussion in this chapter include changes in the social perception of the water environment, the evaluation and distribution of the cost of managing water as a resource, the repercussions of water politics on socio-economic and spatial development and the reform of the institutional water management framework. This chapter analyses these debates, paying particular attention to the discursive aspects, that is, the ensemble of ideas, concepts and categorizations that attach meaning to physical and social realities.

2 HYDRAULIC POLITICS AS A CENTRAL ISSUE IN THE CONSTRUCTION OF THE MODERN SPANISH STATE

For over a century, one particular water management policy has prevailed in Spain, namely the *hydraulic paradigm* or the *hydraulic mission*. This has been described at length elsewhere (Allan, 2003; Feitelson, 1996; Reisner, 2001). The central axiom of this late 19th-century paradigm was the need to provide an adequate water supply for all social agents who were prepared to use it to develop and modernize the country.

This idea embodied a project for the geographical transformation of the country: the *regeneration, the revival (la regeneración)* of an adverse landscape characterized by aridity and barrenness and the resulting underdevelopment and lack of growth. A landscape able to respond favorably to tenacious human intervention based on sound geographical knowledge, know-how and the collective will. Any hope that the private sector would be able to pool the resources required to promote this project for physical, economic and moral regeneration had been ruled out even before the 1890s. Thus, end-of-century regenerationists entrusted the governments to use the taxpayers' money and lead the country in this grand-scale endeavor.

Joaquín Costa (1846–1911) was one of the main advocates and potent symbols of this broad social movement for modernization through his hydraulic policy (*Política Hidráulica*), in which water engineering would be the substratum for fostering growth and enabling social reform and cultural emancipation.

The specific characteristics and different historical manifestations of the hydraulic paradigm throughout 20th-century Spain have been addressed often by both Spanish (Ortí, 1984; Gómez Mendoza & Ortega Cantero, 1987; Ortega Cantero, 1992; Naredo, 1997; López Ontiveros, 1998) and by foreign authors (Drain, 1995; Swyngedouw, 1999). Neither the country's recent history, nor its present geographical layout can be understood without taking into account the involvement and radical transformation of the water environment.

While, at the end of the 19th century, other industrialized European countries, such as the United Kingdom, France and Belgium, were consolidating colonial expansion overseas, Spanish society was in a state of shock at the loss of its last colonial possessions (Puerto Rico, Cuba and the Philippines) following a disastrous war against the United States in 1898. In the absence of an overseas colonial project as a means for modernization, Spanish elites advocating social and economic reform concentrated their efforts on a national program, involving a radical trans-formation of the country's geography (Swyngedouw, 1999). This vision combined a decidedly political strategy, a call for a scientific-positivist understanding of the natural world, a scientific-technocratic engineering mission and a popular base rooted in a traditional peasant rural culture. It united diverse social and political sectors (reformists, socialists, populists, industrialists and agricultural elites), while keeping the more radical left-wing forces (revolutionary socialists and anarchists) and the traditionalist conservative right at bay. This alliance of reformists, focused on reconstructing the country's hydraulic geography, served the twin purpose of uniting hitherto excluded political sectors into a powerful coalition, while addressing social conflicts by seeking to prevent radicalization (Swyngedouw, 1999).

Yet, in view of the limited prospects of private investment, the regenerationists regarded the State as the only body capable of generating the investment funds required to mobilize the country's water resources. They pushed through the necessary reforms in the face of strong and sustained opposition from the traditional oligarchies. At the same time, this reformist route was able to secure the support of part of the old elites, as it did not threaten their fundamental rights as landowners and defended rural power against the rising tide of the urban-industrial elites and proletariat.

Furthermore, hydraulic politics, hailed as the top priority by the country, played an important role in the social legitimization of the State. Reformism advocated *hydraulic regeneration*, whereby the State would take center stage to organize the necessary socio-spatial transformation. After the failed attempts to initiate reform during the first few decades of the 20th century, it was to provide a substra-tum on which the later Falangist (Spanish fascist wing) ideology would thrive. The regenerationist agenda was, in fact, not to materialize on a grand scale until after the Spanish Civil War, during the Franco regime (1939–1975). Although the coalitions of social agents, the objectives and the means were to change over time, the geographical basis for modernization remained the constant guiding principle. The vision of modernization based on *hydraulic regenerationism* became the lynch pin of progress and development in Spain until the end of the Franco era (Swyngedouw, 1999).

This hydraulic regenerationism advocated by Joaquín Costa coincided with an *intellectual regenerationism*, symbolized by the movement of poets, writers and thinkers known as the Gen-eration of 1898 (after the year of Spain's defeat in the Cuban War—1898). This movement redis-covered, both aesthetically and sociologically, the underdeveloped regions of arid inland Spain (Castilla), whose only prospect of future emancipation laid in embracing hydraulic politics. The 'hydraulic desire' of these arid lands became the leitmotiv of much of the Generation of 1898's literature (Unamuno, Azorín, Baroja, Macías Picabea and others).

3   IMPLEMENTATION AND CRISIS OF THE TRADITIONAL HYDRAULIC PARADIGM

While the turn of the 19th century was characterized by the dynamic development and expan-sion of the discourse of hydraulic regenerationism, the 20th century was to be characterized by the slow and tortuous implementation of the geographical project proposed by the regeneration-ists. In spite of the 1902 and 1933 plans for hydraulic engineering works and the creation of river basin authorities (*Confederaciones Hidrográficas*) from 1926 onwards, the hydraulic project failed in its attempt to quell the outbreak of social unrest that eventually led to the Spanish Civil War (1936–1939). Following the Civil War, hydraulic politics became an instrument of control, national integration and autarchic development under the Franco regime. Although the rhetoric of the original social objectives was maintained, the emphasis gradually shifted towards the discourse of reconstruction and 'national' development (Ortega Cantero, 1979; Swyngedouw, 1999).

In 1959 there was a break with the largely inward-looking development model followed in the 1940s and 1950s. Preparations for European Community membership, the opening up of the economy and the progressive incorporation of Spain into the international community required a renewed intensification of the country's resources to meet the demands of fiercer international competition. Its competitive advantages in the production of fruit and vegetables and the booming tourist industry on the Mediterranean coast led to an increase in the demand for water in the driest areas of the country. In the 1970s, work began—still in a context of social consensus on the water policy model to be followed—on the first major water transfer between river basins, from the River Tagus basin to the River Segura basin (see Chapter 19).

Following the death of Franco (1975) and the restoration of democracy (1978 Constitution), democratic Spain had to rise to the challenge of meeting the demands placed on it by rapid growth in a context of increasing integration in Europe and economic globalization.

At the end of the 1980s, discussion began on new water management policies designed to meet the growing demand for water, primarily by means of massive water transfers between river basins. This was the main thrust of the draft National Hydrological Plan (NHP) presented in 1993. Significantly, the introduction to the plan, advocating what it called the National Integrated Water Balance System (SIEHNA), ended by stating in the truest hydraulic regenerationist rhetoric that the plan heralded "a new era in which Costa's old dream can, at last, come true ... the Ésera[1] and many other Éseras will flow over the skin of Spain and their clean waters will be its lifeblood, its dew, its gold, the path to collective liberation and wealth" (Ministerio de Obras Públicas, Transporte y Medio Ambiente 1993). Chapter 19 reviews the details of the 1993 NHP.

Yet, contrary to both the first quarter of the 20th century and the Franco years, a notorious absence of a 'hegemonic' project has been the main feature of the water policy arena over the last two decades, for reasons that are now reviewed.

4   FACTORS INVOLVED IN THE CONTINUITY OF THE MODERNIST HYDRAULIC PROJECT

This situation, characterized by the coexistence of contradictory ideas and the absence of a clearly dominant strategy, is the result of tension between factors of continuity and factors of change. Both have cultural, social and economic dimensions.

First, at the deepest-rooted cultural level, the mainstay of the resistance and continuity of the hydraulic model in Spain up to the present day is the perception of water, a perception that Spain shares with other countries on both sides of the Mediterranean. The water environment is perceived as a hostile medium, with spectacular swings between sometimes intermittent torrential water flows and extremely low water levels that coincide precisely with the hottest season of the year.

The hostility of this water environment, however, can be transformed through human intervention into splendor and beauty. Hence, human-made hydraulic landscapes (the *huertas* or orchards, *domesticated water*) are viewed as ideal images of the power of hydraulic engineering. The positive image of *domesticated water* as a basic feature of development overrides the negative image of the impact on natural aquatic media, that is generally linked to changeability, uncertainty and danger. In broad terms, Spanish society does not have a clear idea of a natural water environment against which to assess the excesses of transformation, when the traditional technologies of local irrigation systems (which actually created those highly appreciated human-made landscapes) were substituted by the great works of modern hydraulics.

As late as 1996, the former mayoress of an important Spanish city and former Minister for Culture in the Spanish Government backed the idea that a new reservoir planned for a valley, which for once was well conserved and covered by national and international environmental and landscape protection standards, "would be positive for biodiversity, as it would contribute to maintaining green areas in the city". The concept of nature underlying these statements—clearly representative of the model of the link between nature and society at the heart of the debate—is perhaps an extreme vision of nature as a social construction founded on the high social appreciation of

---

[1] The Ésera is the river that runs through the birthplace of Joaquín Costa.

dominant *domesticated water* in Spanish society (see the review of the transformation of the natural landscape in Chapter 3).

In the case of the reservoir mentioned above, the mayoress' opinion is even shared by the locals affected by the flooding of their lands, who look kindly on the creation of an artificial lake in their vicinity, although they will not benefit from the hydraulic resources created as a result. Apart from the jobs that would be generated as a result of its construction, it is understood that the infrastructure will allow for the development of leisure activities related to the reservoir. Note, however, that in other areas affected by plans for reservoirs, such as Santaliestra (Huesca), Castrovido (Burgos) and Genal (Málaga), there are signs of growing opposition.

Secondly, another factor of continuity in the hydraulic paradigm is the ongoing perception of *geographical imbalances* as major obstacles to development and well-being. This confirms and strengthens the idea of the transformation of nature, primarily the hydrological system, as a key feature of any modernizing political program. Once the resources of the driest regions have been exploited to the limit and the technology for long-distance transport has been developed, the solution of *hydrological imbalances* between river basins becomes an all-important objective (again, Chapter 19 reviews the major interbasin transfers). In fact, this topic holds the key to the debate in the most recent period of the hydraulic paradigm's development and is conditioned by processes of changes of scale in shaping decision-making bodies, an issue that is discussed in greater detail below.

Thirdly, continuity can also be explained by the ongoing process of transformation of the actors that have controlled the country's hydraulic policy to date, that is, the *hydraulic policy-making community* (O'Riordan & Jager, 1996). This is a tightly-knit, fairly autonomous, highly stable policy network consisting of the main stakeholders involved in the working definition of the hydraulic paradigm: the main agricultural organizations, construction and hydroelectric power companies, the main water management authorities and the engineers (Pérez Díaz et al., 1996; Giansante et al., 2002).

The cohesion of the community was guaranteed by economic interests, the homogeneity of technical criteria and the fluency of contacts within and through the public administration. The inclusion of new and the exclusion of other traditional stakeholders caused by changes of scale in the distribution of power, the fragmentation of prevailing interests and democratization, as well as new emergent values and social aims, have still not succeeded in completely undermining the strong cohesion of interests within this powerful group. As in other countries affected by the hydraulic paradigm, groundwater specialists, who have been mainly excluded from the policy-making community, have severely criticized the predominant hydraulic policy model, characterized by a lack of knowledge of and respect for aquifers and their resulting mismanagement (Llamas, 1988).

The panorama of continuity factors is completed by an ever-increasing demand for water for irrigation purposes. This still constitutes the main drain on resources (about 80% of total demand, see Chapter 5 for a more detailed description of sectorial water demands). On account of the social, cultural, landscape and even ecological role attributed to it by society, it also conserves a great deal of the social *legitimacy* it has traditionally enjoyed (Moyano, 2003).

Over the course of the last decade, there has been a significant increase in the total area of irrigated land, mainly as a result of private initiative. Private initiative has, in fact, taken over from the public sector in the development of new irrigation systems in the last few years (Estevan, 2008). This does not mean that the State has abandoned its role in large-scale hydraulic works, as can be inferred from the predominance of public initiative in the construction of reservoirs. At present, 87% (39,175 Mm³/year out of a total of 45,034 Mm³/year of total water resources available in Spain are reservoir-regulated surface waters (MIMAM, 2000). Note, however, that the role of groundwater in recent irrigation projects is on the increase (see Chapters 14 and 15).

5   PROCESSES OF CHANGE TOWARDS A REFLEXIVE SOCIETY MODEL

In spite of the factors of continuity mentioned above, some processes of change have eroded the cultural foundation of the hydraulic paradigm in recent years. Such processes are, in fact, the regional expression (with specific features at a local level) of large-scale phenomena operating on a worldwide scale (del Moral *et al.*, 2003). These drivers of change include the evolution of values regarding nature which, although occurring later than in other countries (change in the *myths of nature*, according to Douglas & Wildavsky, 1983), are now permeating Spain.

From the mid 1970s onwards, *insalubrious marshy areas*, which were then in the process of drying up, began to be timidly conceived as *wet zones* of great ecological value. The rivers, which had to be channeled and, if possible, diverted away from populated areas, became spatial resources of great potential, especially for urban design and the image and promotion of cities (Chapter 3 reviews the ecological impacts of these transformations). These new values are connected to other features of the *culture of the reflexive society* (Allan, 2003), which contributed to undermining some of the conceptions of the hydraulic paradigm.

Challenging the modern rationalist attitude, the roots of the ecological school of thought, which finds fertile ground in the *reflexive society*, are characterized by two basic precepts. While everything is interrelated, not everything is reproducible or interchangeable. As a consequence, value and impact are (almost) impossible to measure.

Therefore, a hydraulic project does not end with its planning, execution and operation, as modern reasoning would have us believe. Instead, "it is the source of an endless stream of impacts which, on being removed from their origin, grow and cross over into other disciplines and scales, where local matters are contaminated by global issues, and the present is tainted by the future" (Riesco, 1999). One of the reasons for the slow transition of water policies in Spain is, precisely, an inability to quickly and easily accept the new values of nature that might lead to the visions of the *modernist hydraulic model* being rejected.

This progressive but difficult expansion of environmental awareness is reinforced by tendencies towards market mechanisms and the application of criteria of economic rationality to public investments. Nobody doubts that much of the pressure exerted on the aquatic environment would be alleviated, albeit perhaps sometimes traumatically, by applying the principle of cost recovery included in the European Union's Water Framework Directive (WFD, see Chapter 16, and Chapter 13, for its economic impacts).

This is the reason why there has been some convergence between the supporters of liberalization and ecologists in their opposition to the traditional paradigm based on *state paternalism*, inherited from the old project of reform and social cohesion through the public-funded hydrological rebalancing of the country (Gómez Mendoza & del Moral, 1995). On the one hand, a system that recovers only a small part of the replacement cost of public investment made in dams and canals is condemned. Additionally, the substantial extent of water regulation already in operation—over 50% of renewable resources—places many projects in a clearly marginal position on account of a decline in their output. On the other hand, a distinction has to be made between economic demand, based on willingness to pay, and physical demand, based on stated needs (Naredo, 1997).

The general rejection of the paternalistic and interventionist system in force has actually led to some consensus about the introduction of market instruments having positive effects on the management system as a whole, although it may only affect a small part of resources. The buying and selling of water rights would act, in the opinion of experts, as a mechanism for revaluing water as a scarce commodity. It would also bring the economic dimension to users' minds, making them think in terms of opportunity costs and levels of marginal productivity in water use. The deployment of a system of exchanges among users would help to avoid water shortages in places situated near irrigation areas. In this way, transfers between extremely distant regions would take second place as a solution to local water shortages. For example, 'aberration' is the word used to describe the proposed transfer project "from the headreaches of the Tagus to supply the municipalities of La Mancha and their protected nature areas by means of a unified supply network, which has been planned without taking into account local resources and infrastructures that are supplying vast irrigated areas very inefficiently: for instance, it takes a metric ton of water to grow a kilo of corn" (Naredo, 1999).

The liberalization of water allocation has been the topic of heated debates, especially concerning the effect on social and territorial equity, as well as on the environment and the landscape.

On account of historic tradition and a distinct institutional framework, there are various legal forms of buying and selling water in operation in the Canary Islands. Despite its well-known pitfalls (Aguilera, 2002), this institutional framework has a positive effect on the water economy and the development of residual water purification and re-use. However, even the most productive agriculture cannot compete with urban uses, mainly coastal tourism. Agricultural landscapes are therefore under threat. With this experience as a backdrop, left-wing political parties, as well as associations of small-scale and medium-scale farmers, and some ecological groups, have criticized the introduction of market mechanisms (learn about the regulatory framework in Chapter 10). Nevertheless, even these sectors consider that making the system of concessions more flexible (preferably through water banks under administrative control) is a good idea because of its potential for reducing irrational uses and minimizing social unrest in the course of a transition to more sustainable models.

Another factor undermining the hydraulic paradigm is the crisis in financing for public works, caused by sustained efforts to reduce the public deficit and a decline in European Union structural and cohesion funds. The *2000 Agenda* (the European Commission financial schedule for this period) cut these sources of funding by 25% for the 2000–2007 period. Also the accession of Central and Eastern European countries to the European Union has added uncertainty about Spain's eligibility for such funds. This idea has been clearly expressed in recent times as "the historical commitment on the part of state budgets to set up irrigation infrastructures no longer makes sense" (MIMAM, 2000, p. 839).

Furthermore, the strength of Spanish irrigation (significant increase in land area, solid social legitimacy and strong presence of the sector as a political pressure group) contrasts with the reality of sectorial disunity and an uncertain future. An increasing number of studies (Consejería de Agricultura y Pesca, 1999 & 2003) highlight the great differences in productivity, job creation and efficiency in the use of resources from one area to another (see more details in Chapter 6). Noteworthy also is the discrepancy between the 1.2 million hectares of newly irrigated land included in the river basin hydrological plans for 2012 and the 0.2 million foreseen for 2008 in the National Irrigation Plan, which was drawn up in the same year as the others were passed (1998).

This discrepancy is due, in part, to the varying representation of the Autonomous Communities, agricultural lobbies and other pressure groups under the umbrella of the competent river basin authorities that tend to favor the greatest possible involvement in public investment and hydraulic resources. On the other hand, the Ministry of Agriculture, responsible for irrigation planning, is more realistic and sensitive to the ever-increasing pressure of the World Trade Organization, the successive reforms of European Union agricultural policy and the irreversible fall in the number of people actually working on the land (a decline of over 40% in the last decade) (Pérez Picazo, 2003).

Attempts have been made to find a solution to this paradox by subordinating the river basin plan projections to "the programs, deadlines and projections established in the National Irrigation Plan (NIP, *Plan Nacional de Regadíos*) in force at any given time" (Royal Decree 1664/1998 approving river basin water plans). The NIP is, in fact, a planning instrument that is not considered in the State legal framework, whereas, on the contrary, basin hydrological plans which should be drawn up in conjunction with the "any other plans that are likely to affect them" (article 38.4 of the Water Act)—are highly formalized, normative instruments. This has given rise to a peculiar situation, which has been described as "the progressive substantiation of a plan [the Irrigation Plan] without a specific legal basis" (Embid Irujo, 1999, p. 92).

All these changes are interrelated and influenced by the transformation of the hydraulic policy-making community discussed above. The process is similar to what has happened in other countries. Fissures have appeared in the previously close-knit traditional community. Interest groups are divided and promote their interests more cautiously and prudently than they did before. Underlying this new situation are two structurally intertwined trends.

Firstly, there has been an increase in the number of agents involved in political deliberation: the *policy-making community* has become an *issue network* that is larger and less integrated than it was before. There are new actors that operate on a much more open, less stable public stage, and do not all agree on the problems at issue and the means by which they should be tackled (O'Riordan & Jager, 1996). This process has been caused by four interacting transformations in the institutional framework of political life: i) the restoration of a democratic regime, the subsequent development of the logic of competition among political parties, ii) the activation of public opinion and, very importantly, iii) the change in the political and territorial structure of the Spanish State, primarily the emergence of the State of Autonomies (a near federal model, see the institutional description in Chapter 12) and, iv) the growing power of local councils. These changes contributed to the appearance of new social movements and new representatives of local opinion and interests (Pérez Díaz *et al.*, 1996).

Secondly, the increasingly important role played by the global scale (filtered, in the case of Spain, by the European scale) and the parallel rise in power of the regional/local scale in policy making—already implied in the above-mentioned changes—should be emphasized as being a decisive factor in the amendment of the traditional water policy arena. Globalization is rapidly introducing new conditioning financial factors (World Trade Organization), political factors (growing importance of EU legislation, especially the recent WFD) and cultural factors (dissemination of extra-Mediterranean values as criteria for the evaluation of water policy). Yet, the growing role of regional and local water policy bodies makes the regional and local scale increasingly important as a privileged arena for confrontation and the struggle for social support and political legitimization (Sauri & del Moral, 2001).

In view of the tumultuous history of the NHP, from the 1993 draft to its approval in 2001, and its partial repeal in 2004, it looks as if the politicians responsible for the initial proposal were making a grave mistake when they stressed that, "It is not a problem that has to be discussed among autonomous regions, it is not a problem of the political division of the State, it is a question of the physical organization of the Spanish State" (Borrell, 1993). On the contrary, the political reality has proved to be more important than the physical reality: interbasin transfers, when they take place within a single autonomous community, are less (although still very) conflictive than the distribution of water within a single river basin but between different autonomous regions. As Pérez Royo (1999) said, "if it had occurred to any of the members of the constitutional assembly [the Spanish Parliament that established the present-day structure of Autonomous Communities] or to any of the commentators on the Constitution in 1978 to predict that in 1999 we would be where we are now with regard to the structure of the State, he would have been considered a visionary ... And anyone who fails to see this has no business in politics".

6   FINAL REMARKS

The effects of an important change in the way water issues are addressed have made themselves felt in Spain over the last decade. In view of the dynamics of the situation, it is risky to venture any definitive assessment of the final outcome of the changes. Suffice it to say that the climate change issue, which is likely to have crucial repercussions on Spanish hydrology, has only just come onto the water debate agenda in Spain (Iglesias *et al.* in this volume). It is likely, however, to have far-reaching implications in the medium term.

With regard to operational considerations, the new water policy discourse comes up against one powerful obstacle in Spain: agricultural interests and the values of the agricultural sector still hold great sway over Spanish society. Furthermore, there has been much talk about the negative effects of cost recovery on different social sectors, especially agriculture. However, the latest reforms are bringing about a significant change in this situation.

The repeal of plans for the Ebro transfer constituted a historical milestone in Spain's long-standing hydraulic policy (see Chapter 19). The arguments raised to justify abandoning the plans were compelling: "the 2001 NHP", said the decree repealing the plan, had "significant and serious deficiencies" regarding fundamental economic aspects (overstatement of benefits, systematic underestimation of costs, failure to explain price structure), environmental aspects (lack of measures to protect affected rivers, lack of measures to ensure the safety of protected species, ineffectual approach to salinity problems) and technical aspects (inaccuracy of studies on actual availability of water to be transferred) (MIMAM, 2004).

The partial repeal of the NHP coincided with the implementation of the new WFD. The WFD calls for such a change in water policy logic and addresses the issues surrounding it. Concepts such as restoring the ecological status of water, responsibility for the cost of water resources, incorporation of coastal water management and active social participation are the pillars of the new legal framework. This is a change of strategy that entails a profound transformation of objectives, procedures, routines, professional profiles and type of social agents involved.

In support of, and beyond this new legal framework, there is an extensive process of reflection on water in the field of social, economic and environmental science, a rapid change in water treatment and purification technology and an important process of social awareness and mobilization in defense of waterscapes as the expression, in many cases, of the territory as a whole, and living conditions. As a result, water policy is perhaps the sector in which the shift towards sustainability strategies is most clearly formulated and has most specific technical and legal resources.

In view of the system's inertia, though, everything points to competition for water resources continuing to increase in the near future in metropolitan and coastal areas between urban, new and increasing environmental uses and agricultural demands. This process may also expand to the inland areas of the large river basins and across different irrigation zones with respect to their location, the types of crop grown, productivity and infrastructure efficiency.

The debate and conflict surrounding water quality will, very likely, come increasingly to the fore, as will the opposition of local/regional communities upon seeing their natural heritage spoiled for the sake of the development of alien production activities. In such a context, efforts to make the present concession system more flexible, develop mechanisms to reallocate water resources among users, progressively reduce administrative water supply allocations for irrigation and introduce incentives consisting of reduced rates for water-economizing practices are bound to succeed. How equitable the outcome is, what kind of economic efficiency is achieved, what the social consequences are and how the new structure of the power relationship for water will be forged will all depend on the definitive institutional framework in which the transition to the new management model takes place.

REFERENCES

Aguilera Klink, F. (2002). *Los mercados de agua en Tenerife* [Water Markets in Tenerife]. Bilbao, Spain: Bakeaz.

Allan, T. (2003). IWRM/IWRAM: a new sanctioned discourse? *Occasional Paper 50*, SOAS Water Issues Study Group, School of Oriental and African Studies, King's College London, University of London. www.wca-infonet.org/cds_upload/1061476217996_water_policy.pdf

Borrell, J. (1993). El Mundo, 16 January.

Consejeria de Agricultura y Pesca. (1999) and (2003). *Inventario y caracterización de los regadíos de Andalucía.* [Inventory and characterization of Andalusian irrigated areas] Seville, Spain: Dirección General de Desarrollo Rural y Actuaciones Estructurales, Junta de Andalucía.

Douglas, M. & Wildavsky, A. (1983). *Risk and culture: an essay on the selection of technological and environmental dangers.* Berkeley, CA: University of California Press.

Drain, M. (Ed.). (1995). *Les conflits pour l'eau en Europe Méditerranéenne.* [Water conflicts in Mediterranean Europe] Montpellier: Université Paul-Valéry.

Embid Irujo, A. (Ed.). (1999). *Planificación hidrológica y política hidráulica (El Libro Blanco del Agua).* [Water planning and policy (The White Paper on Water)]. Seminario de Derecho del agua de la Universidad de Zaragoza. Zaragoza: Confederación Hidrográfica del Ebro and Civitas.

Estevan Estevan, A. (2008). *Herencias y problemas de la política hidráulica española.* [Inheritance and problems of Spanish water policy]. Bilbao: Bakeaz.

Feitelson, E. (1996). The implications of changes in perceptions of water in Israel for peace negotiation with Jordan and the Palestinians. In Allan, J.A. and Radwan, L. (edits.) *Proceedings of the European Seminar on Water Geography. Perceptions of value of water and water landscapes*, London, SOAS, University of London, 17–22.

Giasante, C., Aguilar, M., Babiano, L., Garrido, A., Gómez, A., Iglesias, E., Lise, W., del Moral, L. & Pedregal, B. (2002). Institutional Adaptation to Changing Risk of Water Scarcity in Lower Guadalquivir. *Natural Resources Journal*, 42(3): 521–563.

Gómez Mendoza, J. & Ortega Cantero, N. (1987). Geografía y Regeneracionismo en España [Geography and *Regneracionismo* in Spain]. *Sistema* 77: 77–89.

Gómez Mendoza, J. & Moral Ituarte, L. del (1995). El Plan Hidrológico Nacional: criterios y directrices. [The National Hydrologic Plan: criteria and directives] In *La planificación hidráulica en España* [Water Planning in Spain], edited by Gil Olcina A. and Morales Gil, A. Alicante: Caja de Ahorros del Mediterráneo, 331–378.

López Ontiveros, A. (1998). El regadío, salvación de la patria y fuente de felicidad, según los congresos nacionales de riegos (1913–1934). [Irrigation as the nation's salvation and source of happiness, according to the national irrigation congresses (1913–1934)] *Demófilo. Revista de Cultura Tradicional de Andalucía* 27: 7–64.

Llamas, M.R. (1988). *El agua subterránea como recurso económico, ecológico y como agente geológico.* [Groundwater as an economic and ecologic resource, and as a geologic agent] Revista de la Real Academia de Ciencias Exactas, Físicas y Naturales, Madrid.

MIMAM (2000). *Libro Blanco del Agua en España* [The White Paper on Water in Spain]. Madrid, Spain: Ministerio de Medio Ambiente.

MIMAM (2004). Real-Decreto Ley 2/2004, de 18 de junio, por el que se modifica la Ley 10/2001, de 5 de julio, del Plan Hidrológico Nacional [Amendment of the National Hydrological Plan Act (Royal Decree 2/2004, 18th June 2004)]. Madrid: Boletín Oficial del Estado 148.

Ministerio de Obras Públicas y Medio Ambiente. (1993). *Plan Hidrológico Nacional. Memoria.* [National Hydrologic Plan. Background studies]. Madrid.

Moral, L. del, Van del Werff, P. Bakker, K. & Handmer, J. (2003). Global Trends and Water Policy in Spain. *Water International* 28(3): 358–366.

Moyano, E. (2003). La nueva cultura del agua: discursos, estrategias y agentes sociales. [New Water Culture: discourses, strategies and stakeholders]. In *III Congreso Ibérico sobre gestión y planificación del agua. Ponencias* [Third Iberian Congress on Water Management and Planning. Proceedings], edited by Arrojo, P. and del Moral, L. Zaragoza, Spain: Institución Fernando el Católico, 547–556.

Naredo, J.M. (1997). *La Economía del Agua en España.* [New Water Economics in Spain] Madrid: Fundación Argentaria.

Naredo, J.M. (1999). El agua y la solidaridad. In *Ciudades para un futuro más sostenible. El Boletín de la Biblioteca*, nº 11. http://habitat.aq.upm.es/boletin/n11/ajnar.html

O'Riordan, T. & Jaeger, J. (1996). Social Institutions and Climate Change. In *Politics of Climate Change: A European Perspective*, edited by O'Riordan, T. and Jordan, A. London: Routledge, 65–105.

Ortega Cantero, N. (1979). *Política Agraria y Dominación del Espacio.* [Agricultural Policy and Spatial Dominance]. Madrid: Editorial Ayuso.

Ortega Cantero, N. (1992). El Plan Nacional de Obras Hidráulicas. [The National Plan of Water Works] In *Hitos históricos de los regadíos españoles* [Historical landmarks of the Spanish irrigated agriculture], edited by Gil Olcina, A. and Morales Gil, A. Madrid: Ministerio de Agricultura, Pesca y Alimentación, 335–364.

Ortí, A. (1984). Política hidráulica y cuestión social: orígenes, etapas y significados del regeneracionismo hidráulico de Joaquín Costa [Water policy and social issues: origins, stages and meaning of Joaquin Costa's hydraulic *regeneracionismo*]. *Agricultura y Sociedad.* 32: 11–107.

Pérez-Díaz, V., Mezo, J. & Alvarez-Miranda, B. (1996). *Política y economía del agua en España* [Water Policy and Economics in Spain]. Madrid: Círculo de Empresarios.

Pérez Picazo, M.T. (2003). Aspectos económicos de la planificación hidrológica [Economic aspects of water planning]. In *III Congreso Ibérico sobre gestión y planificación del agua. Ponencias* [Third Iberian Congress on Water Management and Planning. Proceedings], edited by Arrojo, P. and del Moral, L. Zaragoza: Institución Fernando el Católico, 233–252.

Pérez Royo, J. 1999, *El País,* 18 December.

Reisner, M. (2001). *Cadillac Desert. The American West and its disappearing water*. London: Pimlico.

Riesco Chueca, P. (1999). La traza de lo medioambiental en la cultura contemporánea [The mark of the environment on contemporary culture]. *Argumentos de Razón Técnica* 2: 137–152.

Sauri, D. & del Moral, L. (2001). Recent developments in Spanish water policy. Alternatives and conflicts at the end of the hydraulic age. *Geoforum* 32(2001): 351–362.

Swyngedouw, E. (1999). Modernity and Hybridity: *Regeneracionismo*, the Production of Nature and the Spanish Waterscape, 1890–1930". *Annals of the Association of American Geographers* 89(3): 443–465.

CHAPTER 10

Water sector regulation and liberalization

Gaspar Ariño Ortiz
Department of Public Law, Autonomous University, Madrid, Spain

Monica Sastre Beceiro
Ariño & Asociados Law Firm, Madrid, Spain

1 INTRODUCTION

Like other sectors such as electricity, gas, telecommunications, infrastructure and transport, water regulation is an area of public law currently undergoing an extensive legislative transformation. The hydraulic sector reform was consolidated by the Water Act 46/1999 (1999 Water Act, hereafter) dated December, 13th, which amended the 1985 Water Act. Its possibly most controversial reform was the creation of a so-called "water market", overwhelmingly acclaimed in many sectors and cautiously accepted by others.

In light of this changing situation in sector legislation, it is worth considering how far water regulation has come and where it is going. The issue goes back a long way in Spain. If there is one example studied and followed by foreign legislators, this would have to be the water industry, where our long-established institutions erected authentic legal monuments that stood the test of time.

Even these institutions, however, have been unable to survive in times of change and uncertainty. Over the last hundred years, many of society's standards have changed: the economic revolution, technological progress and, above all, the enormous transformation of society as a whole by modern urban and industrial life slowly cracked the foundations of our magnificent 1879 Water Act. A new era was to begin in 1985. Yet, many of the old act's principles are everlasting, as they are inherent in human nature. As this chapter argues, some inaccuracies in the 1985 Act were due to neglect for the 1879 Water Act's guiding principles. The 1999 Water Act, which endeavored to make amends by introducing a water use transfer agreement, was only a partial solution due to the limits imposed by the so-called, although fictitious, "water market", which has never really existed (as opposed to the "black market", which obviously does exist).

2 HISTORY OF WATER REGULATION IN SPAIN: SIX STAGES

2.1 *Stage One: The 1879 Water Act's pro-market approach*

Spanish water legislation had traditionally been that of a country with scarce water resources, in which public authorities were granted substantial powers by the institutions of the Kingdom of Aragon, always at the forefront of our water laws. Note, though, that our legislation was, at the same time, pro-market: it encouraged private initiative and respected the appropriation of rights, guaranteeing progress and wealth. For instance, according to the old 1879 Water Act, rainwater and water from springs and streams, amongst other sources, was considered private property; landowners were able to dam, build deposits and channel water on their own property and were also entitled to "vested ownership" of groundwater brought to the surface on either public or private land (Menéndez Rexach, 1986).

Another indication of the philosophy behind our old water law is the principle of open private initiative with respect to licensing, prescription of rights and exploitation, promotion of private initiative in public works and the allocation of surface water through *order book licenses* by forecasting future public service needs, especially those relating to water supply to the general public and energy demand.

All of the above was designed by the State to urge society to act, accompanied by a set of legal instruments granting "water rights" to different organizations and allowing them to transact and make use of the commodity. The system was driven by the market.

2.2   *Stage Two: The 1985 Water Act*

As opposed to the 1879 Water Act, the 1985 Water Act introduced a substantial degree of interventionism. It included positive features in relation to previous legislation, such as the formulation of principles of water and environmental protection; unity of management and decentralization; establishment of a comprehensive water quality protection system by prohibiting unauthorized use and granting the authorities the power to enforce and punish offenders, as well as the "water levy". In our view, though, the 1985 Water Act was flawed for a number of reasons:

– It declared groundwater to be in the public domain, which was neither "necessary nor sufficient" for better water management (Llamas & Custodio, 1985). Even so, the Authorities have been able to control groundwater by making it public, although they still do not have an acceptable notion of its volume or current use (see Chapters 14 and 15). As Llamas aptly puts it, "licensed use hardly reaches 100,000, which means that over 90% of actual water is obtained illegally" (Llamas *et al.*, 2001). The consequences are illegal use, over-exploitation of reserves and black markets.
– The 2001 NHP Act attempted to regulate the situation, granting license holders of private water reserves a non-renewable extension of three months as of the date upon which the National Hydrological Plan came into force to apply for Regional Water Classification. Once this period had expired and the order had not been complied with, no water reserves could be classified as private unless declared as such by means of an irrevocable court order (not subject to appeal). Owners of wells that are not registered or catalogued would therefore have to seek a court decision declaring the water's domain, by providing documented proof of alternative registration, such as a Property Registry Certificate (Moreu Ballonga, 2001; Del Saz, 2001).
– It established compulsory planning schemes based on water use assignment that, as of 1985, would be administrative decisions. The foreword of the above Act had already warned that "Regional Hydrological Plans and the National Hydrological Plan (...) would be the basis for all hydraulic administration activities relating to water, the choice of different alternatives for possible hydraulic projects, and resource and concession assignment and reservation". Nobody questions the soundness of some degree of strategic and informative planning, but it should not predetermine all use rights (Ariño, 2001).
– It replaced the private initiative permitted in the 1879 Water Act by public initiative. The 1985 Water Act, therefore, blocked the possibility of water markets, which, although to a limited extent, existed under the 1879 Act, taking away incentives.
– It set out the principles of free water, classifying all water as public property, and comprehensive water planning. In effect, the provisions of the 1985 Water Act make a distinction between three types of levies—(a) occupying public hydraulic domain, (b) flow control and a regulating levy and (c) water use fee (see Chapter 13)—, applicable to the beneficiaries of water regulation schemes or other hydraulic works, respectively. All these levies are charged not based on water use but to the beneficiaries of the water domain or the regulation infrastructure. Tariffs should be imposed on water use only. This would make its value dependable on the volume of water used, water quality, and on whether or not water is consumed and, in this case, based on economic profits (as in the case of nuclear plants).

2.3   *Stage Three: 1999 amendment of the 1985 Water Act*

The 1999 amendment had negligible effects on the identified errors in the traditional system. They can be summarized as follows:

- *Maintenance of binding and comprehensive planning*: The new law upheld the concept of water planning contained in the 1985 Act. In other words, it still conceived planning as the omniscient, dominating and binding focal point of all water policy.
- *Change in the financial and economic regime of the use of public hydraulic domain*: The levy on the occupation or use of public hydraulic domain (natural watercourses, lakes and lagoons, as well as artificial dams on public waterways) was increased to also cover its exploitation, by means of the tax entitled "levy on the use of the public hydraulic domain".
- *Licensing agreement: an inefficient market*: The 1999 reform introduced the water rights transfer agreement, which is analyzed below in more detail.

2.4   *Stage Four: The adoption of Law 10/2001, the National Hydrological Plan*

The Law of NHP of 2001 envisaged the transfer of 1050 hm^3 of the Ebro delta to the following regions: 190 hm^3 to the internal basins of Cataluña, 315 hm^3 to the Jucar basin, 450 hm^3 to the Segura basin, and 95 hm^3 to the South basin of Almería (see Chapter 19).

   To further avoid the expansion of irrigation in the recipient basins' beneficiaries of the transfer, the Act forbade the use of water transferred for new irrigated acreage, or the expansion of existing ones in areas affected by the transfers. Water transferred would be allocated according with the following criterions: 1) to access the use of water transferred users had to have concessions or any other sufficient proof that the right to exclusive use of water duly registered in the water catchments, 2) transferred water would not cause negative environmental changes, 3) prior to the use of any transferred water, the RBA could develop other rules meant to ensure that the only urban users and existing irrigated acreage would end up using transferred volumes.

2.5   *Stage Five: Law 62/2003, of december 30, which transposed into Spanish law the Water Framework Directive 2000/60/EC*

Law 62/2003 amended the 2001 Consolidated Water Act, to incorporate into Spanish law Directive 2000/60/EC, laying down a framework for Community action in the field of water policy (the WFD). The amendment's main objectives were basically those of the WFD. It also foresaw (1) the creation of a registry of protected areas, and defined the proceedings for public participation, (2) a reform of the economic and financial regimes introducing the principle of cost recovery for services related to water management; (3) a more precise regulation municipal for works of general interest; (4) the declaration of urgent occupation of certain properties affected by water works, particularly those related to infrastructure works for water transfers authorized by Article 13 of Law 10/2001, of July 5, the National Water Plan, and establishing the transitional regime of water from the reservoir Negratín to Cuevas de Almanzora (in Almería).

2.6   *Stage Six: The change of government in march 2004*

The *Water Management and Use Program* (AGUA) marked the start of this phase. The resulting redirection of water policy first became evident with the passing of Royal Decree—Act 2/2004, dated June 18th, which revised the 2001 NHP Act. The basic content of the law was to repeal the inter-basin transfer from the Ebro River to the Júcar, Segura and South basins. This was replaced by a series of projects to be carried out in the Mediterranean basin, consisting of desalination, re-use, purifying and upgrading of hydraulic infrastructure, many of which were listed in Annex II of the NHP Act.

The intended reforms in water legislation can be summarized as follows:

- All desalinated water is placed in the public domain, regardless of whether it comes into contact with public domain hydraulic elements or not after being produced.
- Implementation of Water Exchange Centres in the areas of the Segura, Júcar and South basins, subject to prior authorization by the Cabinet of Ministers (see Chapter 13).
- Transfer of control of effectively consumed or used water from the river basin authorities (RBAs) to the Ministry of the Environment.
- Mandatory RBA reviews evaluating the availability of resources, whatever the volume, in the case of regional planning projects and initiatives carried out by Autonomous Communities (ACs) that affect water ecosystems and result in greater demands for water.
- Envisaged modification of the levy on the use of public domain hydraulic assets that are used for protection, improvement and re-generation. The amendment consisted of a fixed levy on the volume of surface or groundwater actually consumed, called the "ecotax". The levy rate was to be from 0.001 €/m^3 to 0.016 €/m^3 depending on the type of water use. However, the proposal was eventually withdrawn by the Government as a result of the objections raised by the National Water Council's Standing Committee, given that the water ecotax would have cost the Spanish irrigation sector 30 million euros per year.

3   REGULATION OF THE SO-CALLED "WATER MARKET" IN SPAIN

The transfer of water use rights is nothing new under Spanish water legislation, as public auctions of water have taken place for centuries between landowners of Levante (Valencia and Murcia region). In this eastern region, groundwater has been bought and sold for centuries (when it was privately owned) at public auctions and managed by irrigation communities (Maestu, 1997). At the present time, following the 1985 Act, a change in ownership or use of water involves a *conversio tituli* of the right to consume private water under administrative license, leading to loss of its privateness and inclusion in the public hydraulic domain (Díaz Lema, 1999; Ménedez Rexach, 1998). This is because Transitory Provision 2 of the Water Act establishes that owners of the right to use groundwater registered either in the Water Registry or in the Private Water Catalogue that wish to modify the conditions or type of use must apply for the appropriate license.

3.1   *The water market in the Canary Islands*

There is no single water market in the Canary Islands; nor is there even an Island market. There are as many markets as productive areas that require water to run their operations. Such markets operate mainly on the islands of Tenerife, Gran Canaria and, to a lesser extent, La Palma. Exchanges represent between 20–50% of available resources, in other words, around 150 Hm^3/year, with a yearly value ranging from €60,000 to €90,000 (Jiménez Suárez, 1997).

There are two types of privately managed water markets on the major islands: shares and water. The share market involves the free transaction of shares owned by the irrigation communities. Irrigation community shares are therefore bought and sold as if it were a kind of water stock exchange. The share market operates via brokers. When someone is interested in purchasing water shares in an irrigation community as security or to gain access to reserves of water for irrigation, this person contacts a water broker (trader or marketer) and buys one or more such shares in the irrigation community for an assessed price. The broker usually contacts the person in charge of the irrigation community to establish how the water can be delivered. The share market reached its peak in the 1960s and 1970s with a volume of 5,000 shares per year, worth 6 million euros.

The water market (as a liquid element) operates in both the private and public sectors. The private sector market was generated by the need to supply urban consumption. Initial supply was controlled by the "Heredamientos"[1] and used for agriculture. Hydraulic companies in the last century purchased water from the "Heredamientos" and either sold it to town councils for distribution or distributed it themselves. The seller was normally the "Heredamiento" or a community of assets in charge of the sale of water not used by its members. The main buyers were, and still are, town councils, industrial and tourism entrepreneurs and new farmers. It is well known that the Canary Islands went from being a farming society to a services society with completely different water needs and uses.

Like all other operators, municipal authorities that purchase water are normally subject to market rules. They purchase water from brokers, although they may, in times of drought, expropriate (or confiscate) water for emergency supply to the community.

The institutional market is another type of market. It is public and was set up in 1960 by Gran Canaria Island Council and obtained water from reservoirs that was then sold to farmers. At the present time, it has a license for 5 dams, moving a volume of 10 Hm^3/year, and sells water annually at a fixed price. In 1997, this price was 0.36 €/m^3.

Tenerife Island Council set up its own organization, "BALTEN", with rights for 13 dams, 350 km of pipes, fourteen concessions for surface water, three wells and a desalination plant. The organization buys water in winter and sells it in summer, for the purpose of operating on the market. In addition, town councils now sell purified water resulting from their access to waste water networks.

The Canary Island market system has had positive effects, such as the promotion of competition and water saving practices by companies and communities. The developed system was highly efficient, bearing in mind that water has never been scarce in areas of very little rainfall, even in drought periods. Note also that the authorities have not made substantial investments in hydraulic projects on the islands (as is the case on the mainland), as private capital was there to do the job. In Tenerife, for example, there is a complex transport network which connects the entire island hydraulically. This makes it easy to send water from any area of the island to another at a lower altitude. Many canals were built with private capital through the above-mentioned share procedure. Only when the profitability of the networks was doubtful, normally due to high costs relating to surveying difficulties (such as the case of the Garafia-Tijarafe canals and Barlovento-Fuencaliente), did the Central Government have to intervene with totally State-funded public works.

It is also true that the market has operated with very little transparency and public information, according to customs that were not legally protected, as well as subject to some operator monopolies. This made official regulation necessary. This regulation has made the market and its operators more transparent. The lack of monitoring of conveyance networks and connections resulted in some abuse with respect to unjustified reductions in transported volume as well as inflated prices, as this is the only part of the water market entertaining a monopolistic temptation. There have been attempts by some town councils to buy water directly from owners, thus excluding brokers. However, they encountered too many barriers due to the fact that canals cross different properties, and brokers blocked operations. This has led some users to construct private networks in order to avoid transport abuse (Aguilera Klink, 2000). Finally, there has been a lack of concern for the environment in the Canary Island water market (non-sustainable use of aquifers) (Aguilera Klink, 1994; Maestu, 1997).

3.2 *Water market regulation in mainland Spain*

Exchanges within the same basin
In mainland Spain, it was the 1999 amendment to the 1985 Water Act that institutionalized surface water markets by making the rigid licensing system more flexible. The Preamble stated that *"in order to increase the efficiency of water use, it is necessary to make the current licensing*

---

[1] "Heredamientos" are owner associations that were later regulated as joint stock companies. They are historically linked to the division of land following the Conquest and successive transfers of land and its associated spring water by inheritance. Land and spring water were later separated, during the 17th and 18th centuries. This is where the name "Heredamiento" (inheritance), referring to water and then to privately owned wells and galleries comes from.

system more flexible by introducing a new water rights transfer agreement, thus enabling optimum social use of such a scarce resource".

However, articles 67 to 72 of the 2001 Consolidated Water Act established so many limits and precautionary measures on the water market that it has never really existed. The NHP background documents admit that after one year's experience with licensing agreements, *"the system (transfer of rights) established under the recent water law reforms is not working as intended, despite the demonstrated existence of potential buyers for urban supply"* (MIMAM, 2000). Apparently, the two Ebro River Delta Irrigation Communities (both sides of the river) put part of their licensed water up for sale (300 Hm³) for supply to Barcelona in September 2002. The proceeds of the sale were to be divided between the Irrigation Communities (20% to 30%), and the remainder invested in Ebro Riverland projects. However, the sale did not take place. The truth is that the limits established by the 2001 Consolidated Water Act on water transfer agreements are excessive and must be reviewed. The water rights licensing agreement is subject to numerous precautionary measures and limits, such as:

– *Time Limit*: maximum duration until the assignor's license for private use expires.
– *Destination Limit*: assignment to another licensee or entitled organization of an equivalent or higher category as per the order of preference established by river basin planning or in accordance with the Water Act (Art. 60, 2001 Consolidated Water Act). Under exceptional circumstances and provided it is in the interests of the general public, the Ministry of the Environment may authorize licenses in a different order of preference (Article 67.2).
– *Quantitative Limit*: "the annual volume subject to the license may not exceed what is actually used by the licensor", where "regulations shall govern the calculation of this annual volume, based on the volume used over a specified number of years, adjusted as the case may be to the target allowance established under the river basin plan and proper use of water, and may never exceed the assigned volume" (Article 69). This poses the problem of how to accurately determine the rights assigned relating to the following three measures: 1) the rights granted according to title, 2) volume actually used and 3) the target allowance, based on the principles of efficient water use.
– *Subject to prior discretionary administrative authorization* by the RBAs, there are four options:
  • Authorization of the license
  • Refusal of the license, based on a reasoned decision in the following cases: "when it has a negative effect on the river basin resource use system, third party rights or environmental reserves, or does not comply with the requirements specified above, where stakeholders have no right to compensation whatsoever". The generic term of reference "negative" can potentially lead to arbitrary refusals that are difficult to control.
  • RBAs can also exercise preferential acquisition rights over the volume to be licensed during this period of time, thus putting an end to all types of private use.
  • Administrative silence then applies. "Considered as authorized, without effects between the parties thereto until such time, after one month as of notification to the RBAs, provided they do not object when licenses are granted to members of the same consumer association, and after two months in all other cases" (Article 68.2).
– *Formal Limits*: the following formal requirements exist.
  • "Licenses must be written and a copy of the agreement sent to the RBAs and consumer associations to which the licensor and licensee belong within fifteen days of signature" (article 68.1 2001 Consolidated Water Act).
  • "Should the license be granted to and by consumers for irrigation purposes, the contract must specify the land subject to non-irrigation by the licensor for the term of the contract, as well as the land to be irrigated by the licensee and the assigned volume" (Article 68.1). In addition to formal implications, this *prohibits dealers* or *traders*. As opposed to gas or electricity, water trading or re-sale (water licenses acquired under agreement) is unjustifiably forbidden.
  • RBAs must register water licenses in the Water Registry (Art. 68.4, 2001 Consolidated Water Act).
– *Physical Limits*: use of infrastructure. Negotiated third party access (TPA) is established in relation to hydraulic infrastructure belonging to third parties, and regulated TPA, with official tolls, when the regulating body is a RBA.

– *Political Limits*: licenses for the use of infrastructure connecting areas subject to different river basin plans may only be authorized if they come under the National Hydrological Plan or other specific laws relating to a particular diversion scheme. This means that diversion is still a political operation, which is a serious mistake.
– *Limits on open pricing*: according to Art. 69.3 of the 2001 Consolidated Water Act, "regulations may determine maximum price limits" for water licenses. Competitive pricing is substituted by administrative intervention.
– *Market organization limits*: the Act provides for setting up Water Exchange Centers at the initiative of the Cabinet and at the request of the Minister of the Environment (in the cases foreseen in Articles 55, 56 and 58, 2001 Consolidated Water Act). In such cases, RBAs are allowed to buy and sell water rights. However, these Water Banks are subject to public contract procedures ("acquisition and disposal of water rights carried out under this section must comply with the principles of publicity and open bidding in accordance with the procedures and selection criteria established under applicable regulations"). This is absurd, costly and difficult to implement. "Selection criteria" imply central planning, as opposed to pricing criteria or open agreement between parties (Ariño & Sastre, 2004). To date, Water Exchange Centers have only been authorized exceptionally for a particular river basin. This was the case of the Sinclan de Calasparra aquifer in the Segura River basin, used to cover a water deficit in the Canales de Taibilla Local Corporation, which supplies 76 towns in Murcia, Alicante and Albacete, many of which suffered annual water restrictions of up to twelve hours a day.

In sum, the water market is subject to all kinds of restrictions. Real water prices reflecting economic worth and therefore facilitating efficient use of such an essential commodity are highly unlikely to come out of such regulations.

Exchanges across basins
The regulatory framework of exchanges across basins is much more complex, and includes a number of provisions. Among the most significant ones are that a 3-month review period is mandated, instead of just one month for intra-basin exchanges, and the required approval of the Ministry of the Environment. And yet, the Royal Decree—Law 15/2005, approved urgent measures for the regulation of the transactions of rights to the water use. It allowed for the utilization of existing infrastructures basins between the reservoir of the Negratin and Almanzora's Cuevas, as well as the aqueduct Tagus-Segura which will be used for the implementation of rights transfer agreements. The economic-financial regime applicable to these transactions was established with singular criterions regulating the use regime of the corresponding infrastructures. Likewise, in agreement with the Royal Decree—Law 3/2008, of April 21, of exceptional measures to guarantee the supply of populations affected by the drought in the province of Barcelona, it allowed for construction of an inter-basin conduit from the Water consortium of Tarragona and the system Ter-Llobregat (CAT-ATLL) for the rights transfer agreements.

The experience during 2005, 2006, 2007 and 2008 shows a clear inclination of the Ministry to renew all needed permissions for inter-basin exchanges and remove most obstacles by means of annual Royal Decree—Laws, that the 2001 Consolidated Water Act had envisioned to regulate. In the four cases, the recognition of special drought circumstances in the recipient basins has been invoked to provide rationale and support to the removal of the red-tape and burdensome administrative procedure which the Law had established.

Water Banks or Exchange Centres
In cases of drought and other exceptional circumstances, the 2001 Consolidated Water Act foresees the setting up of "Exchange Centres of Concessions" (called "Water Bank"), as proposed by the Ministry of Environment. The RBAs can offer for tender the acquisition of water use rights and offer them to other users. In 2004, the government authorized the constitution of Exchange Centres in Segura, Júcar and Guadiana, and in 2008 in the Guadalquivir. In the latter case, the

centre will be closed down when the Program A.G.U.A. becomes fully operative (some economic information of these exchanges is reported in Chapter 13).

4   WATER SUPPLY TO THE GENERAL PUBLIC

4.1   *Regulatory framework for water services*

Regulation of the supply of drinking water and sewerage services in Spain must be defined according to the different levels at which they are provided. They are basically wholesale and retail water services.

Bulk or wholesale services
The wholesale services cover large projects involving regulation, piping, purification and storage activities that enable the supply of water to so-called head deposits by means of outlets at dams, rivers and wells designed to take water from underground deposits, or, in the case of desalination, simply from the sea.

This first tier, or level of activity, is determined by the need for large and strategically important infrastructure projects requiring substantial investment (dams, sewage plants, extensive channeling and, generally, large investments that enable a wholesale approach to water). The fact that such activities are specifically linked to State Authorities necessarily requires the presence of the Public Administration, which must have some control or influence on how services are managed. Even so, semi-public companies could play a key role in this sector. This would strike a balance that would enable first a sufficiently attractive return on private capital and adequate levels of safety and, second, participation by public authorities not only in funding but also in running the services, without detriment to the management control of the private company partner or partners.

At this level of activity, there is no reason whatsoever for exclusive rights or monopolies. Water today is not only a natural product, but is also a potentially "industrial" product (that can be "produced"). If there were an organized "bulk water market", it could lead to a greater possible supply of water for public projects at potentially lower prices. In this way, for example, water could be transferred from the Rhône River (in France) to the Ebro. However, the entire European Union would have to liberalize their markets so as not to breach the community principle of reciprocity.

Retail services
Head installations and reservoirs also supply consumers via smaller and more diverse piping systems. These systems comprise the so-called retail supply to subscribers. Retail services may be divided again in two levels (metropolitan and municipal services). These services should be managed locally due to the proximity to consumers. Water supply and sewerage networks often call for substantial investment, similar to that required for water intake (construction of dams, wells, etc.). In addition, the calculation of consumption, invoicing, payment, network conservation and maintenance take much more managing that bulk services.

Needs at this level focus mainly on the fact that considerable management is required: small-scale but constant projects; installations and services repair, numerous administrative and payment collection staff, as well as the ability to rapidly respond to needs that only the private sector is able to do. At this retail level, private operator participation must be channeled through what is known in Spain as "indirect management systems", which are referred to below.

4.2   *Management systems of supply to the general public*

In Spain, there are several organizational systems for supplying water to the general public.

Municipal systems
In municipal systems, water is supplied to the general public by the town council, either *directly* (as such or by setting up a decentralized town council-run organization or management company), or

indirectly "in conjunction" with private entrepreneurs and companies (businesses or associations) that are incorporated specifically for the purpose.

Association systems: Local corporations, consortiums and metropolitan areas

Services can also be provided by municipal systems acting through associations, a system designed to compensate for the insufficiencies resulting from the strict territorial limits of municipalities and to provide the quality of service demanded by modern day society.

This system includes *local corporations,* which are voluntary associations of town councils (formed by official agreements) incorporated for the purpose of jointly carrying out projects and providing authorized services such as water supply to the general public. An example is the Pamplona Region Local Corporation, which manages the integrated water and sewage cycle.

In *local corporations,* municipal bodies join together with other public authorities, or private non-profit organizations, to provide water supply services by setting up an independent organization. A typical example of such an organization is the "Consorcio de Aguas de Bilbao-Bizkaia" (Bilbao-Bizkaia Water Consortium) comprised of forty-three town councils, the Basque Government and Vizcaya Council.

Regions are groups of similar town councils, combined for geographic or historical reasons, normally with similar social and economic activities.

Metropolitan areas are local bodies comprised of large urban town councils that have economic and social ties, requiring joint planning and coordination of some services and projects.

State models, the "Los Canales de Taibilla" local corporation

The "Canales de Taibilla" Local Corporation was set up to supply drinking water to the cities of Cartagena, Murcia, Lorca and Alicante, as well as other towns in the southeast region. Despite its name, it is not a true association of town councils, but rather a State institution (self-governing body). Instead of providing integrated and comprehensive water supply services, it is only involved in bulk supply (intake, regulation, purification and piping or transport) to head deposits. Therefore, it does not have the jurisdiction or competences of a town council authority regarding municipal supplies. It also deals with the supply of water to municipal networks linked to deposits, where the retail distribution phase of municipal services takes over, after having invoiced at set rates (Morales Gil & Vera Rebollo, 1989).

Regional systems, Community of Madrid

By virtue of the exclusive rights entitling Autonomous Communities to carry out "public works", as well as "projects, construction and exploitation of hydraulic developments" (Art. 148.1.4th the Spanish Constitution), some phases of water supply have been regionalized (intake, adduction, bulk supply, basic networks), including those classed as being in the interests of the region and thereby excluding town councils.

This is the case of the Autonomous Community of Madrid, which assumes control of water adduction and purification through an organization called Canal de Isabel II (see Chapter 13 to learn about its water pricing criterions).

4.3   *Forms of private participation in water supply to the general public*

The participation of private entities in providing water supply services to the general public can be achieved through the following systems of indirect management.

Licensing

Licensing is a form of contract between Public Administration agencies and private individuals, in which the former grants the latter the right to manage a public service at their own risk, normally accompanied by an investment by the licensee (works and installations) that is recovered by charging a fee for services provided. It may also include an equity contribution from the public authority concerned. In any case, the public authority maintains internal management and control of the ways and means of providing the service. Services are provided on behalf of the Public Administration agencies, which therefore manage and are responsible for all services rendered.

The Public Administration performs its tasks in different ways, such as: 1) the power to manage by issuing instructions and orders, 2) power to guide and monitor company activities, 3) power to fix rates, although licensees are able to make proposals to the authorities. The authorities are obliged to ensure that the license is economically balanced at all times, and 4) the right to revoke a license as a result of a serious breach of a licensee's obligations. Licenses may last for a maximum of 75 years. Even so, the term is normally much shorter (10 to 35 years).

In recent years, there have been numerous examples in Spain of town councils having changed from direct water supply management systems via municipal companies to indirect management systems via licensing, including, since 1996, the Oviedo, Toledo, Jaén and La Laguna Town Councils. Licensing in Spain currently represents around 29% of the population, as opposed to 23% in 1995.

Semi-public companies

This system exists when public authorities own part of the equity—in their own name or through a public body—in a private company that manages public water supply services. Management by a semi-public company has two essential traits: firstly, the presence of private capital that normally assumes company *management*, and secondly, the compulsory *time limit* on control (with a maximum term of 75 years), meaning that services and installations revert to the titleholder upon expiry.

Company shares (both public and private) are normally balanced. Furthermore, the key to the successful operation of semi-public companies does not depend on the level of public and private shares, but on the combination of multiple elements: individual initiative and public control, profit and quality of service, public funding and private management, shareholders and consumers, apart from maintaining this delicate balance for the entire duration of the company.

Private participation in a semi-public company may take place at incorporation or at a later date in water supply and waste management companies that are fully owned by the town council and later wish to privatize part of their share capital by selling shares by public tender.

We consider this system of management of water supply to the general public to be positive, as the creation of semi-public companies leads to private operators contributing finance, industrial experience and management efficiency to public services. In effect, semi-public entities involve a combination of public and private (financial, technical and human) resources. In Spain, this type of indirect management accounts for 7% of the population (5% in 1995).

Leasing

The main purpose of leasing is to exploit municipal installations, the use and exploitation of such being the service subject to agreement. The municipal authorities are responsible for the services provided.

More specifically, this type of management consists of a private company managing a service according to the terms of tender. The management company receives a fixed fee for m^3 invoiced, as previously stipulated under contract, and undertakes to assume the financial risks involved in exploitation. The council assumes the funding obligations. Installations are owned by the council and leased to the private company. The maximum term of lease agreements is 75 years, however they normally last for much less (5 to 10 years). Lessees are obliged to keep premises and installations in perfect repair, only use them for the agreed purposes, pay for all necessary repairs and return them upon expiry of the lease term in the same condition as they were delivered. Fees are determined in order to balance income and expenses, the latter including a fee charged by the management company for services rendered.

In population terms, lease agreements make up for approximately 8% of the Spanish market (7% in 1995).

The above goes to show that in population terms the management of water supply and sewage services using private capital (including semi-public companies) represents around 44% of the Spanish market. The most important private entities operating in Spain are AgBar, FCC-Vivendi and Bouyages. These three groups account for 90% of the Spanish market (17 million inhabitants of a population of 40 million).

The distinction between bulk and retail services is also applicable to the waste water sanitation and collection services. Therefore, the most appropriate management models are also worthy of the same kind of analysis. There is a wholesale level of water treatment, drainage and recycling related to river basins and the potential secondary or tertiary uses of recycled water (gardens, fountains, parks, street cleaning, etc…), as well as a urban retail level, which basically involves maintaining drainage systems in healthy conditions. In theory, distribution and sewerage should go together, as in Great Britain.

In relation to the above paragraph, we would argue that there are no reasons why treated water cannot be sold for secondary or tertiary use.

5 CONCLUSIONS

We conclude by saying that we are on the verge of a new era, with greater private sector participation in the water and hydraulic projects industries moving in two directions: 1) increasing transformation of public water supply and recycling from direct municipal management systems to indirect systems involving semi-public companies and licensing (in Toledo, Oviedo, Granada, Vigo and elsewhere), and 2) faced with the crisis in traditional public funding of infrastructure, more and more private funding systems are being used to finance hydraulic projects, ranging from traditional public works licensing agreements to administrative contracts on total price payment systems and different types of project funding systems.

Furthermore, the 2001 Consolidated Water Act represents a shift towards greater levels of private initiative in groundwater management and the reassigning of water use by means of license transferability. The creation of even a limited water market will enable improved allocation of hydraulic resources. Nevertheless, until now, water license transfer agreements have had, as far as we know, very little practical effects.

Finally, we are moving towards water management systems with increasingly more room for private participation. The key to change lies in opening public services up to competition and market pressure, deregulation and, above all, changing the course of such regulation. As long as water is public domain, it will continue to be subject to public control and conservation, based on the responsibility of guaranteeing general public interests, in particular those related to supply and environmental protection. It therefore becomes necessary to find a happy medium between public and private management of water and hydraulic projects. In this respect, we all have a role to play in finding the way to make the famous third way a reality.

REFERENCES

Aguilera Klink, F. (1994). *Proyecto Ardka* [Ardka Project]. Madrid, Spain: Dirección General de Calidad de las Aguas. MOPTMA.
——. (2000). El papel económico de las aguas subterráneas en Canarias [The role of groundwater in the Canaries]. Papeles del proyecto Aguas Subterráneas [Groundwater Project Series Documents]. Serie B-La Economía de las Aguas Subterráneas en España [Series-B- The economics of groundwater in Spain. Madrid, Spain: Fundación Marcelino Botín.
Ariño Ortiz, G. & Sastre Beceiro, M. (2004). Regulación del Agua [Water Administration]. In *Principios de Derecho Público Económico* [Principles of Economic Public Law], edited by G. Ariño Ortiz. Granada, Spain: Editorial Comares, 34–56.
Ariño Ortiz, G. (2001). La economía del agua en España [Water Economics in Spain]. *Revista del Instituto de Estudios económicos* 4: 45–78.
Díaz Lema, J.M. (1999). La cesión de derechos de uso del agua en el Proyecto de Ley de 7 de mayo de 1999, de Modificación de la Ley de Aguas de 2 de agosto de 1985 [Water Rights exchange mechanisms in the Bill of Law 7, May 1999, of the Reform of the Water Act of August 2, 1985]. Privatizaciones y liberalización de servicios [Privatization and Services Liberalization]. Madrid, Spain: Universidad Autónoma de Madrid.

Jiménez Suárez. (1997). *Personal communication.*

Llamas, M.R. & Custodio, E. (1985). *El Proyecto de Ley de Aguas. Informe Científico-técnico* [The Water Act Draft. Technical-Scientific Report]. Madrid: Spain, Instituto de Estudios Económicos.

Llamas, M.R., Fornés, J.M., Hernández-Mora, N. & Martínez Cortina, L. (2001). *Aguas subterráneas: retos y oportunidades* [Groundwaters: Challenges and Opportunities]. Madrid, Spain: Fundación Marcelino Botín and Mundi- Prensa.

Maestu, J. (1997). Dificultades y oportunidades de una gestión razonable del Agua en España: la flexibilización del régimen concesional [Difficulties and Opportunities of Rational Water Management in Spain: the liberalization of the water rights system] *In La economía del agua en España* [Water Economics in Spain]. Madrid, Spain: Fundación Argentaria, 24–38.

Menéndez Rexach, A. (1986). *El Derecho de Aguas en España* [Water Law in Spain]. Madrid, Spain: Editorial Ministerio de Obras Públicas y Urbanismo.

——. (1998). Consideraciones jurídicas sobre el mercado del agua [Legal Considerations on Water Markets]. Jornadas *INTERREG II Sobre Legislación, Planificación y Gestión del Agua* [Interreg II Workshop on Legislation, Planning and Water Management] Málaga, Spain, October 5–8.

MIMAM (Ministerio de Medio Ambiente) (2000). *The National Hydrological Plan* [Spanish Water White Paper] Madrid, Spain: Ministerio de Medio Ambiente.

Morales Gil, A. & Vera Rebollo, F. (1989). *La Mancomunidad de los Canales del Taibilla* [The Mancommunity of the 'Canales del Taibilla]. Alicante, Spain: Real Academia Alfonso X El Sabio and Universidad de Alicante.

Moreu Ballonga, J.L. (2001). Los problemas de la legislación sobre aguas subterráneas en España: posibles soluciones [Legal problems related to groundwater in Spain: possible solutions]. Papeles del proyecto Aguas Subterráneas [Groundwater Project Series Documents]. Serie D-Aspectos jurídicos de las Aguas subterráneas [Series D-Legal Aspects of Groundwater]. Madrid, Spain: Fundación Marcelino Botín.

Saz, S. del. (2001). ¿Cuál es el contenido de los derechos privados sobre las aguas subterráneas? [What is the content of private rights on groundwater] Papeles del proyecto Aguas Subterráneas [Groundwater Project Series Documents]. Serie D-Aspectos jurídicos de las Aguas subterráneas [Series D-Legal Aspects of Groundwater]. Madrid, Spain: Fundación Marcelino Botín.

CHAPTER 11

The foundations and principles of modern water law

Antonio Embid Irujo
Department of Public Law, University of Zaragoza, Spain

1 THE FOUNDATIONS OF MODERN WATER LAW: THE SPANISH CONSTITUTION (1978) AND EUROPEAN LAW

There have been two major institutional events in the evolution of Spanish water law as we know it today: the Spanish Constitution and Spain's accession to the European Union.

a. Apart from being Spain's key piece of legislation, the Spanish Constitution also marks the beginning of the most fruitful and creative period for Spanish water law. The Constitution's provisions redefining the public domain (art. 132), its unprecedented concern for the environment (art. 45), and the structural influence that the new territorial organization of the State was to have on the legal system as a whole (through arts. 148.1.10 and 149.1.22) provide the juridical foundations on which later water legislation was developed and eventually came into force.

b. Spain became a Member State of the European Union on January 1st 1986, and this was also the date that Spain assumed the EU's package of measures concerning water resources. From this date onwards, she was to be not only a recipient, but also a partner in the creation of a European legal system concerning a matter that was constantly growing in complexity and ever more difficult to interpret. This set of European regulations has influenced Spain's national law in the field of water quality and the environmental assessment of hydraulic schemes. As a result, the country's national regulations in this area are directly or indirectly offshoots of the European regulations. Indeed, the latest example of this creative drive is Directive 2000/60/EC of the European Parliament and Commission of October 23rd, which establishes a Community framework for action in the area of water (hereafter referred to as the WFD, see Chapter 16). This Community Directive was transposed into Spanish law by Act 62/2003 of December 30th, modifying the Consolidated Water Act, passed by Executive Order 1/2001 of July 20th, and these are the regulations currently in force in Spain (Tirado Robles, 2004).

2 WATER AS A RESOURCE: PUBLIC AND PRIVATE OWNERSHIP

The significance of article 132 of the Constitution is primarily illustrated by the redefinition of the water as public property, with profound implications for subsequent legislative initiatives on water. And this is substantially how art. 132 should be understood as a provision that is remarkably unique in the field of comparative constitutional law. Art. 132 declares certain resources, including any which could be associated with the generic concept of the maritime-terrestrial public domain, as being public property. The declaration of these resources as being part of the public domain by a State Act was to open the way for one of the most important provisions of the 1985 Water Act concerning water resources, which in practice allowed the inclusion of all continental water resources in the hydraulic public domain. (This 1985 Act was to be integrated along with others into the 2001 Consolidated Water Act, that is currently in force.)

Article 2 of the 1985 Water Act listed what were to be considered the public hydraulic domain assets or resources, declaring all continental water resources to belong to the public domain. It was based on the principles of a hydrological cycle (*cf.* 1.2). It also abolished the judicial distinction of water resources based on their physical appearance (rainwater, subterranean or surface water).

They were now to be referred to as continental waters with no further classifiers (Alcaín Martínez, 1994; Del Saz, 1990; Moreu, 1996).

Even so, the 1985 Act was not a radical break away from inherited principles. For instance, it granted holders of private water rights the choice to opt, within a set time limit, to continue under a private ownership system or else to switch over, under advantageous conditions, to a public ownership system. The obvious implication of such a switchover was that their status changed from being owners to being concessionaires of public water resources.

These measures were connected to other legal provisions that placed further restrictions on private owners' rights, while expanding the room for public intervention from several viewpoints. The possibility of acquiring private rights to use waters by positive prescription was eliminated, and the maximum period of usage rights was limited to 75 years. Concessions for water use could no longer be granted in perpetuity. In return, landowners with private use rights were compensated with usage rights for rainwater and water from wells or springs.

Constitutional Court Decision 227/1988 of November 29th ratified the legitimacy of the above-mentioned constitutional provisions. Essentially, it settled that the above provisions could by no means be declared an expropriation in content or nature in spite of allegations by some contenders claiming that, since there was no specific mention of compensation, this type of decision breached art. 33.3 of the Constitution.

Another innovation, of minor importance, by the 1999 Water Act (that amended the 1985 Act), was the extension of the property of the hydraulic public domain to "water resulting from sea water desalination when, once outside the plant's gates, it is incorporated into any of the above elements" (the elements into which desalinated water is "incorporated" are continental waters or natural water flows). Assimilation then is the sole criterion determining the desalinated water's status as public property. Within the plant's facilities, this water belongs to the private domain. It can then be commercialized by desalination companies. This 1999 Water Act was also incorporated into the 2001 Consolidated Water Act. The last mention of private ownership in Spanish legislation—apart from the continued ownership of some subterranean and spring waters—is the private ownership of water courses in which rainwater occasionally flows and ponds on private property forming part of these estates or even entered separately in the Land Registry, a situation which has been extended to lakes and tarns.

Yet, there are still management problems related to water ownership. In actual fact, the problems with some aquifers in Spain are a result of a combination of disproportionate claims from private individuals, together with inadequate policing by the River Basin Agencies (RBAs) (see Chapters 14 and 15).

3   WATER AS A NATURAL RESOURCE: THE PROTECTION OF WATER

This is one of the aspects that have undergone the greatest change over the last twenty years, as far as the normative situation is concerned, and will continue to do so in the near future. With regard to this question, note the steady but growing effect of article 45 of the Spanish Constitution. This article prescribes the rational use of natural resources (including water), stipulating penal consequences for breach of this principle. This "rational use" of water is also present throughout the 2001 Consolidated Water Act and forms part of the European Union's environmental policy, being specifically included in the EU Treaty (*cf.* art. 174.1).

Indeed, private individuals' use of water must be "rational", and the public decision on allotment of use will need to be governed by the criterion of rationality. This principle should be present in projects that involve better use of the resource from the point of view of conservation. Without wholly rejecting the resource's consideration as being simply productive, Spanish water law shows a marked trend towards the conservation of water resources and related ecosystems. In the case of special water projects, specific processes for evaluating environmental impact are required before they can be approved.

European law makes a decisive contribution to the implementation of these principles with an endlessly growing body of directives. Reading these directives is hard work and sometimes disheartening for any jurist who loves clarity and method. Transposing them into national law and enforcing them is equally hard. EU Court of Justice decisions have continuously penalized countries for non-compliance with these directives, and this is probably the best proof of the fact that European Commission authorities take monitoring very seriously. Another factor having a decisive impact on the application of European law is the European funding required to develop water works through funds specifically conceived for countries like Spain, such as the EU Cohesion Fund or Regional Development Fund.

This European influence has been fundamental in some respects. First there is Directive 91/271 concerning urban wastewater treatment (transposed to Spanish law by Executive Order 11/1995 of December 28th), which was the origin of a systematic policy of heavy investment in this field by the State and the Autonomous Communities (ACs) (Setuaín Mendía, 2002).

Most of these initiatives are promoted by the ACs, which for these purposes make use of an assorted but relatively standard legislation. ACs that do not have their own river basins have established sectorial water authorities. These authorities are responsible exclusively for water policy and often for the management of a specific levy, usually known as a *canon de saneamiento* (wastewater treatment tax). This levy is meant to recover part of the investment and exploitation costs (and here, the word '*canon*' does usually cover all the exploitation costs) of these infrastructures.

Additionally, environmental protection of the aquatic ecosystems still has a long way to go. The principles of the applicable act, which establishes a zoning system aiming to protect surface waters (policing and water rights), as well as aquifers (perimeters of protection) and surface reservoirs, appear to be satisfactory. Yet law enforcement is still flawed, and the situation is not especially favored by the different types of authority granted to the various Administrations. The powers of the ACs (regional planning, environment, town planning), the town and city councils (town planning) and the RBAs do not seem to be clearly laid out in practice. Whereas Constitutional Court decisions often state the need to foster cooperation among the layers of the administration, this is still quite deficient in practice.

One example of this lack of cooperation is the entire regulation concerning floodable zones. In view of the damage done to lives, property and farms as a result of regular flooding, it is clear that much more needs to be done in this area.

4   HYDROLOGICAL PLANNING AND THE CONCESSIONS SYSTEM

The management of water use is tied in with the existence of any water law. In this respect, the most notable thing in the historical evolution of Spanish water law is the fact that water management has taken on aspects of planning, as one of the central provisions of the 1985 Water Act is hydrological planning (Embid Irujo, 1991).

Today, hydrological planning can be perfectly described in legal terms, and hydrological plans are complex documents that specifically serve the purpose of regulation. A vivid expression of their relevance is that Spanish water law is today a collection of acts and regulations, plus the Hydrological Basin Plans approved by Executive Order 1664/1998.

With regard to national water planning, the 2001 NHP Act approved the National Hydrological Plan (NHP), intended by the 1995 Water Act to be the key element of hydrological planning (Embid Irujo, 1993). This regulated the transfer of water from the Ebro basin to various basins along the Mediterranean coastline. During its drafting and even after its enactment, the 2001 NHP Act ignited an unprecedented political/legal dispute in Spain. There were street demonstrations as well as appeals to the Constitutional Court and proceedings brought before the European Commission claiming that this regulation was in breach of EU law. In the end, the 2004 NHP Act amending the NHP repealed the Ebro transfer regulation. Act 11/2005 suppressed any reference to the Ebro transfer.

Owing to its novelty, complexity and, at the same time, success, the conception and the ups and downs of hydrological planning in Spain is one of the most interesting points in contemporary Spanish water law. The 1985 Water Act marked the turning point in a water policy tradition that looked upon hydrological plans as being simply a collection of hydro works and is now finally moving towards making hydrological plans into legal regulations. In addition, the drafting process enshrines the principles of openness and transparent cooperation among institutions and stake-holders. The importance of hydrological planning is based on the fact that is binding upon RBAs, and the mandate that before new water rights are granted they must be checked for compliance with the plan. Also existing concessions may be reviewed if they are incompatible with plans.

Based on planning, the acquisition of private water rights falls within the system of conces-sions. According to the law now in force, use rights can only be granted either according to *ex lege* principles, applying to rainwater and water originating from groundwater and springs only, and by concessions. Yet under some circumstances, particularly in some transfer regulations, the legislation provides for a sort of "mixed" form of acquisition, where a specific act details uses and final purposes. In the case of interbasin transfers, for instance, a law has been enacted to explicitly regulate the allotment of the transferred volumes among beneficiaries (*cf.* art. 1 of Act 18/1981 of July 1st, concerning water resources in Tarragona, previously quoted in this section in reference to transfers).

A water concession is a sectorial manifestation of the typical domain concession that materi-alizes in a real property right (registrable and transmittable) for the concessionaire. It is usually acquired by means of a public bidding process and is granted at discretion for a fixed time and for uses specifically designated in the concession document. By no means may a concession be granted, if it is not compatible with the hydrological plans in force at the time.

5   WATER WORKS AND THE EVOLUTION OF THE WATER ACT

Water works and projects take center stage in Spanish water legislation. Interbasin transfers are a clear sign of their importance. This book provides evidence that, without complex works and projects, "natural water systems" would provide largely insufficient water services. This social need is one of the hallmarks of Spanish water law, a basis that has led to exceptional juridical institutions. It is also a source of significant controversy surrounding many of these institutions.

The historical analysis of Spanish law shows a marked shift away from a complete reliance on private initiatives and entrepreneurs in the 19th century, towards the *regenerationists'* appeal for State intervention, which evolved into the 1911 Water Works Act, whose economic and financial foundations were eventually integrated into the 1985 Water Act.

It also raises organizational issues, which are typical of Spanish law. These organizations, such as the basin bodies set up in 1926 under the name of River Basin Agencies, have expanded hugely due to their excellent conception and beneficial effects. One of their main tasks was to develop water projects within a self-governing users' system and counting on permanent support from the Spanish Government (Martín-Retortillo, 1960; Fanlo Loras, 1996).

The most recent juridical issue is essentially administrative, as either the State or the ACs can exercise authority. State intervention is indicated for the so-called "works of general public interest" (*cf.* their justification in art. 149.1.24 SC). The State is empowered to develop such works even in the territory of the intercommunity hydrographical basins, just as the Constitutional Court stressed in Decision 227/1988, following the best doctrine (Martín-Retortillo, 1992).

An understanding of what "works of general public interest" means is vital (Embid Irujo, 1995). For a long time, no regulations have been established to attempt to deal with this problem and the regional governments have contended for State intervention in order to carry out works of general public interest. There is a perfectly logical explanation for this apparent paradox: a declaration of work of general public interest is accompanied by State funding and regional government treasur-ies are in permanent crisis.

Despite the juridical controversies, the development of water works has experienced some "privatization" in more recent times as a result of Act 13/1996. The word "privatization" refers to the Act's provisions for setting up State Corporations (*Sociedades Estatales*) to build and/or operate water works. Although incorporated with 100% State funds, these corporations behave in accordance with the viewpoint of private law (Embid Irujo, 1997; Malaret, 1998).

State Corporations raise economic and juridical issues. They have been generously endowed with public funds. At the same time, they develop an activity within the private law system that has previously been "extracted" from the initial "natural" authority of the RBAs. This implies then the loss, to a large extent, of the fundamental function that presided over the creation of the RBAs in 1926.

At one important level Spanish water law currently appears to have one permanent concern. This is the relationship between the environment and water works. Often this relationship has been at the heart of highly complex legal battles. The case of the Itoiz reservoir in Navarre is a very representative case of such conflicts. The project was not granted full legitimacy until the Constitutional Court issued a favorable ruling. Contenders even filed a case against the project before the European Court of Human Rights, but the appeal was rejected in 2004.

Currently, the 2004 NHP Act and Act 11/2005 foresee the construction of a large number of water works. Once again, many are surrounded by controversy on the grounds of poor economic rationality or significant environmental impacts.

6   THE FLEXIBILITY OF THE CONCESSIONS AND THE SETTING UP OF WATER MARKETS

The reform of water legislation under 1999 Water Act introduced into Spanish Law a regulation that has been dubbed "the water market". It is actually a very complex regulation governing a number of different institutions, such as the contracts for the transfer of the water use rights, regulations governing the use of state infrastructure for materializing such contracts, the organization of public centers for the exchange of water use rights and, finally, the connection of all the above with the problem of the inter-basin transfers.

The official doctrine underlying the 1999 Act is the need to reform the system of concessions to make it more flexible. This, it was said, would be a way of helping to overcome the damaging effects of the periodic droughts suffered in Spain. The arguments expressed in the recent past in favor of the reform claimed that facilitating contracts for the transfer of water use rights between private right-holders would make some water works unnecessary. Since water resources would be channeled "naturally", by market forces alone, from the socially and economically less profitable towards the more profitable uses, new works would become redundant.

The regulation allows concessionaires or holders of use rights to enter into contracts to assign water use rights. To this end, the 1999 Water Act repealed the appurtenance of land and water rights, allowing land other than stated in the concession to be irrigated. This is a key legal provision, since it allows a potentially large volumes of water to enter the market, given that irrigation accounts for 80% of the water used in Spain.

On the whole, the market regulation is intended to avoid the obvious weaknesses that have been identified with water markets in other countries. For instance, while the Act lets the parties agree on the transfer terms, the Act carefully lays down a number of preventive measures to ensure that the contracts do not lead to increased consumption levels (see Chapter 10).

Additionally, the Act provides for maximum prices to be set that may be stated in the contracts. While this may be excessively intrusive, it is one way to counteract the general public's negative perception of speculation with water rights that were granted to holders for free. The regulations also provide for an administrative authorization, whose content is governed by the applicable rules, of the contracts for the assignment of water use rights. However, there is a system whereby permission is granted automatically within very short deadlines, which can in practice be interpreted as an abandonment of the powers that should be inherent to domain ownership.

During the last drought (2005–2008), water market exchanges have alleviated conditions of those regions and basins (Segura and South) where waters scarcity was most severe.

7   ADMINISTRATIVE AND POLITICAL ORGANIZATION

A good part of Spanish water law is synthesized by the institutional set-up of RBAs. They are a clear reference to a model of administrative organization based on the principle of user participation to the extent that the term "River Agencies" and its inherent content, seems to be something essential, and naturally linked. And something as simple as this took Spain through historical evolution in which the organizational formulae previously tried and tested under the Upper Aragon Irrigation Act (1915) and in the area that was later to come under the Ebro Agency were highly relevant.

Of course, models necessarily evolve over time, and the composition, internal organization and functions of the agencies today probably have very little in common with the model as originally conceived in the 1930s and with practices used over the following years. In particular, the growing concern of not only of traditional users but also of the general public about water issues can be seen everywhere. It is then essential to get them involved in the different bodies and branches of the RBAs. This is true to the point that some of such bodies are inconceivable without their participation.

Similarly, representatives from the scientific and technological community should also be given a role in providing a permanent link between water policy, which is ever more dependant on scientific aspects and advances, and the possibility of channeling those concerns directly into the bodies that have to adopt or report decisions.

The membership of the new Administrations created as a result of the 1978 Constitution of the RBAs is also essential. In actual fact, the ACs, whose frank and loyal co-operation with the Spanish Government within the RBAs and, generally, the exercise of all water-related functions, is the key to the success of water policy. Evidently, there is still a lot to be done at this level, and conclusions still remain to be drawn from two undeniable facts. They are: firstly, the permanent value of the river basin as the arena for structuring administrative competence over water matters and, secondly, the essential and primarily territorial consideration afforded to water policy. This would be something much wider and more open than what one might think, were they considered merely as water supply policies.

Note, however, that we have come a long way in all this time. On the one hand, we have seen the consolidation of the organizations that were set up basically as a grouping of private interests and which later took on board the management of the hydraulic public domain. On the other hand, RBAs have survived and kept on working despite the tremendous impact of the new territorial organization of the State as a result of the 1978 Constitution.

To sum this issue up, the positive points made above do not, however, rule out the need for the modernization and reassessment of the RBAs. To a large extent, they do not have capacity enough at present to meet the demanding requirements imposed on them by the legal framework. Nor do they receive much assistance from other governmental agencies. There is no doubt that the budget available to the agencies and their staffing levels are severely limited.

Another key component within the Spanish institutional framework are the user communities. They can be considered another accomplishment of Spanish water Law. The legal definition of user communities is very appropriate to their potential and real functionality. The decision to structure collaboration in the field of domain management and hypothetically contradictory interests around these public law corporations (which therefore hold public powers) is completely appropriate. It can only be hoped that in those cases where user communities should have been set up by means of legal mandate, the water administration will encourage or demand their creation. They should be given the powers and responsibilities necessary for the best possible exercise of their important functions.

8   ECONOMIC ASPECTS: COST RECOVERY PRICES AND TARIFFS

After the enactment of the 1911 Water Works Act, an economic and financial system was created under the water projects umbrella. Under the previous legislation, this system was legally weak, but has now been substantially structured under arts. 111 ff. of the 2001 Consolidated Water Act and its subsidiary regulations. This does not in any way mean that the 2001 Water Act and further regulations contain the entire economic and financial system applicable to water. In fact, the dispersion of the regulations brings with it a number of problems relating to the exact comprehension of the existing system, though it conforms to understandable principles, including the very structure of the State's territorial organization.

Water law refers to two charges, a regulation levy and water tariffs that provide for the periodic payment by users of the amounts that the State Administration has allotted to the construction of reservoirs for the regulation of the water supply (regulation levy) or other infrastructures (water tariffs). The juridical system applicable to these charges seems to have become very finely tuned over time, with the amounts payable being shared out among the different users of the infrastructure for which they are applied in proportion to consumption and specific uses. Even so, this juridical system cannot be seen as a whole as a system for the recovery of the entire investment made by the State, because some of this infrastructure is subsidized, and the State accounts for some costs. This is based on the fact that the infrastructure concerned also fulfills functions that are in the public interest, such as defense against flooding.

Water law also includes an environmental charge called 'effluent control levy'. This charge is related in a way to other environmental charges (water treatment charges) regulated by the Autonomous Communities. Their objective of generating proceeds meant to finance the construction and operation of water treatment and purification facilities. Looking at their configuration and the specific use of the proceeds for the above purposes, these are clearly ecological charges. Some Autonomous Communities, such as Catalonia (which, remember, has a basin of its own), have a more complex system of charges that takes into account the cost of not only the water treatment infrastructure but also the supply infrastructure. This should probably serve as an example to be followed by the rest of Spanish law if it is to fulfill the principles of art. 9 of Directive 2000/60 (WFD) on the cost recovery of public investment in water supply for different beneficiaries, which are to be implemented by December 31st, 2010 at the latest.

The future evolution of the economic and financial system will be marked, above all, by the real impact of the WFD, and the recovery of public costs that the supply of water to private consumers brings with it. In this matter, we must start with a clear statement of the principle of recovery contained in art. 9. This really is much more complex when taking an overall view of unclear clear precept stipulated in art. 111 bis of the 2001 Consolidated Water Act. All of this suggests that the adequate justification by States of the motives for the exceptional nature of cost recovery would allow the general rule to be evaded. But the key question is, precisely, that the European Commission or, eventually, the European Union's Court of Justice must consider the "adequate justification" to be sufficient. This means that it is not enough simply to resort to the traditional claim of Spain's exceptional climate in order to gain special treatment, but it must be demonstrated in detail, case by case, exactly what that exception consists of and its extent quantitatively.

9 FINAL REMARKS

An examination of the historical evolution of Spanish water law up to the present day shows that we are clearly at a moment of transition. This is a transition from a coherent, self-contained and well-constructed juridical structure, which has, however, some markedly obsolete institutions. It is the end of an era, of a way of tackling water problems, but the new era has not yet completely dawned.

And as in all periods of transition, it is difficult to say which is now the stronger current, the weight of tradition, or the winds of change. Some social and economic stakeholders are clearly looking in the new direction that, to them, is perfectly clear. Yet, others only find security in clinging to a juridical structure that they think is changing too quickly, especially bearing in mind that the 1879 Water Act was in force for a whole century.

Further, social and territorial heterogeneity prevails. What is seen in some regions as an inescapable necessity is viewed in others as a milestone from the past. What for some social groups is a testimony to their very existence, to their identity, and the only guarantee of a reasonably secure economic future, for others is an unjustifiable load that leads to immobility and inefficiency. The road that some Autonomous Communities traveled years ago with foresight, others are only just glimpsing today.

But a few forceful ideas have begun to be accepted by all: the primordial value of the environment, of the conservation of resources, of demand management, of the use of economic means for the management of a scarce resource, of training in new technologies, and of the need to increase investment in research and development into water. Water works have come taken down from the top of the altar where they have always been revered. They have not been completely banished either from thought or action, and they probably never should be. But there is no doubt that today very few people think that a mere provision of funds for water works will solve the water problems of this country or in any other country.

These are very exciting times. And from the point of view of institutional change, they are genuinely promising for stakeholders. The ground covered by Spanish society in this field in the twenty-four years since the 1985 Water Act came into force has been simply spectacular. It is to be hoped, nevertheless, that the transition, like Spain's political transition, will be peaceful, and that the necessary changes can be achieved without throwing overboard all that is, in the final analysis, most important to a lawyer when the historic journey comes to an end. And that, in Spain, is the value of water law. The tradition of respect for the law, the value that is attached to the fact that we have a Water Act that is accepted as a kind of structured, systematic code offering solutions for the many potential situations and problems and upon which everyone accepts that the inevitable conflicting interests must be resolved.

REFERENCES

Alcaín Martínez, E. 1994. *El aprovechamiento privado del agua y su protección jurídica* [Private water use and its legal protection]. Barcelona, Spain: Bosch.

Carpi Abad, M.V. 2002. *Aprovechamientos hidroeléctricos: su régimen jurídico-administrativo* [Hydropower uses: its legal and administrative regime]. Valladolid, Spain: Lex Nova.

Del Saz, S. 1990. *Aguas subterráneas, aguas públicas (El nuevo derecho de aguas)* [Groundwater, public water (The new water law)]. Madrid, Spain: Marcial Pons.

Embid Irujo, A. 1991. *La planificación hidrológica. Régimen jurídico* [Hydrological planning. Legal regime]. Madrid, Spain: Tecnos.

———. 1993. *El Plan Hidrológico Nacional* [The National Hydrological Plan]. Madrid, Spain: Civitas.

———. 1995. *Las obras hidráulicas* [Water works]. Madrid, Spain: Civitas.

———. 1997. *Gestión del agua y medio ambiente* [Water management and the environment]. Madrid,———
—. 2001. *El derecho de aguas en Iberoamérica y España: cambio y modernización en el inicio del tercer milenio* [Water Law in Latin America and Spain: change and modernization at the beginning of the third millennium]. Madrid, Spain: Civitas.

Fernández Torres, J.R. 2000. *La Política Agraria Común. Régimen jurídico de la Agricultura Europea y Española* [The Common Agricultural Policy. Legal Regime of European and Spanish Agriculture]. Pamplona, Spain: Aranzadi.

Fanlo Loras, A. 1996. *Las Confederaciones Hidrográficas y otras Administraciones hidráulicas* [River Basin Authorities and other Water Agencies]. Madrid, Spain: Civitas.

Malaret, E. (1998). "Las sociedades para la construcción de obras hidráulicas" [Companies for the construction of water Works] in Embid Irujo, A (dir.) *El nuevo derecho de aguas: las obras hidráulicas y su financiación* [The New Water Law: water works and its financing] Civitas, Madrid: 97–145.

Martín-Retortillo, S. 1960. *De las Administraciones autónomas de las aguas públicas* [On the Autonomous administrations of public water]. Seville, Spain: Instituto García Oviedo.

———. 1963. *La Ley de aguas de 1866. Antecedentes y elaboración* [The 1866 Water Act. Precedents and Drafting]. Madrid, Spain: Ediciones Centro de Estudios Hidrográficos.

———. 1966. *Aguas públicas y obras hidráulicas* [Public water and water projects]. Madrid, Spain: Tecnos.

———. 1992. Competencias constitucionales y autonómicas en materia de aguas [Constitutional and Autonomous Community competences over water issues]. *Revista de Administración Pública* 128: 23–84.

Moreu, J.L. 1996. *Aguas públicas y aguas privadas* [Public waters and private waters]. Barcelona, Spain: Bosch.

Sanz Rubiales, I. 1997. *Los vertidos en aguas subterráneas. Su régimen jurídico* [Groundwater effluents. Their legal standing]. Madrid, Spain: Marcial Pons.

Setuaín Mendía, B. 2002. *El saneamiento de las aguas residuales en el ordenamiento español. Régimen jurídico* [Wastewater treatment in the Spanish legal system. Juridical regime]. Valladolid, Spain: Lex Nova.

Tirado Robles, C. 2004. *La Política del agua en el marco comunitario y su integración en España.* [Water policy in the European Community framework and its integration in Spain]. Cizur Menor, Spain: Thomson-Aranzadi.

CHAPTER 12

Institutions and institutional reform in the Spanish water sector: A historical perspective

Consuelo Varela Ortega
Department of Agricultural and Social Sciences, Universidad Politécnica de Madrid, Spain

Nuria Hernández-Mora
Environmental Consultant, Madrid, Spain

1 INTRODUCTION

In modern institutional economics, institutions have a much broader context that mere organizations or administration agencies. Institutions in this ample vision are defined as a set of rules and property rights that govern the behavior of individuals and determine their actions in such a way that these actions affect not only those individuals but also other members of society (Ostrom, 1992; Eggertsson, 1990; Bromley, 1989). In this context, water institutions are conceived as an integrated system that includes the legal framework (water law), the policy regime (water policy) and the administrative or organizational arrangements (water administration). All these elements operate interactively and cannot exist independently from one another. Water organizations have the precise task of executing, implementing and enforcing the legal and policy provisions (Saleth & Dinar, 2004).

Several key elements determine the establishment, performance and evolution of water institutions. These are, namely, water-related factors such as scarcity or drought spells and path dependency, determined by historical traditions and other factors that initiate outside the water sector, like decentralization trends in nation-wide policy reforms (Saleth & Dinar, 2004). Following these arguments, the evolution of water administration institutions in Spain has been determined largely by historical traditions as well as by top-down policy reforms and political will.

Other chapters in this book treat in detail both water laws and water policies, two of the three pillars of the Spanish institutional setting. This chapter aims to complete the analysis by focusing on water management organizations. Following this introduction, the second section of the chapter gives a general panorama of the organizational institutions that govern water management in Spain. The third section looks at the origins, tasks and roles of water management organizations. The fourth section explores the performance of water institutions in the light of the Water Act currently in force, as well as their capacity to address existing technical, societal and political challenges. The final section includes some concluding remarks.

2 THE SPANISH INSTITUTIONAL FRAMEWORK FOR WATER RESOURCES MANAGEMENT

The political and administrative model of modern Spain has shaped the recent statutory evolution of water management institutions to adjust to the new decentralized regionally-based territorial organization. At present, Spain's 17 Autonomous Communities have regional governments with large political, economic and administrative competences. Regional governments have powers in matters of land use planning, the environment, agriculture, forests and other natural areas and are involved both directly and indirectly in the administration of water resources. For the management of water resources, Spain has been divided into 15 river basins or water planning districts, defined in the 1985 Water Act as 'the territory along which waters flow to the sea in a network of secondary water courses that converge into a principal and unique river bed' (title II, article 14 of the 1985 Water Act). River basins are governed by the correspondent River Basin Authorities that have distinct administrative character.

The Ministry of the Environment and Rural and Marine Affairs[1] is the national authority on water resources and exercises its powers through the Office of the Secretary of State for Rural Affairs and Water. Its functions are, namely, the drafting of national regulations, the preparation, monitoring and review of the National Water Plan, the assistance to River Basin Authorities in River Basin Water Plans, the reporting to the European Commission on the progress toward the implementation of the Water Framework Directive (WFD) and the international coordination between River Basin Water Authorities.

The National Water Council (*Consejo Nacional del Agua*) is an advisory body ascribed to the Ministry of the Environment. The Council issues recommendations on projects and plans that impact public water resources and are national in scope. These include, among others, National and River Basin Water Plans; agricultural, industrial, energy or land use plans with a significant impact on water resources; or any water management issue that affects more than one river basin. It is made up of appointed representatives of the national and regional governments; the River Basin Authorities; representatives of professional and economic organizations with interests directly tied to water use (energy, agriculture, commerce, water supply, local governments and environmental interests); and representatives of the scientific and environmental community. Representatives of the national and regional governments usually hold a majority of seats. As a result, the Council's reports are usually supportive of official plans and only have minority dissenting opinions issued by its more independent members (environmentalists, scientists and sometimes others).

As Figure 1 shows, in addition to the Ministry of the Environment and Rural and Marine Affairs (MERMA) and the River Basin Authorities, many other levels of government have water resources-related responsibilities. The policies of the European Union, through its directives and

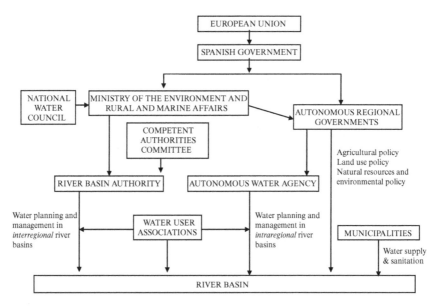

Figure 1.   Institutional framework for water decision making in Spain.
Source: Own elaboration.

---

[1] The Ministry of the Environment and Rural and Marine Affairs was created in April 2008 as a result of the merger of the Ministries of the Environment and of Agriculture, Fisheries and Food, which were independent until that date.

regulations on matters of environmental protection, agriculture, water management and financial assistance, have a direct impact on river basin management.

Natural resources, agricultural policies and land use planning are the primary responsibilities of the Autonomous Communities. The MERMA (until April 2008, the national Ministry of Agriculture) is responsible for setting national policy guidelines and act as a conduit and negotiator in the European Union agricultural and environmental policies. For the most part, however, agricultural policies are set at regional level. Autonomous governments are also responsible for setting environmental policies in accordance with national and European guidelines and standards, for developing regional land use policies and for the management of natural and protected areas.

Finally, local municipalities are responsible for waste water treatment and management and for guaranteeing urban water supply. They are also responsible for urban land use planning.

3   WATER MANAGEMENT ORGANIZATIONS

3.1   *River Basin Authorities*

The origins of River Basin Authorities
The administrative organizations in charge of the management of water resources in Spain have a solid historic tradition, and are autonomous River Basin Authorities. The first Spanish Basin Authority was created in 1926, marking a decisive step towards a model of water administration embracing all water uses within a river basin. The 1985 Water Act brought all these organizations under the general title of River Basin Authorities (*Organismos de Cuenca*) extending their powers and enhancing the role of users in decision-making. The River Basin Authorities are defined as authorities established under public law which bear most of the administrative weight of water management in the river basins with a high degree of organizational, functional and economic autonomy.

When the river basin runs through more than one Autonomous Region, the River Basin Authority is affiliated to the Central Administration (Department of Waters and Coasts at the Ministry of the Environment)[2] and is called *Confederación Hidrográfica* (literally, 'Hydrographic Confederation'). In these cases, the law, adapting the precepts of the WFD, determines that a Competent Authorities Committee—made up of representatives of the Central Administration, of the affected Autonomous Communities and of local governments, and presided by the President of the River Basin Authority—is to coordinate policies and measures that aim to achieve the goals of the WFD in each river basin and to report on progress made. Where the river basin lies within a single Autonomous Region, it is managed by the regional government through the Water Agency of the Autonomous Community, an institution that varies from one region to another, but is always subject to the principles laid down in the national Water Act.

Whether the administration is inter- or intra-regional, the management unit is the river basin and management is integral (that is, dealing with both surface and groundwater resources), multi-sector and participative, with financing shared between the users and the State (MIMAM, 2000).

Function, tasks and roles of River Basin Authorities
The 2001 Water Act—which consolidated the 1985 Water Act and its 1999 reform—and its developing statutes, determines the organizational structure of River Basin Authorities. According to the law, the role of the Authorities is to develop, monitor and revise basin management plans; to administer water and related public resources; and to develop the necessary public infrastructure works to develop their mission. The Authorities are made up of two complementary organizational structures: the participatory boards and the administrative and technical services, responsible for providing information, implementing decisions and day-to-day operations (Figure 2). The user participated boards illustrate the principles of user participation and inter-governmental integration in water management that inform the Water Act.

---

[2] See Figure 1 in Chapter 2 for the distribution of the Spanish territory into Autonomous Communities and Figure 2 for the geographical location of the River Basins in Spain.

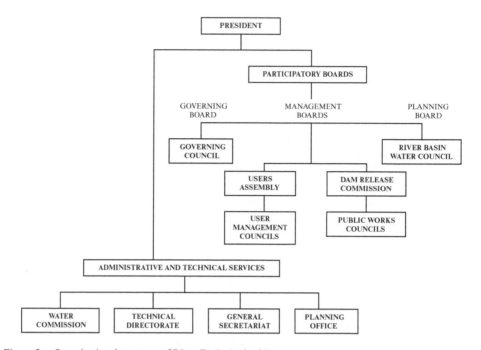

Figure 2.   Organizational structure of River Basin Authorities.
Source: Own elaboration with information from Duero, Guadiana and Tajo Basin Authorities.

The roles, responsibilities and composition of the different participatory boards and councils are statutorily determined.

- The *Governing Council (Junta de Gobierno)* is responsible for approving action plans and budgets for the Authority; propose and approve aquifer management plans; encourage and approve the creation of water-user communities; and propose periodic reviews of the Basin Hydrologic Plan to the Water Council. Its membership includes representatives of the Authority, the national government and the autonomous governments within its boundaries. At least one third of its members represent water users.
- The *Users Assembly (Asamblea de Usuarios)* coordinates the management of hydraulic works and water resources throughout the basin. It is responsible for electing user representatives to the Governing Council, the Dam Release Boards, and the Water Council. Its members are all the user representatives that form the User Management Councils of the basin.
- The *Users Management Council (Juntas de Explotación)* have the same responsibilities as the User Assembly but only in sections of the basin where uses are particularly interrelated. They are made up of representatives from the administration, water supply companies, irrigator associations, hydroelectric companies and industrial users, in proportion to their volume of water use (Iglesias & Moneo, 2005). Aquifer Management Councils have also been formed for some aquifer systems with overuse problems. Although the Water Act is committed to conjunctive use of surface and groundwater resources, the aquifer management councils that exist in different river basins are independent from the surface Management Councils, even if they cover the same geographic area.
- The *Dam Release Commission (Comisión de Desembalses)* is responsible for proposing water release regimes from the various dams in the basin. It is made up of representatives from user associations, the Ministries of Agriculture and Industry, and the National Electric Consortium. It is headed by the Authority's president.

- The *Public Works Councils (Juntas de Obras)* are formed at the request of the future beneficiaries of a proposed hydraulic infrastructure with the purpose of keeping them informed of the progress in the construction.
- Finally, the river basin *Water Council (Consejo del Agua)* is the planning arm of the basin authorities. Their role is to debate and approve the Basin Hydrologic Plan and its periodic reviews. It also has an advisory role to the Governing Council on water management related issues. It is made up of representatives of the national and regional governments within the boundaries of the basin, staff of the Basin Authority, and at least one third representatives from the user associations appointed by the User Assembly.

The administrative and technical services of River Basin Authorities are responsible for day to day management. They also provide technical and administrative support to the participatory boards. Their functions and role are also statutorily determined.

- The 1985 Water Act merged the *Water Commissions (Comisarías de Aguas)* that existed in each basin with the Hydrographic Confederations to form the River Basin Authorities. The responsibilities of the Water Commissions today are very broad. They include the review and granting of water use and discharge permit applications and of permits related to any actions that may affect the 'hydraulic public domain'[3] (*dominio público hidráulico*). They are also responsible for managing the Registry of Public Waters and the Catalogue of Private Waters, where all water uses must be registered. Among their functions is the dealings with water users associations. The Commissions are also responsible for monitoring, control, enforcement and data gathering activities.
- The *Technical Directorate (Dirección Técnica)* is responsible for the development, construction, inspection and management of hydraulic works projects built with public funds. It provides administrative and technical support to the Dam Release Commission and executes its water release proposals; and develops water use fees proposals.
- The *General Secretariat (Secretaría General)* is responsible for the management of financial, accounting and personnel issues. It also provides legal advice and support to the Authority. It holds the Secretary position in the Governing Council, User Assembly and River Basin Water Council.
- The *Planning Office (Oficina de Planificación)* is responsible for gathering the necessary information for the development, follow-up and revision of the Basin Hydrologic Plans. It is also responsible for the preparation of water use plans for intensively used aquifers that have legally been declared overexploited or are experiencing salinization or overdraft problems. The Planning Office also provides technical support to the River Basin Water Council.

3.2   Users' participation in water management: Water Users' Associations

The role of users in water management finds its most prominent and emblematic example in Spain in the numerous irrigation associations that exist throughout the country. Crafted along an extended historical tradition, the Spanish irrigation associations emerged beyond medieval times and their origins are rooted deeply in the Roman and Arab civilizations. The organization of irrigation associations was not clearly defined, historically, in the Spanish Legal texts as the ancient norms governing water distribution were part of Common Law rules. These norms emerged from informal institutions such as brotherhoods, guilds, unions and boards, and were often based on verbal agreements passed on from generation to generation (del Campo, 1999; Bolea, 1998).

The legal status of irrigation associations is explicitly defined, for the first time in a legal text, in the 1866 Water Act, which stated clearly that all surface waters were of public entitlement. Further reiterated by the 1879 Water Act, the law affirms that a communal interest for the collective use of public waters requires a commonly shared administration. In the case of a collective use of water for irrigation, it required necessarily the establishment of an irrigation community, considering

---

[3] The legal term 'hydraulic public domain' is used to refer not only to public water resources, but also to the river beds, lake beds, wetlands, and the floodplain and shoreland areas surrounding public waters (MIMAM, 2000).

the deeply-rooted historical tradition in which different types of associations and unions had safeguarded a good administration and distribution of common waters.

The 1985 Water Act strengthened the statutory character of users' associations which are considered independent public administration bodies with administrative competences. The law states that the users of public waters from the same outlet or concession, or of the same aquifer system, should form a Users' Community (*comunidad de usuarios*) which in the case of irrigation water will be named an Irrigators' Community (*Comunidad de Regantes*, equivalent to the commonly used Water Users Associations, WUAs). Alongside, the 1985 Water Act reinforces the public character of all waters and therefore includes, for the first time in the Spanish water legislation, groundwater within the hydraulic public domain. This results in the fact that groundwater users, also for the first time, constitute water users associations and become an active part of the overall water management regime. In fact, the new legislation extended considerably the concept of water user including other interests and uses beyond irrigation and incorporating representatives of water supplies for municipal consumption.

The by-laws of the Water Users Associations are drawn up and passed by the users themselves with the requisite that they comply with the basic statutory guidelines of the Water Act and are approved by the River Basin Authority. In general terms, these statutes regulate the organization of the irrigators and establish the norms for distribution and control of the irrigated water, and also penalties. They also regulate the use and maintenance of the hydraulic systems: the payment in equitable proportions of the cost of use, maintenance, repairs and improvements, as well as the corresponding fees and charges. In addition, each Irrigators Community is responsible for the resolution of any problems that might arise within its irrigation district. Thus, the Community must resolve arguments which, if presented to the administration, would paralyze the legal and administrative services of the River Basin Authority.

The basic guidelines of the statutes constitute a framework model that includes four organizational structures: the president, the general assembly, the government council and the water jury. *The president*, elected in the general assembly, is the legal representative of the Irrigators' Community. *The general assembly* is the supreme body of the Association and it exercises most of its powers, such as the elaboration of the Association bylaws, the election of the president, the admission of new members, the approval and execution of irrigation works and the application all legal provisions. The assembly includes all water users, that is, irrigators and municipal and industrial users as well. However, the decision capacity is not based on 'one man-one vote' and votes in the assembly are proportional to the ownership of irrigated land in the case of the irrigators and to the relative importance of the correspondent water use in the case of the other type of users. *The government council*, in charge of the Association management responsibilities, is composed of the president, a varied number of voting members, the secretary and the treasurer. *The water jury*, composed by a president and a selected number of members, is responsible of the resolution of conflicts among members, the establishment of infraction penalties, sanctions, compensations and the correspondent obligations (Bolea, 1998).

The Water Users Associations take part in the participatory boards of the River Basin Authorities. They are active stakeholders seeking to safeguard their own interests in various matters such as the guarantee of the water volumes granted in the administrative concessions, the effective use of public funds for investment in irrigation works, the equitable establishment of tariffs and levies for the repayment of these public investments and the demands of publicly-funded irrigation modernization programs.

At present there are more than 7,000 Water Users Associations in Spain. They are responsible for the administration and operation of around 2.6 million ha, which represent 70% of the total 3.4 million ha of irrigated lands in the nation. The remaining 1 million ha are operated by individual farmers (MAPA, 2001). Water Users Associations are remarkably diverse. Their structure and operational modes are determined largely by different factors such as historical tradition, size, origin, water source, irrigation system, production technology and the characteristics of their members.

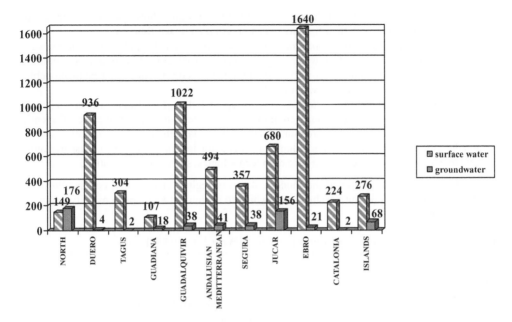

Figure 3.   Distribution of water users associations by basin.
Source: For surface water: MOPTMA (1994); for groundwater: adapted from Llamas *et al.* (2001).

Figure 3[4] shows the distribution across river basins of Water Users Associations using surface water (Merino, 2001; MOPTMA, 1994) and groundwater (Llamas *et al.*, 2001[5]).

With respect to surface water Irrigators' Communities, large basins with an extended irrigation surface, such as the Ebro and Guadalquivir (625,000 ha and 310,000 ha respectively) have the largest number of Communities. They also have the bigger communities, both in size and number of members. In fact, in the Ebro basin, where irrigation has a long historical tradition, over 100 Irrigation Communities have more than 500 members and about 250 associations in this basin operate more than 500 ha each. In contrast, in the Guadalquivir and Duero basins, large Irrigation Communities are not so numerous. As irrigated surface in these basins is about half of the Ebro, large Irrigation districts are only one third of their Ebro counterparts (Merino, 2001).

3.3   *Types of Irrigators' Communities*

Attempts to group the Spanish Irrigators' Communities into a coherent typology are numerous and reflect their exceptional variety (Ramos & Merino, 1998; Bolea, 1998; Sumpsi *et al.*, 1998; Campo García, 1999; Merino, 2001), evidencing that 'there is no one best way to organize irrigation activities' (Ostrom, 1992, p. 41). However, a simple straightforward classification can be drawn from the origin of irrigation communities as follows (Ramos & Merino, 1998):

a. *Traditional or historic irrigators' communities*. This group includes those communities created before 1866, some as far back as the Middle Ages. They are usually located in river floodplains and use low-efficient traditional irrigation technologies, such as flood or gravity irrigation, characterized by high water distribution losses. These communities own the irrigation infrastructures they use and are responsible for their maintenance. The 1985 Water Act

---

[4] Data is shown for the 11 planning districts or basins that existed in 2001. Some of these have been subdivided into one or more planning districts, with management responsibilities in some cases handed over to the Autonomous Communities.
[5] 1999 data. Includes groundwater associations of public nature organized according to the 1985 Water Act.

acknowledged the existence of these historic communities and allowed them to continue to operate in accordance with their traditional customs and laws (see Table 1).

b. *Irrigators' communities of public initiative.* These were formed by the public water administration to manage the large surface water irrigation networks developed in conjunction with the building of large dams and reservoirs throughout the 20th century. They rely heavily on public subsidies, low water use fees and water use tends to be inefficient.

c. *User associations of private initiative.* This group includes groundwater user associations developed by private farmers using groundwater. They tend to use water more efficiently as they bear the full cost of construction, operation and maintenance of the irrigation system and therefore pay a higher price per volume of water extracted. In consequence, these irrigators tend to adopt modern water-saving irrigation technologies, such as drip irrigation.

The 1985 Water Act, imprinted the modern statutory regime of irrigation communities and therefore constitutes a relevant baseline upon which to sketch out a comprehensive typology as shown in Table 1.

Table 1.   Typology of irrigators' associations based on the Water Act of 1985 (1985 WA).

Typology criteria	Type of irrigation association	Characteristics
Date of establishment	Prior to 1866 Water Act	• Structured and organized according to their historical tradition based on written rules or Common Law rules • Their operation must comply with the constitutional principles of representation and democratic structure of the 1985 WA
	After Water Act of 1866	• Structured according to the statutory obligations of the 1985 WA • Their operation must comply with the constitutional principles of representation and democratic structure of the 1985 WA
Mode of establishment	Voluntary	• Based on the private initiative of the users
	Compulsory	• Imposed by the River Basin Authority based on public interest for ameliorating water use in a given area
Number of members	Normal number	• Ruled by their approved by-laws • By-laws must be approved by the River Basin Authority
	Reduced number (< 20)	• Ruled by elementary and simple agreements • Agreements must be approved by the River Basin Authority
Types of members	Ordinary associations	• Formed by individual irrigators
	General associations	• Formed by different groups of associations that take the water from the same public riverbed using a common water outlet
	Central unions	• Formed by all types of associations, ordinary or general, that take the water from the same public riverbed using different water outlets
Type of water source	Surface water	• Surface water irrigators' associations of (public domain)
	Groundwater	• Groundwater irrigators' associations (can be private or public domain)
	Mixed waters	• The same irrigators' association can use conjointly surface and groundwater
Type of water entitlement	Public water entitlement	• Ordinary irrigators associations with public administrative legal status
	Private water entitlement	• Groups of irrigators that lack public administrative legal status and are structured as simple common good properties or private societies

Source: Own elaboration based on Bolea (1998).

3.4   *Types of groundwater users' communities*

Groundwater users associations are also extremely diverse. Their size and organizational complexity varies from a few members using the same well, to General Users Communities that include thousands of individual irrigators, municipalities and individual irrigators' associations. In spite of these differences, according to Carles *et al.* (2001) all groundwater users' communities can be classified into two categories, according to their goals and objectives. One category would include those associations whose objective is the common exploitation of a well or group of wells. These are named *associations for the collective management of irrigation networks* and, to a large extent, they operate like surface water irrigation associations, dedicated primarily to the distribution of water among their members. But in contrast to those, they pay for all drilling, installation, operation and maintenance costs.

The other category includes the *associations for the collective management of* aquifers, a group of user communities that comprise all or a majority of users within an aquifer. In addition to pursuing their own interests, that is, maximizing the private utility in the exploitation of the resource, they also contribute to its conservation and serve therefore a social goal.

Today, there are hundreds of groundwater users' associations in Spain. Some are organized according to the 1985 Water Act and are therefore of public nature (Figure 3). Conversely, there is no official count of the numerous groundwater irrigators associations that are organized and operate under private law. Often called well associations, they are organized as private business entities for the exploitation of a well or group of wells. However, the public or private nature of the groundwater users' associations is not determinant of their activities (Llamas *et al.*, 2001). Although in general private users associations tend to focus on maximizing economic returns, there are instances of private associations whose goals are broader and related to the long-term sustainability of the resource itself. At the same time, some of the public associations are not concerned with managing their water resources in a sustainable manner. Rather, they were originally private companies formed for the common exploitation of a well and became public entities in order to be eligible for public subsidies for irrigation modernization programs, but have not substantially modified their goals or the way they operate.

4   PERFORMANCE OF WATER MANAGEMENT ORGANIZATIONS

The performance of Spanish water management organizations needs to be evaluated in light of their success in implementing and enforcing the new rules and regulations emanating from the Spanish institutional reform in water resources management. In the 1980s and 1990s, the main guiding principles of this reform included the establishment of the river basin as the basic planning and management unit; the introduction of economic principles as a water management goal; conjunctive management of surface and groundwater resources; participatory management; and the adoption of environmental goals, including water quality protection. Starting in 2003, the implementation of the WFD has further transformed the goals and structure of Spanish water management institutions. Indeed, the WFD firmly establishes good ecological status as a primary policy goal relying again on economic rationality and public participation as basic management instruments. It is still early to evaluate the impacts of the legal and organizational changes that are taking place and which are still underway. Overall however, and while significant improvements have been made, much work remains to be done.

Other chapters in this book have discussed in some length the challenges and limitations of the implementation of these guiding principles. We will therefore only refer to some of them briefly here.

In terms of the participatory management framework established by the 1985 Water Act, representation in Basin Authority boards is to a large extent still limited to permitted users. Representatives of non-consumptive uses such as environmental or recreational interests are either not included or, when they are included, they only have a symbolic presence on these boards. Table 2 shows the composition of User Assemblies in different Basin Authorities. Three of them are shared river basins that run through more than one Autonomous Region (Ebro, Guadiana

Table 2.   Composition of Users' Assemblies in different River Basin Agencies.

	Ebro	Guadiana	Tagus	Catalonia
Central government	2	2	2	–
Autonomous governments	14	5	7	–
Local administrations	–	–	–	8
Basin management agencies	4	4	5	–
Users	**397**	**113**	**85**	**25**
– Water supply & sanitation	50	15	24	3
– Irrigation	232 (58%)	81 (72%)	39 (46%)	4 (16%)
– Livestock	–	–	–	4
– Hydroelectric uses	46	7	22	–
– Industrial uses	–	8	–	4
– Other uses	48[1]	2	–	–
– Environmental	–	2	–	6
– Recreational	–	–	–	4
Labor organizations	–	–	–	4
Neighborhood associations	–	–	–	1
Consumer protection associations	–	–	–	2
Scientists/academics		–	–	7
Total	**417**	**124**	**100**	**47**

[1] Includes nuclear plants, thermoelectric plants, aquaculture, individual irrigators and other permitted uses.
Source: Modified and updated from Llamas *et al.* (2001).

and Tagus) and thus managed by River Basin Authorities affiliated to the central government. The Catalan Water Agency (*Agencia Catalana de l'Aigua*) manages river basins that are entirely within Catalonia's territory. The composition of the User Assembly is significant because they appoint user representatives to other participatory boards and councils in numbers proportional to those that sit in the Assembly. The table indicates the number of representatives from each level of government and each water use and are statutorily determined. For instance, in the three shared river basins there are representatives from the central government, as well as representatives from all the Autonomous governments that share the river basin. Additionally, the number of representatives of each use is proportional to their relative importance in terms of indirect calculations of total amount of water consumed (hectares irrigated in the case of agriculture, number of inhabitants, in the case of municipal water supply, and kilowatts/hour produced in the case of hydroelectric plants). Therefore irrigation interests usually predominate (58% of seats in the Ebro Basin Authority, 72% in the Guadiana Basin, and 46% in the Tagus Basin), as they use a proportionately larger share of the resource. This system of representation does not take sufficiently into account the socioeconomic importance of different water uses.

This situation is strikingly different in the Council for Sustainable Water Use (*Consejo del Uso Sostenible del Agua*), the equivalent body to the User Assembly in the Catalan Water Authority (*Agència Catalana de l'Aigua*). The Catalan agency manages river basins entirely within the Catalan Autonomous Community, which includes the metropolitan areas of Barcelona and Tarragona and their industrial belts. User representation in this case is more balanced and diverse, in an effort to incorporate the views of all users, stakeholders and the public at large into the decision-making processes. The Catalan model could serve as a guide for future reforms.

Barreira (see Chapter 17) addresses the legislative rationale behind the changes in participatory modes that are needed to fulfill the requirements of the WFD. Indeed, starting in 2004, different Water Authorities have undertaken participatory processes to accompany the elaboration the new basin management plans to be submitted to public review in 2009. These processes have so far operated independently of the existing participatory bodies of the Basin Authorities that have been described above. As Hernández-Mora & Ballester (2008) discuss, while the participatory processes underway have served to open up the deliberative space to more interests outside of permitted uses, many have so far failed to truly modify planning and management goals and procedures. To a large extent, Water Authorities are simply complying with legislative requirements by holding public meetings at different times throughout the planning process. Although some basins have been more active than others in launching the stakeholder-participatory programs required by the WFD management plans (e.g Segura, Guadiana and Ebro) these are still not fully implemented. A few significant exceptions are the Catalan Water Authority (www.gencat.cat/aca), and the processes undertaken by the Cantabrian Office for Water Participation (www.ophic.es). Both have designed comprehensive and meaningful participatory processes for each river basin within their territory.[6]

It remains to be seen what impact this public input is going to have in the final basin management plans. However, it is important to note that the new planning process is clearly resulting in greater transparency and ease of access to information that was not readily accessible to the public in the very recent past, as a visit to the websites of all water authorities clearly demonstrates.

A second area in need of improvement is the protection of the quality of Spanish water resources. Once again, the implementation of the WFD has brought about significant changes in this area, most importantly and for the first time, a broad inventory of the ecological status of Spanish surface and groundwater resources. While the absence of baseline information in most basins is making it difficult to have a clear picture of the situation, results so far indicate that a large proportion of surface and groundwater bodies are at risk of not achieving the goal of good ecological status by 2015 (see Chapter 14 for the situation of groundwater resources). Pollution from industrial and urban discharges continues to be a major problem with a majority of industries lacking discharge permits, and many urban centers either lacking or having insufficient wastewater treatment services. As for non-point source pollution, there is a general lack of information and an absence of comprehensive plans to deal with the problem. Early estimates indicated that over 40% of Spanish rivers were polluted or very polluted, and between 40% and 70% of Spain's 1000 dams are heavily eutrophied (Prat *et al.*, 2000). Ongoing studies and reports by Water Authorities are proving these early estimates accurate.

A third area of concern, also with a direct impact on environmental and water quality protection, is the situation of the hydraulic public domain. Encroachment on floodplain and shoreland areas due to uncontrolled urban and suburban development, degradation of these environments due to inappropriate land uses, and the continued destruction of wetland ecosystems, are resulting in water quality deterioration and environmental degradation. In order to address these issues, the Ministry of the Environment launched in 1993 the LINDE Project. The goal of the project is to physically identify and delimitate the boundaries of the hydraulic public domain and develop comprehensive plans for its protection. However, the program has suffered from significant cost overruns and is still incomplete (Villarroya, 2000). Another more recent initiative, is the National Strategy for River Restoration, launched in 2007 by the then Ministry of the Environment. It is an ambitious multi-faceted approach to river restoration through research, outreach, grassroots involvement, river and shoreline restoration, and protection of sensitive and ecologically valuable areas. While it is still early to evaluate its results, some critics already point to insufficient public funds and political commitment to the strategy that can jeopardize its success.

Along the same lines, the continued increase in illegal irrigation, often relying on the uncontrolled exploitation of groundwater resources (Llamas *et al.*, 2001), continues to be a significant problem in Spain. In the region of Andalusia, for instance, total irrigated area increased by 54,000 ha between 1998 and 2002 (Vives, 2003), and in Murcia, illegal irrigation increases by 5,000 ha per year, according to some estimates (Esteve, 2002). In some aquifers, such as those located in the

---

[6] Parallel to the official participatory processes currently underway, several European-funded projects are using participatory methodologies in some basins in Spain with the active involvement of the stakeholders. These are aimed to develop participatory scenario building dynamics for water management and climate change (see for instance Varela-Ortega *et al.*, 2008), risk assessment and socioeconomic and water vulnerability (Varela-Ortega and Blanco, 2008, Varela-Ortega, 2009), or identify major management and planning issues in a basin (Martínez Santos *et al.*, 2007).

Upper Guadiana Basin, the problem of illegal irrigation and the associated overexploitation is so significant, that the central government launched the Special Plan for the Upper Guadiana in 2008 with a total estimated budget of 3 million euros for the period 2008–2027, in order to reduce water consumption and restore the Ramsar-catalogued associated wetland ecosystems. The Plan is an example of a comprehensive publicly-launched initiative that seeks to balance environmental and socioeconomic goals and maintain rural livelihoods in the area as a response to long-lasting social conflicts. It provides specific actions such as the purchase of water rights from the irrigators by the official Water Rights Exchange Center (see Chapters 10 and 13), the closing-up of illegal drillings as well as reforestation and rainfed farming measures. The full implementation of this comprehensive and ambitious program is subject to uncertainties regarding budget limitations and the irrigators' willingness to sell, on a permanent basis, their water concessions at the publicly set prices (CHG, 2007).

A common theme in the performance evaluation of River Basin Authorities is their inability to enforce existing rules and regulations. Several reasons can help explain this situation. A basic limitation is the absence of comprehensive inventories of existing uses. The 1985 Water Act created the National Registry of Public Waters and the Catalogue of Private Waters, to centralize the information on all existing water uses and facilitate the processing and inscription of new water use applications. The registration process has been very complex and, in some cases, riddled with conflict. Today, over 20 years after the Act was passed, and in spite of substantial public investment, it is still incomplete. In 2001, the Spanish Ministry of the Environment launched the Alberca Project, a renewed effort to complete the user rights registration process, while trying to remedy some of the limitations of previous efforts, by creating a unified technological and database management approach nationwide. While progress is being made, the registry is still not completed.

In addition to a lack of comprehensive information and basic data networks, Basin Authorities are limited by insufficient financial, technological and staff resources to handle the responsibilities taken over by the reform of the 1980s and 90s (Llamas *et al.*, 2001; Hernández-Mora, 2002). This situation has only been aggravated with the new planning, monitoring and management responsibilities that derive from the WFD. Given these limitations, Basin Authorities' staff dedicate a considerable amount of time to processing and managing external contracts, thus losing motivation and technical expertise (MIMAM, 2000).

5   CONCLUDING REMARKS

Water management in Spain must balance competing environmental, social and economic demands on water resources. The traditional Spanish water management model, based on comprehensive hydrologic planning, is inadequate in the face of growing uncertainties with respect to the future evolution of water resources demand and supply. While a significant modeling and projection effort is underway, the impact of climate change on the future availability of water resources is uncertain. On the other hand, future water demand will also be affected by demographic and socio-economic changes, the reform of the European Union Common Agricultural Policy, and changes in social preferences. Perhaps most significantly, the requirements of the WFD are determining a new direction for water resources management.

In many respects, the Spanish institutional framework for water resources management is ideally positioned to tackle these challenges. Basin Authorities with their participatory management boards, water user associations and Basin Hydrologic Plans, already constitute a solid basis from which to work. In addition, the merger of the Ministry of the Environment and the Ministry of Agriculture into a unique ministerial department may facilitate the coordination and implementation of the EU water and agricultural policies that are being progressively interrelated across different administrative competences. In fact, the Spanish basin authorities are entailed to fulfill the ecological requirements of the WFD and to assure water services to all users as stated in the national legislation. Alongside, as the CAP has been progressively including environmental requirements into its farm programs over the last years, it has specifically defined water management as one of the new environmental challenges in the newly approved reform. Therefore, under this mew ministerial setting, a coordinated and synergy-seeking implementation of these policies in a transparent, accountable and public-participatory manner is one of the challenges facing the Spanish national and regional administrations (Varela-Ortega, 2009).

However, the WFD demands a fundamental change in the way water resources planning and management is understood (Estevan & Naredo, 2004). In the Spanish water management tradition, water was conceived as a resource that needed to be harnessed and distributed to meet an ever-growing demand. The WFD, on the contrary, understands the river basin as a complex ecosystem that needs to be managed through the consideration of interrelated hydrological, ecological, socioeconomic, technological and institutional variables. In order to undertake this transformation, water management must become more flexible, adaptive and participative. To enable this transition, a first step for Agencies will be diversifying their staff covering an ample variety of disciplines.

Another necessary reform, already under way in several Basin Authorities, is the improvement of the data collection and information networks. Information also needs to be made easily available to improve transparency and accountability. Basin Authorities must continue the effort to develop comprehensive information and outreach programs.

Finally, a key reform is the transformation of participatory processes in water resources decision making by including other stakeholders beyond water users and integrating them into the decision making process of River Basin Authorities beyond the public participation processes that are a part of current planning efforts. Only then will water resources in Spain be managed in a way where the public interest is truly at the forefront of policy and management decisions.

REFERENCES

Bolea, J.A. (1998). *Las Comunidades de Regantes* [Irrigators' Communities]. Zaragoza, Spain: Comunidad General de Usuarios del Canal Imperial de Aragón.

Bromley, D.W. (1989). *Economic Interests and Institutions: The Conceptual Foundations of Public Policy.* Oxford: Blackwell.

Campo García, A. del. (1999). *Las Comunidades de Regantes en España y su Federación Nacional* [Irrigators' Communities in Spain and their National Federation]. Madrid: FENACORE and Ministerio de Agricultura, Pesca y Alimentación.

Carles, J., García, M. & Vega, V. (2001). Gestión Colectiva de las Aguas Subterráneas en la Comunidad Valenciana [Collective Groundwater Management in the Valencia Autonomous Region]. In *La Economía de las Aguas Subterráneas y su Gestión Colectiva* [Groundwater Economics and Collective Management], edited by Hernández-Mora, N. and Llamas, M.R. Madrid, Spain: Mundi-Prensa and Fundación Marcelino Botín., 291–323.

Confederación Hidrográfica del Duero: www.chduero.es, May 2005.

Confederación Hidrográfica del Guadiana: www.chguadiana.es, May 2005.

Confederación Hidrográfica del Guadiana (2007). Plan Especial del Alto Guadiana; documento de síntesis, CHG, Ciudad Real, Spain. (http://www.chguadiana.es/corps/chguadiana/data/resources/file/PEAG/0_DOC_SINTESIS.pdf).

Confederación Hidrográfica del Tajo: www.chtajo.es, May 2005.

Eggertsson, T. (1990). *Economic Behavior and Institutions.* New York, U.S.A: Cambridge University Press, New.

Estevan, A. & Naredo, J.M. (2004). *Ideas y propuestas para una nueva política del agua en España* [Ideas and Proposals for a New Water Policy in Spain]. Bilbao, España: Bakeaz.

Esteve, M.A. (2002). *Implicaciones Ambientales de la Gestión del Agua en las Cuencas Receptoras del Trasvase Ebro-Júcar-Segura, Especialmente en las Tierras del Sudeste Ibérico* [Environmental Implications of Water Management in the Receiving Basins of the Ebro-Júcar-Segura Water Transfer, Especially in Southeastern Spain]. Proceedings: III Congreso Ibérico Sobre Gestión y Planificación de Aguas. Sevilla, Spain, November 2002.

Hernández-Mora, N. & Ballester, A. (2008). Participación pública en el proceso de elaboración de los planes de cuenca en España. (Public participation in the water management plans planning process in Spain). 6th Iberian Congress for Water Planning and Management. Proceedings. Vitoria Spain, 3–7 December 2008. (www.fnca.eu/congresoiberico/documentos/c0417.pdf).

Hernández-Mora, N. (2002). Groundwater Management in Spain: Local Institutions for Collective Management of Common Pool Resources: An Analysis of Three Cases from La Mancha. Master's Thesis, Gaylord Nelson Institute for Environmental Studies, University of Wisconsin-Madison.

Iglesias, A. & Moneo, M. (2005). *Drought Preparedness and Mitigation in the Mediterranean: Analysis of the Organizations and Institutions.* Revue Options Mediterranéennes. CIHEAM, Centre international de Hautes Etudes Agronomiques Méditerranéennes, Paris.

Llamas, M.R., Fornés, J., Hernández-Mora, N. & Martínez Cortina, L. (2001). *Aguas subterráneas: retos y oportunidades* [Groundwater: challenges and opportunities]. Mundi-Prensa/Fundación Marcelino Botín, Madrid: 1–529.

Merino, A. (2001). *Análisis Estratégico de las Comunidades de Regantes. Estudio de Casos en Aragón* [Strategic Analysis of Irrigators' Communities: Case Studies in Aragon]. Doctoral Dissertation. Facultad de Ciencias Económicas y Empresariales. Universidad San Pablo-CEU, Madrid.

MAPA (Ministerio de Agricultura, Pesca y Alimentación) (2001). *Plan Nacional de Regadíos* [National Irrigation Plan]. Madrid, Spain: Horizonte-2008.

Martínez-Santos, P., Varela-Ortega, C. & Hernández-Mora, N. (2007). Making inroads towards adaptive water management through stakeholder involvement: the NeWater experience in the Upper Guadiana basin, Spain. Proceedings of the International Conference on Adaptive and Integrated Water Management. Coping with complexity and uncertainty (CAIWA). Basel, Switzerland, (12–16 November) (on line: http://www.newater.uos.de/caiwa/basel.htm).

MIMAM (Ministerio de Medio Ambiente) (2000). *Libro Blanco del Agua en España.* [White Book on Water in Spain]. Madrid, Spain.

MOPTMA (Ministerio de Obras Pública, Transportes y Medio Ambiente) (1994). *Catálogo General de las Comunidades de Regantes* [General Catalogue of Irrigators' Communities]. Madrid, Spain: MOPTMA, Secretaría General Técnica.

Ostrom, E. (1992). *Crafting Institutions for Self-Governing irrigation Systems.* San Francisco, USA: Institute for Contemporary Studies Press.

Prat, N., Munné, A., Rieradevall, M. & Bonada, N. (2000). La Determinación del Estado Ecológico de los Ecosistemas Acuáticos en España [Determination of the Ecologial Status of Aquatic Ecosystems in Spain]. In *La Aplicación de la Directiva Marco del Agua en España: Retos y Oportunidades* [The Application of the Water Framework Directive in Spain: Challenges and Opportunities], edited by Fabra, A. and Barreira, A. Instituto Internacional de Derecho y Medio Ambiente, 47–81.

Ramos, J.L. & Merino, A. (1998). Las Comunidades de Regantes y la Nueva Política del Agua: Los Problemas de la Acción Colectiva [Irrigators' Communities and the New Water Policy: Problems of Collective Action]. Paper presented at I Congreso Ibérico Sobre Gestión y Planificación de Aguas. September 1998, Zaragoza, Spain.

Saleth, R.M. & Dinar, A. (2004). *The Institutional Economics of Water. A Cross-Country Analysis of Institutions and Performance.* Washington, DC: World Bank—Cheltenham and UK, Edward Elgar.

Sumpsi Viñas, J.M., Garrido, A., Blanco Fonseca, M., Varela-Ortega, C. & Iglesias Martínez, E. (1998). *Economía y Política de Gestión del Agua en la Agricultura* [Economics and Politics of Water Management in Agriculture]. Madrid, Sapin: Mundi-Prensa.

Varela-Ortega, C. (2009). The contribution of stakeholder involvement for balancing ecological and human systems in groundwater management. *Ecology and Society*, Special Issue "Implementing participatory water Management: recent advances in theory, practice and evaluation" (under revision).

Varela-Ortega, C. & Blanco, I. (2008). *Adaptive capacity and stakeholders' participation facing water policies and agricultural policies.* Proceedings of the XIIth Congress of the European Association of Agricultural Economist–People, Food and Environments: Global Trends and European Strategies. Ghent, Belgium (26–29th August, 2008). http://www.eaae2008.be/

Varela-Ortega, C., Esteve, P., Blanco, I., Carmona, G. & Hernánez-Mora, N. (2008). Storylines and conceptual model (key drivers and water visions) in the Guadiana river basin. EU integrated project SCENES (Water Scenarios for Europe and for Neighboring States) (project number, GOCE 036822). European Commission, Brussels (www.environment.fi/syke/scenes)

Villarroya, C. (2000). *Inscripción de aprovechamientos en el Registro de Aguas. El Proyecto ARYCA* [Inscription of Water Uses in the Water Registry. The ARYCA Project]. Curso Sobre Gestión Administrativa de Recursos de Agua Subterránea [Seminar on Groundwater Resources Administration]. Ministerio de Medio Ambiente, Madrid, July 2000. 25 p.

Vives, R. (2003). Economic and Social Profitability of Water Use for Irrigation in Andalucia. *Water International* 28(3): 326–333.

CHAPTER 13

Trends in water pricing and markets

Alberto Garrido
Department of Agricultural Economics and Social Sciences, Technical University of Madrid, Spain

Javier Calatrava
Department of Firm Economics, Technical University of Cartagena, Spain

1   INTRODUCTION

Over the last two decades water policy in Spain has witnessed profound changes in the way water services are priced at all levels and in all sectors. Levels are defined by the intermediary governmental agencies or private agents that intervene to provide water services, ranging from abstraction out of water bodies to wastewater treatment and disposal.

The era of modern Spanish water legislation began in 1985 with the enactment of the Water Act that replaced the old 1879 Water Act (see Chapters 10 and 11). The 1879 Act was to be accompanied by the 1911 Irrigation and Land Reclamation Act. Jointly they granted very generous economic conditions for irrigators that benefited from the State water projects. Agricultural users were asked to repay less than 50% of the project costs, with 25-year reimbursable loans at an interest rate of 1.5%. Overall, farmers' charges ended up being about 10 to 20% of the development costs. Operation and maintenance costs have seldom been covered in full.

Spanish water policy experienced a radical departure from this state of affairs with the enactment of the European Union Water Framework Directive (WFD) in December 2000 (see Chapter 16). While the WFD foresees that Member States will implement water charges pursuant to the concept of 'cost recovery prices' by 2010, Spain's present systems of charges are generally a long way from complying with this. This chapter reviews the history of Spanish pricing policies, and conjectures what impact the WFD's pricing criterion is likely to have. It also covers the most recent experiences with water market exchanges.

2   GENERAL PRINCIPLES IN THE APPLICATION OF WATER PRICING POLICIES IN SPAIN

There is a stunning variety of water pricing in Spain. Even so, there is a general or canonical model, which stems from the application of the 2001 Consolidated Water Act. We first discuss this general model, and then go on to review other pricing systems that can be found in Spain. Actual tariffs across sectors and regions will be reported in the next section.

2.1   *The general model*

Chapter 11 details the principles and definitions of the four levies and charges stipulated in national jurisdictions. Figure 1 is an outline of the general model that applies only to surface waters, as most groundwater developments are legally private.

First, users that occupy, utilize or make use of the public hydraulic domain will be charged a levy that is meant to protect and improve the state of this domain. This levy is charged on the use of land, river beds and river flows, but not on water consumption. Second, an effluent control levy is set at 0.012 €/m³ for urban sewage, and at 0.03 €/m³ for industrial wastewaters,

131

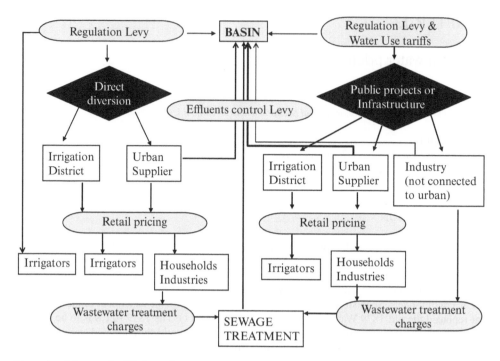

Figure 1.   Scheme of tariffs for surface water users.

although this may vary depending on the pollution levels of the discharged effluents. The third charge is the 'regulation levy' (*canon de regulación*) and is meant to compensate the administration for the investment costs associated with water regulation works and to cover their operation and maintenance costs (O&M). And the fourth and last charge is the 'water use tariff' (*tarifas de uso*). This levy is meant to compensate the administration for the investment, operation and management costs of specific infrastructures, including main canals, inter-basin transfers, diversion schemes and any other elements that pertain to specific projects. The 1999 Water Act (see Chapter 11) introduced a factor ranging from 2 to 0.5. This factor would be applied to increase or decrease the levies and tariffs charged to irrigators whose actual consumption was above or below reference consumption levels.

The exact figures for the 'regulation levy' and 'water use tariff' are the result of adding together (i) the expected O&M costs; (ii) the administration costs directly attributable to the management of these infrastructures, and (iii) 4% of the actual value of the total investment made to develop the infrastructures, taking into account realistic amortization rates.

In 1997, almost one third of the total collected proceeds were for a single water work, the Tagus-Segura Transfer (TST), which represents only 2% of all water users (MIMAM, 2000). Further, levies and tariffs are charged on about 90% of the whole country's territory and water diversions, and the grand total averages about 0.005 €/m³.

The actual recovery rates across River Basin Agencies (RBAs) applying this general model differ widely. Generally, RBA costs are borne by each sector according to simple rules. In the Júcar RBA, the State pays a variable percentage for flood prevention services (generally 20%, but for some dams up to 80%) (Estrela, 2004).

Since earlier work conducted in 1998, and pursuant to the application of the WFD, significant progress has been made in refining the analysis and measurement of water service costs. To pave the way for sorting out major conceptual difficulties about the economic analyses of the water sector, the Júcar Basin was selected as a pilot study. Contrary to what is established in the WFD, the recovery rates in the Júcar Basin fell between 1996 and 2001 from 70% to less than 40%.

This indicates that the Júcar RBA will have to reinforce its charge collection capacity and increase the tariffs and levies significantly in the future.

2.2   *Regional wholesale and resource pricing initiatives*

There are a few important exceptions to the general rule described above. As regional governments develop their own environmental policies, some of them have specific water charges. Catalan, for instance, charges all final consumers a water levy (*canon del'aigua*), irrespective of the retail agency servicing users and the kind of public-private institutional organization. Rates vary across sectors, and have differing rating structures. Farmers are exempt from the levy. Most regional governments have enacted similar levies.

2.3   *Wholesale supplies to the irrigation sector*

As Chapter 12 reviews, apart from supplying water to irrigators, irrigation communities play a major role in water management both at the RBA and district levels. Most irrigators are benefici-aries of regulation and specific works and, as such, are charged both 'regulation levies' and 'water use tariffs'. Others who abstract the water directly from the surface water bodies must pay the 'regulation levy' to the RBAs. Groundwater developers are not usually obliged to pay any levy, because most groundwater resources are still under private ownership (see Chapters 14 and 15). They are not registered with the River Basin Authorities and have paid for the drilling of the water wells and their O&M out of their own pockets.

Cost recovery rates in the agricultural sector vary across basins. In the Júcar Basin, the irriga-tion sector should contribute 50% of the RBA proceeds. However, in relative terms, irrigators pay about 25% of what urban users pay for one cubic meter (Estrela, 2004).

2.4   *RBAs and urban supply*

Urban retailers, be it private or governmental, are also charged regulation levies and water use tariffs. In Spain, 78% of all urban uses are generally served from surface waters, but sur-face sources provide only 30% of the water supply in cities with a population of less than 20,000. In addition, 82% of the volumes served to urban customers of any kind are diverted directly from the water bodies by the water suppliers, the remaining 18% being provided by water wholesalers.

Exactly what water service costs are included in household water bills varies across cities and regions. Less than half of municipalities include sewerage costs in the bill, and about 67% include sewage disposal and treatment costs. Tariffs are approved by the Public Service Commission (regional government), the municipalities or by both.

The latest water industry association (AEAS[1]) survey reported that about 80% of the served population pays full cost recovery prices (AEAS, 2004). But about one third receives subsidies to cover operation and management costs, and about 46% are granted capital subsidies mostly for projects related to wastewater collection and sewage treatment facilities. We now review the cases of the two largest metropolitan areas in Spain.

Case 1: Barcelona
The case of the Metropolitan Area of Barcelona (Catalan) is an exception to the general rule. Considered one of the most institutionally complex systems in Spain, final customers are charged a tariff that results from the adding up four components. The first is the wholesale supply service which is provided by *Aguas del Ter y del Llobregat* (ATLL), a public company owned by the Cat-

---

[1]The Spanish Association of Water Supply Companies and Sewage Treatment Providers (AEAS, 2004) sur-veys most water companies, representing more than 56% of the population, and all important cities except Bilbao.

alan Regional Government. As ATLL diverts resources from the natural water bodies, it is subject to the 'water levy' established by the Catalan Government (Catalan Water Agency). In addition, ATLL supplies water to the concessionary companies that retail about 60% of all water served to the final consumers. This is the second cost component, which is in turn divided in two tariffs (an O&M tariff, and an investment tariff). The third component is charged to most households in Barcelona by the water company, Aguas de Barcelona (AgBar). The fourth component is the sewage fee that is charged to consumers to cover wastewater collection and treatment costs.

The regulatory regime of the public-private system in the Catalan inner basins is governed by regional jurisdiction, and, as such, its water tariffs structure is approved by its government. Urban water rates in Barcelona are among the highest in Spain.

Case 2: Madrid water systems

In the case of the capital city of Madrid and more than one hundred municipalities, a regional water company called *Canal de Isabel II* (CYII) is responsible for all water cycle activities. With its water works and infrastructures, CYII diverts, stores, conveys, distributes and retails water to all connected customers, including households, industries and commercial customers. It also collects and treats wastewater from most serviced municipalities. Since CYII runs all infrastructures, its rates are meant to recover all O&M costs, and the unsubsidized segment of capital investments. Lastly, since CYII develops projects and works that provide services to the Tagus RBA, it has special financial agreements for the components of specific facilities. CYII's customers pay complex tariffs with increasing block rates, and, since 2005, summer-winter variations. Sewage systems and wastewater treatment plants are the competence of Madrid's municipality. There has always been some overlap between the regional government and the city council.

2.5   Groundwater sources

Generally, no water charges or levies are charged on groundwater users, although users must have access rights to make use of water resources. This means that groundwater users only incur extraction and distribution costs. In Catalan, however, groundwater users (excluding irrigators) are charged the above regional levy also charged to surface water users as well.

2.6   Historical users

In Spain, historical users are those who can provide evidence of having used water before major modern infrastructures were built. Although some of these may date back to medieval times, it is difficult nowadays to isolate their abstraction patterns from the regulation services provided by the water works. Thus they are, in principle, subject to the 'regulation levy' charged by the RBA.

However, a number of traditional water associations that are owners of their water works fought a legal battle for exemption from the regulation levy. Their legal case was based on the claim that their use was independent of the basin's regulation works, and the administrative courts set a precedent in their favor. This implies that they are presently exempt from tariffs and levies. However, it remains to be seen whether the full application of the WFD results in the need to charge resource and environmental costs to these users.

2.7   Services related to sewage and wastewater treatment and discharge

In addition to the effluent levy charged by the RBA on all point source pollutants, all urban and industrial consumers are charged for wastewater treatment services. These sewage and wastewater treatment tariffs vary significantly across cities, basins and even regions. The collection scheme also varies depending on the organizational structure of the whole system. Some urban water retailers collect the tariff, but it is the municipality that is responsible for providing the service and collecting the charge in some cities. The AEAS survey (2004) indicates that the most critical problems of the sewer systems and wastewater treatment services include the state of the conveyance network, water treatment plants and flood control. This implies that charges for these services vary significantly across cities, depending on the relative obsolescence of the network and the efficacy of the treatment plants. Presently, 95% of the collected wastewater receives a 'secondary treatment', and only 9% goes through a rigorous 'tertiary treatment'. This is a possible explanation for why total household and other urban users' charges in Spain are about half of other OECD countries'.

2.8   *Special water projects and inter-basin water transfers*

One of the most significant water works in Spain is the Tagus-Segura Transfer (TST) (see Chapter 19 for a detailed description). Its economic regime is based on a specific law (Act 52/1980) that set it apart from the doctrine of previous water legislation. First, it foresaw for the full cost recovery of 60% all the project costs; the remaining 40% is not applicable because of the excess capacity with which it was designed. Second, it provided for a surcharge to generate proceeds that would target water works to be developed in the Tagus basin, the area of origin. While this second provision turned out to be insufficient to compensate the area of origin, it set an important precedent for new self-financed infrastructures.

The rationale of this compensatory payment was questioned by the government based on two flaws (MIMAM, 2000). First, although users were charged 'full cost recovery prices', the government diverted a fraction of these proceeds to finance water investments in the area of origin. This practice implies that TST customers do not fully pay for the project costs and the compensatory payments. Secondly, while the spirit of the specific TST Act was to facilitate a mutually beneficial agreement between the area of origin and the recipient basin, it has never worked as such. All in all, TST tariffs for raw water are among the most expensive currently paid in Spain (in the range of 0.09 to 0.12 €/m³). But even here the farmers' tariffs are approximately 50% of the tariffs for urban water supply.

Despite the innovative tariff system applied to the TST project, subsequent interbasin transfer schemes, such as the Guadiaro-Guadalete (1995) and the Tagus-Guadiana (1995) interbasin water transfers, followed the principles of the Water Act, which do not ensure a complete full cost recovery.

2.9   *The Canary Islands*

The economics of water resources in the Canary Islands obeys quite different principles and traditions to those prevailing in mainland Spain. The most populated of the seven islands, Tenerife and Gran Canaria (population 886,000 and 830,000 respectively), are rather arid and mostly volcanic. The Canary Islands have had a specific Water Act since 1990. The absence of surface flows and a traditional water culture that appreciates the importance of water have given rise to numerous institutional arrangements. The private sector has been the major actor in these institutions, though more recently a number of governmental agencies have begun to participate actively in various water services (Aguilera Klink, 2002; see also Chapter 10 for an evaluation of water law, markets and institutions in the Canaries).

3   SECTORIAL WATER PRICING

3.1   *Irrigation*

Farmers pay the 'regulation levy' and the 'water use tariff' to the RBA through the irrigation district (ID), plus an additional tariff to cover the costs of the irrigation district itself (called "*derrama*"). IDs that abstract their water directly and that use publicly developed infrastructures only pay the regulation levy.

Average tariffs paid for irrigation water in areas where water is supplied by RBAs is 0.02 €/m³, except for the agricultural users served from the TST who pay about 0.09 €/m³, whereas areas that use groundwater pay an average of 0.04 to 0.07 €/m³, based on extraction and other O&M costs (see Chapter 14). Groundwater users pay their financial, energy, operation and maintenance costs directly. A recent Júcar Basin study (Estrela *et al.*, 2004) indicates that about 90% of all the expenses related to surface and groundwater are not controlled by the RBA, but by the private sector. In the Júcar Basin, most customers are groundwater users.

These figures are well below the public costs incurred to supply the water. In Andalusia (Guadalquivir, Guadiana and South basins), where irrigation water comes predominantly from surface resources, average water tariffs for surface water are about 0.01 €/m³, with a final cost for farmers, including other water-related variable costs, of 0.04 €/m³, whereas it costs the Administration about 0.12 €/m³ on average to supply the water (Corominas, 2001). The total cost for farmers using groundwater ranges from 0.13 to 0.5 €/m³.

On the other hand, in the region of Valencia (Júcar and Segura basins), where the use of groundwater is very intense and predominant, the total cost of water for farmers ranges from 0.04 to 0.22 €/m³ for surface and groundwater, respectively, with an average of 0.11 €/m³ (García *et al.*, 2004). Examples of farmers' charges for different districts and RBAs are reported in Table 1. Generally, these figures are wholly applicable to 2008.

Most irrigated areas in Spain have opted for one of the following water pricing schemes (MAPA, 2001):

– A fixed per-hectare tariff, calculated as the total costs attributable to farmers divided by total irrigated area. It is the most common option in the traditional districts served from surface resources. In all, fixed rates are applied across 82% of the surface water irrigated acreage in Spain.
– A volumetric tariff, more frequent in districts served from groundwater and/or incurring energy costs, is applied across 13% of the irrigated hectares. Volumetric tariffs are applied (1) as an amount per m³ served; (2) with rates for each time a hectare of land is irrigated, and (3) with rates per hour of irrigating time.

A binomial tariff combining a volumetric rate to cover variable costs with a fixed per hectare rate for investment and management costs, applied across the 5% of the irrigated acreage. Binomial tariffs are more predominant in private and modern publicly developed districts, where metering devices exist and energy costs are substantial.García Mollá (2002), cited by MIMAM (2007) distinguished in the Jucar basin, the following types of water tariffs:

– Two-part: per ha + number of water applications
– Two-part: per ha + duration of turn
– Per ha
– Two-part: per ha + (duration of turn and number of water applications).
– Two-part: par ha + volumetric
– Duration of turn
– Duration of turns and number of applications.

While Table 2's right column provides average cost recovery rates for the principal basins in Spain, very few available studies permit a detailed application of the criterions for cost evaluations and the cost recovery rates. Pérez and Barreiro (2007) performed a detailed evaluation of cost recovery rates at the wholesale level of the Gállego basin (a tributary of the Ebro), applying the Spanish Water Law criterions. Among the most crucial assumptions of their exercise were: (a) capital replacement costs were not taken into account; (b) since canals were more than 25-year old, amortisation rates were set to zero; (c) a cost sharing rate to irrigation of 50% among the remaining sectors (flood control, 12%; hydropower, 32%; urban and commercial, 5%). They came up with a full financial cost-recovery charge equivalent to 0.0077 €/m³, whereas the actual rate is in the Gállego basin 0.00403 €/m³, that is an actual cost recovery rate of 52%.

Table 1. Tariff schemes for several Spanish Irrigation Districts and RBAs.

District	Basin	Type of district (surface in ha)	Irrigation system (F: Furrow S: Sprinkler D: Drip)	Average allotment (m³/ha)	Water supply	Type of tariff	Regulation levy + water use tariff (a)	District tariff (b)	Total (a + b)	Volumetric term (€/m³)	Sources[+]
Guadalmellato	Guadalquivir	Old private (6,129)	F/S	6,000	Surface	Per hectare			150	–	1
Genil-cabra	Guadalquivir	Modern public (15,000)	S/D	4,000	Surface and groundwater	Binomial		85	85	0.025	2
Riegos de Levante (MD)	Segura	Traditional public (4,183)	F/D	5,100	Surface	Binomial			25	0.184*	3
Com. de usuarios de Novelda	Segura	Private, specialty crops (2,056)	F	6,000	Groundwater	Binomial	–	85	85	0.151*	2 & 3
Acequia Real del Júcar	Júcar	Historical (21,736)	F/D	12,600	Surface	Per hectare			211*	–	2 & 3
Canal Cota 220, Onda	Júcar	Modern (2,179)	F/D	7,200	Surface and groundwater	Binomial	3	–	3	0.1*	2
Babilafuente	Douro	Old public (3,570)	F/S	13,000	Surface	Per hectare	66	6	72	–	2
Villalar	Douro	Modern public (510)	S	Unlimited	Groundwater	Volumetric	–	–	–	0.06	2
Bajo Carrión	Douro	Modern public (6,600)	F/S	4,500	Surface	Per hectare	26	12	38	–	4
Daimiel	Guadiana	Modern (19,000)	S/D	Depending on size	Groundwater	Binomial	–	–	5 €/ha + 9 €/well	energy cost	2
Loma de Quinto	Ebro	Modern public district (2,606)	S	5,850	Surface (pumped)	Volumetric	–	–	–	0.034	5
Monegros-Cinca	Ebro	Public (100,000)		10,000	Surface	Per hectare			60	–	6
Canal de Aragón	Ebro	Public modern (105,000)		5,700	Surface	Volumetric	–	–	–	0.05–0.07	6
San José	South	Private, fruit prod (200)	S/D	5,045	Groundwater	Binomial			144 + 36 €/share	0.14	7

All volume-related tariffs (per irrigation turn, per volume or per duration of irrigation) have been converted to per volume amounts.

(*) Includes the cost of labor of the district worker that manages the system.

[+]References: 1. Calatrava (2002); 2. Sumpsi et al. (1998); 3. García (2002); 4. Gómez-Limón et al. (2002); 5. Dechmi et al. (2003); 6. Arrojo (2001); 7. Calatrava and Sayadi (2004).

Table 2.   Payments for irrigation water services in Spain (only in the inter-regional basins), all figures expressed in Euros.

Basin	Groundwater per ha	Groundwater per m³	Surface Per ha Distribution	Surface Per ha ID and basin tariff	Total per m³	Total per ha	Total per m³	Financial cost recovery rates
Douro	500	0.095	19.88	46	0.012	231	0.044	89%
Ebro	829	0.15	49	12	0.011	113	0.02	86.10%
Guadalquivir	744	0.15	101	70	0.035	400	0.081	97.70%
Guadiana	232	0.048	19	102	0.025	188	0.039	54.10%
Júcar	383	0.074	81	16	0.02	283	0.055	85%
Segura	789	0.163	34	151	0.038	463.8	0.096	n.a.
Tagus	541	0.1	36	67	0.02	199.3	0.038	n.a.
Total	**500**	**0.09**	**50**	**56**	**0.021**	**263.5**	**0.051**	**87.10%**

Source: MIMAM (2007).

Since 2001, 95% of the budget devoted to irrigation in Spain is targeted to finance modernization projects, which have reached 1.3 Mha and a budget of 4 billion euros (Barbero, 2005). Beneficiary farmers must pay only 50% of the project's costs, for which they are granted preferential loans. But the process is becoming very costly, as projects have been re-focused to include environmental, structural, technological and land planning/tenancy components. The gains are private in the form of more efficient and productive districts, as well as public in the form of water conservation and reduced pollution. By no means would farmers' full cost-recovery rates suffice to finance such projects. Yet they are praised and uncontested.

3.2   Urban water services

There are remarkable differences in rating structures and tariff levels across Spanish cities and metropolitan areas. Overall, Spanish urban tariffs are in the lower segment of the OECD countries. Spain's average rate is 0.97 €/m³, whereas Australia's is 1.62, Canada's is 0.72, Italy's is 0.67, France's is 2.65, Germany's is 3.88, Japan's is 1.33, the UK's is 2.5, and the USA's is 1.56 (OECD 2003, see also Chapter 8). According to AEAS (2004), the average urban tariff for Spain has an annual growth rate of 5.5%, quite above the inflation rate during 2001–03. The major features of Spanish household and other urban users' water tariffs are described in Table 3. Although the application of block rates is increasing in most cities, large consumers pay lower average prices than households that use less water.

3.3   Industries, hydropower and other users

20% of total urban water consumption corresponds to industrial users. Industries not connected to urban supply services must pay the RBA levies. Each agency applies different criteria to establish the tariff levels. For example, the average water use tariff in the Tagus basin is 0.0161 €/m³, whereas the average regulation levy is 0.0161 €/m³ for industrial and 0.0015 €/m³ for non-consumptive uses. In the Guadiana basin, the average regulation levy is 0.022 €/m³ for industrial uses and 0.0071 €/m³ for non-consumptive uses.

4   EXPECTED CHANGES RESULTING FROM THE APPLICATION OF THE WATER FRAMEWORK DIRECTIVE

The WFD places significant emphasis on the role of economics and pricing in meeting its objectives. Of particular relevance is art. 9, which states that '*Member States shall take account of the principle of recovery of the costs of the water services, including environmental and resource costs, ..., in accordance in particular with the polluter pays principle*'. It goes on to establish that '*Member States shall ensure by 2010 that water-pricing policies provide adequate incentives for users to use water resources efficiently, ... [and that] an adequate contribution of the different water uses, disaggregated into at least industry, households and agriculture, to the recovery of the costs of the water services*' (WFD, pp. 12–13). We can summarize the likely effects of the WFD as follows.

First, cost recovery rates of the water regulation and water use levies will need to be significantly improved. Although the Júcar is the basin that has taken the lead in thoroughly setting up cost recovery rates, the Ministry of the Environment (MIMAM, 1998) foresees a similar situation for the remainder of basins in Spain. It remains to be seen what shape the WFD will finally take when it comes into full force in 2010, but it is clear that basin's levies are likely to double, and perhaps triple, if resource and environmental costs are also included in the definition of service costs. Yet resource cost evaluations indicate that they are extremely variable across basins and wet-dry cycles (Iglesias *et al.*, 2003).

Second, virtually all analyses carried out to examine the effects of higher prices in the agricultural sector predict that it would be hard hit by the strict application of the WFD (see Table 4 for a list of findings about the application of the WFD in the agricultural sector). For the Douro basin, government officials estimated that the additional water cost that would have to be charged to the irrigation sector would be around 0.041 €/m^3 (Berbel *et al.*, 2005). This is below the full cost recovery rates (FCR) estimated by other authors. Present Ministry of the Environment (MIMAM, 2007) estimates of irrigation sector cost recovery rates for the Júcar Basin are 45%, not including environmental and resource costs. The estimates reported in Table 4 show that the application of FCR rates would seriously affect both water demand and farmers' income. Yet, Hernández-Mora & Llamas (2001) contend that groundwater irrigation provides evidence of farmers' greater willingness to pay than results based on mathematical programming models. Perhaps the differences are related to the time frame, with mathematical programming models focusing on short-term impacts and the groundwater irrigation indicating better long-term adaptation.

Third, while the use of groundwater resources is a key factor in sustaining the most profitable irrigated agriculture, the prevailing view is that these users will not be subject to charges or levies. This implies that about 20% of all agricultural users, representing more than 50% of total irrigation output, are likely to be exempt from further charges.

Table 3.   Urban tariffs in Spain.

	Household's consumers (in % of served population)	Industrial and commercial consumers (% of served population)
Fixed rate	Yes (91%)	Yes (90.5%)
Minimum consumption	No (86.5%)	No (88%)
Block-rate structures	Yes (91.5%) of which	Yes (79%) of which
	– 2 blocks: 12.3%	– 2 blocks: 53.8%
	– 3 blocks: 55%	– 3 blocks: 21%
	– 4 blocks: 18%	– 4 blocks: 15%
	– 5 or more blocks: 15%	– 5 or more blocks: 9%
Increasing Block Rates	Yes (96%)	Yes (84%)
With sewerage tariff	With Connection fee: 44%	With Connection fee: 43%
	With volumetric fee: 84%	With volumetric fee: 84%
With sewage treatment tariff	With Connection fee: 71.3%	With Connection fee: 72%
	With volumetric fee: 74%	With volumetric fee: 75%
With incentives for reduced consumption	72%	32%

Source: AEAS (2004).

Table 4.   The effects of the WFD on the irrigation sector.

RBA	Type	Present rate Levels[1] (€ per cm)	Medium	Tariff increase FCR[2]	Results Farm income	Water demand	Other	Sources*
Douro	Per hectare	0.01	0.04	0.06	−40% to −50%	−27% to 52%	Great influence of agricultural policies	1
Guadalquivir	Per ha & Vol	0.01–0.05	0.05	0.1	−10% to −19%	0 to −10%	Same	2
Douro	Per ha & Vol	0.01	0.04	0.1	−10% to 49%	−5% to −50%	Technical response	3
Guadalquivir	Per ha & Vol	0.01–0.05	0.06	0.12	−10% to −40%	−1% to −35%	Technical and crop response	3
Guadalquivir	Per ha & Vol	0.01–0.05	0.03	0.09	−16% to −35%	−26% to −32%	Technical and crop response	4
Guadiana	Per ha	0.005	0.03	0.06	−15% to −20%	−30% to −50%	Technical and crop response	4
Júcar	Per ha, Vol & hourly rates	0.03–0.15	0.06	0.15	−10% to −40%	0 to −40%	Technical response	3
Segura	Per ha, Vol & hourly rates	0.05–0.30	0.10	0.25	−10% to −30%	0 to −10%	Very inelastic demand	3
Guadalquivir	Per ha	0.013	0.04	0.14	−10% to 25%	−10 to −38%	Very inelastic with low availability	5

[1] Equivalent measure.
[2] Full cost recovery rates.
* References: 1. Gómez-Limón and Riesgo (2004); 2. Berbel *et al.* (2004); 3. Sumpsi *et al.* (1998); 4. Iglesias *et al.* (2004); 5. Mejías *et al.* (2004).

Fourth, the WFD assumes that Member States should already be complying with the remaining Directives that are applicable to the water sector. This implies that the cost burden that water users should have to bear is larger than the application of just the WFD. In part, this is due to the fact that many countries, including Spain, have been found in breach of other EU Directives related to the water environment, aquatic ecosystems, drinking and bathing waters, and toxic substances (see WFD, art. 22).

5   WATER MARKETS AND OTHER FLEXIBLE ALLOCATION MECHANISMS

Informal water trading is common in many Spanish regions, especially in the South and East. Temporary leases of water predominate, whereas permanent exchange of property rights is a less common practice. Most exchanges are confined to the local level, depending on the availability and cost of transportation infrastructures, and are usually private agreements among users belonging to one and the same user association.

Permanent exchanges of water rights are commonly linked to groundwater, whereas temporary leases of water relate to both surface and groundwater resources. The different types of water exchanges are not necessarily exclusive but complementary, as they satisfy different users' supply needs. For example, trading in the Canary Islands includes exchanging water shares in the so-called "water communities", year-long water leases, or even spot exchanges at specific times in the agricultural season (Fernández Bethencourt, 2001).

Public water rights to irrigation districts are usually appurtenant to agricultural land. By contrast, private rights to groundwater are usually equity linked to a well that entitles the shareholder to a certain amount of water, irrespective of the amount of land the holder owns. This explains why most permanent exchanges of water rights relate to private rights over groundwater. In horticulture along the Southern Mediterranean Coast, exchanges of shares of groundwater corresponding to either private wells or public pumping concessions are frequent (Calatrava & Sayadi, 2004). Prices along the coast of Granada commonly range between 2,500 and 3,000 euros per share (on average a share is equivalent to about 1,500 m^3/year). Water markets for private groundwater rights are also very active in the Canary Islands.

Informal trading is more common among farmers belonging to the same district. This includes water trading in a strict sense, as well as other types of informal agreements without monetary compensation, such as the exchange of irrigation shifts, based on the customary trust there is among farmers in the same area. They usually consist of leases of water either from surface water supplied by the irrigation district and the RBA or from farmer-owned wells. Albiac *et al.* (2006) report exchange prices for horticulture in the Southeastern Segura Basin in the range of 0.1 to 0.4 €/m^3 and 0.15 €/m^3 on average in years of normal supply. In years of extreme scarcity, prices range from 0.15 to 0.6 €/m^3, and are 0.35 €/m^3 on average. These prices match typical groundwater irrigation costs in the Segura basin, which range from 0.13 to 0.74 €/m^3.

Regarding more formal water markets, the 1999 Reformed Water Act allowed for the voluntary exchanges of water, either directly among trading partners or through publicly-run water banks, all requiring approval from the RBA (see Chapters 10 and 11). There are, however, few experiences, as many users have been reluctant to publicly exchange their water or concessions. In fact, only 35 exchanges of public concessions were authorized in the Segura Basin from 2000 to 2005, amounting to 10 hm^3, less than 1% of total water consumption in the basin. Other experiences are the case of an irrigation district at the Upper Tagus Basin that sold its unused public water concession to the city of Aranjuez (near Madrid), and some cases of concessions sold by farmers in the Upper Guadiana Basin. It seems that farmers are quite wary of formal water markets. The option of public water banks, common in California and Australia, has only been relied upon in the Guadiana, Júcar and Segura Basins, where public "exchange centers" have started functioning with relative success during 2007.

Yet, the drought that started in 2005 forced the Spanish Government to allow inter-basin exchanges of water using previously existing infrastructures, paving the way for a number of major agreements among farmers in the Segura Basin and nearby basins. Public water concessions amounting to 80 hm^3 were bought in both the Lower and Upper Guadalquivir Basin for 2 to 2.4 €/m^3, and transferred using the Negratín-Almanzora inter-basin transfer. Farmers with concessions from the Negratín reservoir also entered into five-year water lease agreements at prices in the 0.15–0.18 €/m^3 range.

Using the Tagus-Segura inter-basin transfer, farmers from the Upper Tagus Basin agreed to transfer 30 million cubic meters annually to farmers in the Segura basin at a price of 0.19 €/m^3. The "Mancomunidad de Canales del Taibilla", a major urban water supplier in the Segura basin, signed an agreement in 2006 with farmers in the Upper Tagus Basin to buy up to 40 hm^3 annually at a price of 0.28 €/m^3. Environment Ministry officials have been active in promoting these agreements, waiving the transportation fees for the buyers, which are equivalent to 0.09 €/m^3.

Other market initiatives have used to tackle the most pressing problems connected to overexploited aquifers (see Chapters 14 and 15). Since the enactment of the 1985 Water Law, which included special provisions to tackle the problem of overexploited aquifers, there have been at least four major initiatives to manage groundwater resources. In short, these were (i) the declaration of overexploited aquifers and the mandate to enforce regulations and implement management plans; (ii) an EU Agri-environmental programme, only applicable to Aquifer 23 in the Guadiana Basin, with subsidies to farmers curtailing their water consumption; (iii) the use of inter-basin transfers, both in the case of the southeast coastal areas and in the Upper Guadiana; and lastly, (iv) The Especial Plan of the Upper Guadiana (PEAG, Spanish acronym), and the creation of exchanging centres in the Segura, Jucar and Guadiana basins (Garrido & Llamas, 2009).

The option to use buyouts of water rights, permanent or temporary, gave a rationale to the establishment of exchanges centres (*centros de intercambio* in Spanish). We will review the different approaches taken in the Jucar and Guadiana. In the Jucar basin, the Offer of Public Purchase (*Oferta pública de adquisición de derechos*, OPA) was targeted to farmers relying on groundwater resources near the Jucar's headwaters. Its objective was to increase the water table levels in Castille-La Manche to ensure that Jucar flows to the Valencia region increase from historical lows. Farmers were given the option to lease-out their rights for one year in return for a compensation ranging from 0.13 to 0.19 cents per m^3, the variation depending on the distance of the farmer's location with respect to wetlands or to the river alluvial plain. The OPA was launched in two rounds, the first with disappointing results in terms of farmers' response while the second had more success. The purchased waters served the unique purpose of increasing the flows, enabling more use downstream in Valencia. But the OPA did not have any specific beneficiaries dowstream, other than the increase of flows.

The OPAs of the Guadiana followed a completely different approach and were meant to address serious problems of overexploitation in the Upper Guadiana. As stated before, the OPA formed part of a more ambitious programme of aquifer recovery, called the PEAG. The Guadiana's OPA made offers to purchase permanent water rights to groundwater, paying farmers €6,000–10,000 per hectare of irrigated land. Note that, since these farmers had seen their allotments reduced in preceding years, what the Guadiana basin was truly purchasing from the farmers was about 1500–2500 m^3/ha, effectively 2–4 €/m^3. The Guadiana basin agency has the objective of 'purchasing' the water rights to 50,000 hectares of irrigated land, and is budgeting €500 million for the whole plan. A marked difference to the Jucar's OPA is that the Guadiana exchange centre will transfer part of these rights to other farmers (growing vegetables) and to the Autonomous Community of Castille-La Mancha. The Guadiana basin will grant fewer rights than it has purchased, allocating the difference to wetlands and to increasing the piezometric levels of the aquifers. One subtlety of the Guadiana scheme is the fact that, while farmers entering the programme must surrender their private rights (honored because they were in the catalogue of private waters before the 1985 water law was enacted), those that gain access to them will be granted 30-year 'concession' rights (which is more attenuated property than the others). So the Guadiana operation had this other dimension that in the long term will imply that the basin agency has more users with 'concessions' than with private rights.

6   CONCLUDING REMARKS

The whole edifice of water services finance and economics is undergoing a profound reform at all system levels in Spain. At the highest level, RBAs, even a narrow interpretation of the WFD principles will necessarily be accompanied by significant percentage increases of the tariffs and levies charged on the diversion water or dumping wastewater into water bodies. A few regional governments, with Catalan at the forefront, have already approved tariffs one order of magnitude larger than in most other inter-regional basins, though Catalan irrigators are presently exempt from these tariffs.

At the basin levels, too, it is too early to predict what portion of the environmental and resource costs will lead to further increases of tariffs and levies. The European Commission has been rather conservative in requesting Member States to add these costs to the rates. In Spain, water banks and exchange systems have been supported from the Ministry of the Environment, which set them in 2005, and included in the set of relevant water management measures established in 2008 by several River Basin Authorities to comply with the WFD. Should voluntary exchanges become a frequent practice in Spain, this will be the test for the countless estimates of water opportunity costs to be found in the literature.

At the retail level, Spain has advanced enormously in establishing complex pricing schemes for households and industries. While users now pay full cost recovery prices, their rates for sewage and wastewater treatment may need to increase substantially to ensure that less polluted effluents are returned to the water bodies. In the agricultural sector, pricing systems lag significantly behind other sectors, though farmers serviced with groundwater have shown a remarkable capacity to deal with scarcity and implement demand management systems.

For the next decade, Spanish water consumers will face increasing water prices, and surface water irrigators will be the ones to experience the greatest increases. Groundwater irrigators are already paying the full cost. The driving forces will be the demand for more reliable urban supplies and the massive restoration needs of most aquatic ecosystems below the 1000 m elevation line. Yet, a few regional governments are becoming active actors in developing and applying new system of charges. But even in this case, farmers are exempt from these initiatives.

REFERENCES

AEAS (2004). Suministro de agua potable y saneamiento en España 2002. VII Encuesta Nacional [Urban water supply and wastewater treatment in Spain. 2002. 7th National Survey]. Madrid.

Aguilera Klink, F. (2002). *Los mercados de agua en Tenerife* [Water Markets in Tenerife]. Bilbao, Spain: Bakeaz.

Albiac, J., Hanemann, M., Calatrava, J., Uche, J. & Tapia, J. (2006). The rise and fall of the Ebro water transfer. *Natural Resources Journal*, 46(3): 727–757.

Arrojo, P. (2001). Valoración de las aguas subterráneas en el marco económico general de la gestión de aguas en España [Groundwater valuation in the general economic framework of water management in Spain. In *La economía del agua subterránea y su gestión colectiva* [The economics of groundwater and collective management], edited by Hernández, N. and Llamas, M.R. Madrid, Spain: Fundación Marcelino Botín and Editorial Mundi-Prensa 3–39.

Barbero, A. (2005). *The Spanish National Irrigation Plan*. In OECD (Ed.). *Water and Agriculture: Sustainability, Markets and Policies*. OECD, Paris.

Berbel, J. & Gutiérrez, C. (2004). (eds). *Sustainability of European Irrigated Agriculture under Water Framework Directive and Agenda 2000* Córdoba, Spain: Sixth Framework Programme, European Commission.

Calatrava, J. (2002). Los mercados de agua en la agricultura y el riesgo económico: Una aplicación en el valle del Guadalquivir [Irrigation water markets and economic risk: an application to the Guadalquivir Valley] Doctoral dissertation, Department of Agricultural Economics, Universidad Politécnica de Madrid, Madrid, Spain.

Calatrava, J. & Sayadi, S. (2005). Economic valuation of water and willingness to pay analysis in tropical fruit production in Southeastern Spain. *Spanish Journal of Agricultural Research* 3(1): 25–33.

Corominas, J. (2001). El papel económico de las aguas subterráneas en Andalucía [The economic role of groundwater in Andalusia]. In *La economía del agua subterránea y su gestión colectiva* [The economics of groundwater and collective management], edited by Hernández, N. and Llamas, M.R. Madrid, Spain: Fundación Marcelino Botín and Editorial Mundi-Prensa, 111–139.

Dechmi, F., Playán, E., Faci, J.M. & Tejero, M. 2003. Analysis of an irrigation district in northeastern Spain: I. Characterisation and water use assessment. *Agricultural Water Management* 61: 75–92.

Estrela, T. (Coord.) 2004. *Júcar Pilot River Basin*, Provisional Article 5 Report Pursuant to the WFD. Valencia, Spain: CH. Júcar.

Fernández Bethencourt, J.D. (2001). El papel económico de las aguas subterráneas en Canarias. [The economic role of groundwater in the Canaries]. In *La economía del agua subterránea y su gestión colectiva* [The economics of groundwater and its collective management], edited by Hernández, N. and Llamas, M.R. Madrid, Spain: Fundación Marcelino Botín and Editorial Mundi-Prensa, 251–267.

García Mollà, M. (2002). Análisis de la influencia de los costes en el consumo de agua en la agricultura valenciana. Caracterización de las entidades asociativas para riego [Analysis of the influence of costs on water use in the Valencian agriculture. Characterisation of the water users association] Doctoral Dissertation. Departamento de Economía y Ciencias Sociales. Universidad Politécnica de Valencia, Valencia.

García, M., J. Carles, J. & Sanchís, C. (2004). Características institucionales y territoriales y su influencia en el costo del agua como input en la agricultura [Institutional and Spatial Characteristics as an irrigation water cost factor]. VII Congreso Nacional de Medio Ambiente [7th National Environmental Congress]. Madrid (Spain), 22–26 November.

Garrido, A. & Llamas, M.R. (2009). Water Management in Spain: An Example of Changing Paradigms. In Dinar, A. & Albiac, J. (Eds.) *Policy and Strategic Behaviour in Water Resource Management*. Earthscan, London, 125–146.

Gómez-Limón, J.A. & Riesgo, L. (2004). The case of the River Duero Basin (Northern Spain). In *Sustainability of European Irrigated Agriculture under Water Framework Directive and Agenda 2000*, edited by Berbel, J. & Gutiérez, C. Córdoba, Spain, Sixth Framework Programme, European Commission, 89–112.

Gómez-Limón, J.A., Arriaza, M. & Berbel, J. (2002). Conflicting implementation of agricultural and water policies in irrigated areas in the EU. *Journal of Agricultural Economics* 53(2): 259–281.

Iglesias, E., Sumpsi, J.M. & Blanco, M. (2004). Environmental and socioeconomic effects on water pricing policies: Key issue in the implementation of the Water Framework Directive. 13th Annual Conference of the European Association of Environmental and Resource Economists. Budapest, June 25–28.

Iglesias, E., Garrido, A. & Gómez-Ramos, A. (2003). Evaluation of drought management in irrigated area. *Agricultural Economics* 29: 211–229.

MAPA (2001). Plan Nacional de Regadíos. Horizonte 2008. [National Irrigation Plan. Range of 2008]. Madrid, Spain, Ministerio de Agricultura, Pesca y Alimentación.

Mejías, P., Varela-Ortega, C. & Flichman, G. (2004). Integrating agricultural policies and water policies under water supply and climate uncertainty. *Water Resources Research*, 40(7).

MIMAM (Ministerio de Medio Ambiente) (2000). *El libro blanco del agua* [Spanish Water White Paper] Madrid, Spain: Ministerio de Medio Ambiente.

MIMAM (Ministerio de Medio Ambiente) (2001). *Plan Hidrológico Nacional*. [National Hydrological Plan] Madrid, Spain: Ministerio de Medio Ambiente.

MIMAM (Ministerio de Medio Ambiente) (2007). Precios y costes de los servicios de agua en España. Informe integrado de recuperación de costes de los servicios de agua en España. Artículo 5 y Anejo III de la Directiva Marco de Agua [Prices and costs of water services in Spain. Integrated Report of the Services' Cost Recovery in Spain. Article 5 and Annex III of the Water Framework Directive]. Ministerio de Medio Ambiente. Madrid.

OECD (Organization for the Economic Cooperation) (2003). *Social issues in the provision and pricing of water services*. Paris, France: OECD.

Peréz, L. & Barreiro, J. (2007). Una nota sobre la recuperación de costes de los servicios del agua en la cuenca del Gállego. [A note about the cost recovery of water services in the Gállego basin] *Economía Agraria y Recursos Naturales* 7(13): 49–56.

Sumpsi, J.M., Garrido, A., Blanco, M., Varela, C. & Iglesias, E. (1998). *Economía y Política e Gestión del Agua en la Agricultura* [The economics and management policies of irrigation water]. Madrid, Spain: MAPA and Mundi-Prensa.

CHAPTER 14

Issues related to intensive groundwater use[1]

Emilio Custodio
Department of Terrain, Cartographic and Geophysical Engineering, Technical University of Catalonia, Barcelona, Spain

M. Ramón Llamas
Department of Geodynamics, Complutense University, Madrid, Spain

Nuria Hernández-Mora
Environmental Consultant, Madrid, Spain

Luis Martínez Cortina
Spanish Technical Institute of Geology and Mine, Madrid, Spain

Pedro Martínez-Santos
Department of Geodynamics, Complutense University, Madrid, Spain

1 INTRODUCTION

Groundwater use in Spain has increased dramatically over the last decades, with total volume pumped growing from 2,000 Mm³/year (million m³ per year) in 1960 to more than 6,500 Mm³/year in 2006. Today, groundwater provides between 15–20% of all water used in Spain, although the proportion may approach 100% in some peninsular areas and the islands. This development is mostly the result of the initiative of thousands of individual users and small municipalities, with scarce public planning and oversight. The result has been intensive rates of groundwater use, in what has been called a silent revolution (Fornés *et al.*, 2005; Llamas, 2003).

Groundwater use in Spain has significant socioeconomic importance, both as a factor of production in agriculture and industry, and as a source of drinking water for over 12 million people, almost one fourth of the total population. Given the importance of irrigation water demand, and in the context of increased competition among the different water demand sectors for limited water resources, recent efforts to improve the quality of data on groundwater use and its economic importance are crucial to inform future water policy decisions. This better knowledge is also needed to meet the environmental requirements of the Water Framework Directive (WFD, 2000, see Chapter 16) and the *daughter* Groundwater Directive (GWD, 2006) of the European Union.

The intensive development of groundwater has brought significant social and economic benefits, although their unplanned nature has also resulted in negative environmental, legal and socioeconomic consequences. While the drivers and conditions may vary, the situation in Spain is

---

[1] The authors are highly indebted to Dr. Alberto Garrido for his significant effort in the edition of this text and previous versions. This chapter is a modification and update of the Report on Spain for the General Report on Groundwater Issues in the Southern EU Member States. This general report was prepared by a working group appointed by the European Academies of Sciences Advisory Council (EASAC), chaired by Prof. M.R. Llamas, of the Royal Academy of Sciences of Spain. The report was presented in the second meeting of the working group held on April 19, 2007 in the headquarters of the ARECES Foundation in Madrid. Parts of the text of Llamas and Martínez-Santos "Intensive use of groundwater: three decades of management failures", formerly prepared to be part of this book, have been used.

similar to what happens in most arid and semiarid countries, both developed and developing ones, as emphasized in the Alicante/Alacant Declaration approved at the International Symposium on Groundwater Sustainability (Ragone *et al.*, 2007).

This chapter presents a critical overview of intensive groundwater use in Spain, with emphasis on economic and institutional aspects. After a review of available data on groundwater use and a brief discussion of the situation in some regions where groundwater is used intensively, the economic parameters associated with this use are considered, focusing on irrigation. After that, the institutional framework for the management of groundwater resources that has evolved from the 1985 Water Act and its later reforms are evaluated. This chapter builds on Chapters 10 and 11, which present and discuss the legal framework of Spain's water sector, and on Chapter 12 which reviews water institutions in Spain with a historical perspective. It provides the technical, economic and political bases for the in-depth discussion of groundwater institutions of Chapter 15.

2   GROUNDWATER RESOURCES IN SPAIN

Spain is well endowed with aquifer formations, which tend to be small and widely spread around the country. Traditionally, only sedimentary, carbonated or volcanic formations of quite high permeability were officially considered aquifers. This meant an aquifer surface area of about 180,000 km², or one third of the country's surface area. Aquifer formations were divided into 411 hydrogeological units. This concept was defined for the first time in the 1985 Water Act, often following administrative rather than hydrogeological criteria. However, many areas with low permeability aquifer formations containing limited water resources were excluded, although they were strategically important resources at the local level.

The WFD introduced the concept of water bodies as the new unit of reference. The spirit of the Directive indicates that water bodies should be considered as subunits of river basins. This is coherent from a management standpoint in order to achieve the environmental objectives of the Directive. These objectives will be compared in each water body with the current status, which must be given in sufficient detail (EC, 2003).

The existing demarcation of hydrogeological units was a starting point for the characterization of groundwater bodies. Areas that were previously classified as having "no aquifers" have now been included following the WFD criteria, that characterize as water bodies those that serve as a drinking water source for more than 50 people or that supply over 10 m³/day.

Under the new classification, 699 groundwater bodies have been identified, covering an area of over 350,000 km², practically 70% of Spain's surface area (MMA, 2006). The size of groundwater bodies varies greatly: from less than 2.5 km² to more than 20,000 km², with an average groundwater body size of about 500 km². A classification was developed in 2005 in order to comply with the requirements of articles 5 and 6 of the WFD and is currently being revised and updated for the elaboration of the new Basin Management plans to be published before the end of 2009 after an extensive public consultation process.

Estimates of the total volume of water stored in Spain's aquifers vary between 150,000 and 300,000 Mm³, depending on the study. However, actual reserves are probably much higher, since the existing calculations (done under the previous definition of hydrogeological units) only take into account the volume stored in the 100–200 m depth range and do not consider smaller aquifer formations (Llamas *et al.*, 2001)—now included in the new definition of groundwater bodies— whose reserves can be significant. In any case, groundwater reserves present a much higher storage than surface water infrastructures, whose full capacity is about 53,000 Mm³, of which only 37,425 Mm³ are annually available for use (MMA, 2007).

From a management standpoint, two hydrogeological concepts are significant: aquifer storage, annual rate of recharge or renewable groundwater resources. The storage of many aquifers usually exceeds the annual natural rate of recharge by one or two orders of magnitude. This has practical implications, particularly important for a country like Spain where evapotranspiration is high and droughts are frequent, and where these reserves can be important to guarantee supply during drought periods.

A mathematical distributed model (SIMPA) was developed to estimate Spain's renewable groundwater resources for the White Book of Water (MMA, 2000). Results show that total renewable resources amount to about 30,000 Mm^3/yr. This amount probably underestimates total resources, since simulations are carried out under natural conditions, certain low permeability areas are ignored, and the model is not able to consider in detail the real behaviour of groundwater (Cruces 1999). Recently, aquifer recharge rates on peninsular Spain have been recalculated through the atmospheric chloride deposition balance (Alcalá & Custodio, 2007). Ultimately, usable groundwater resources are less than renewable water resources due to environmental restrictions, seawater intrusion limitations in coastal areas and islands, and interference with already committed surface water resources.

In spite of their quantitative and strategic importance, groundwater resources have traditionally not been adequately taken into consideration by the Spanish Water Administrations, which have emphasized surface water development, although the situation is improving. From a management standpoint, and given the climatic and hydrogeological diversity of Spain, what is important to develop accurate local estimations of stored and renewable groundwater resources. This is being done for the elaboration of the River Basin Management Plans, required by the WFD and to be completed within 2009 after extensive public consultation processes.

3   GROUNDWATER USE IN SPAIN

Existing information on groundwater uses in Spain is very heterogeneous and often scarce. This is due to two main causes. On one hand, water management responsibilities in Spain are divided between local, regional and national government levels. River Basin Authorities that depend of the national government's Ministry of the Environment, Rural and Marine Affairs (Ministerio de Medio Ambiente, Rural y Marino or MERMA) for the management of shared river basins (those that flow through more than two autonomous regions); Water Management Agencies that depend of Regional Governments for river basins that flow entirely within that region, as is the case in Catalonia, Andalusia, Galicia and the Basque Country in peninsular Spain, and in the Balearic and Canary archipelagos; Regional Governments for the management of protected natural areas and environmental policies; municipalities in issues relative to public water supply and sanitation; and irrigators associations for management and distribution of water among their members. On the other hand, most official statistics about water use for irrigation and urban supply do not differentiate between surface and groundwater sources. This is primarily due to the fact that, until 1986, there were no inventories of existing groundwater uses and no administrative permits other than drilling ones were required to abstract groundwater. This lack of detailed information makes it difficult to make a global estimation of groundwater uses for the entire country.

Table 1 presents an estimation of groundwater use in Spain using data from the 1990s. It indicates that overall groundwater use ranges from 5500 to 6500 $Mm^3/year$, or between 15% and 17%

Table 1.   Estimations of groundwater use in Spain.

Use	T, Total water		G, Groundwater		
	$Mm^3/year$	%	$Mm^3/year$	%	G/T
Domestic supply	5500	15	1000–1500	20	0.18–0.27
Irrigation	24500	65	4000–5000	75	0.16–0.20
Industry	1500	4	300–400	5	0.20–0.25
Hydroelectricity	6000	16	–	–	–
Total	37500	100	5500–6500	100	0.15–0.17

Source: Elaborated with data from MIMAM (internal reports of 2007), MIMAM (2000), MOPTMA-MINER (1994).

of total water use in Spain. However, these percentages vary widely from region to region. For instance, in the Mediterranean basins groundwater can represent up to 75% of all water resources used (MMA, 2007). The volume of groundwater use also increases significantly in times of drought, when surface water resources dwindle (López Geta, 2006).

According to the European Environmental Agency (EEA, 1999), in European countries with sufficient aquifer potential, over 75% of domestic water supply comes from groundwater. In comparison to other European countries and with the exception of Norway, which has very little aquifer potential, in 1999 Spain had the lowest percentage of groundwater used for urban supply: only 19% according to 2007 data from the Ministry of the Environment. This national average is low for a country with the hydrogeological potential and the meteorological characteristics of Spain, where groundwater could play a major role in guaranteeing urban water supply during droughts. However, groundwater as a source of domestic water supply is more important in some particularly arid river basins: 51% in the Andalusian Mediterranean Basins, 49% in the Canary Islands (accounting for desalinization contribution) or 43% in the Júcar (Xúquer) River Basin, and even more in the Balearic Islands. In communities of less than 20,000 inhabitants, approximately 70% of water comes from groundwater sources, whereas the figure is 22% in larger cities (MMA, 2000).

The principal use of groundwater in Spain is for irrigation, as is the case in most arid and semiarid countries. The dramatic increase in groundwater development in Spain has been primarily undertaken by thousands of individual farmers in different regions with very limited public involvement. In some Regions (Castilla-La Mancha, Murcia, Valencia), groundwater is the primary source of water for irrigation. In the Balearic and Canary islands, groundwater is often the only available resource. Approximately 75% of groundwater abstracted in Spain is used for irrigation of around one million hectares, about 30% of the total irrigated area. However both surface and groundwater sources are often used conjunctively to irrigate crops. Groundwater resources allow farmers to guarantee their crops in drought years when surface water resources are not available. Groundwater provides 20% of all water used to irrigate 30% of the total irrigated area.

While seawater, brackish groundwater and waste water desalination and salinity reduction currently contribute to water resources in the islands and the Mediterranean area of the peninsula, it does not significantly change previous figures. Recently, advanced treatment of brackish and polluted groundwater, including membrane technology (reverse osmosis and reversible electrodialysis) are augmenting groundwater use and the reclamation of waste water, a part of which was originally groundwater. Most of this water is for urban water supply, which can pay their relative high production cost, or for highly productive crops, such as those under greenhouses.

When groundwater quality is poor it may be applied to some urban uses, such as gardening, street cleaning and ornamental fountains. In urban areas prone to high water-table problems, drainage water is used for urban uses, as in the Barcelona Plain. There is also an increasing use of treated waste water for irrigation in new areas or to substitute for surface and groundwater resources, or to develop sport facilities such as golf courses. In this last case treated waste water application in new areas may be compulsory.

Artificial recharge of aquifers is rarely done. An exception is provided by the surroundings of Barcelona, where excess irrigation water recharge has been an important term in aquifer water balance. Enhanced river water infiltration by carefully scrapping the Llobregat river bed has been practiced since the 1940's under favourable hydrological conditions. Artificial recharge through deep wells started in the 1950s in the low Besós river area using excess water from a canal, and since the late 1960s in the low Llobregat river area by injecting excess potabilized river water (Custodio, 2009). In 2007 a pilot facility is being operated to control seawater intrusion in the Llobregat's delta deep aquifer by injecting advanced treated waste water, with reverse osmosis salinity reduction, and has now expanded to form a 14 well barrier (Niñerola *et al.*, 2009). Recharge basins, first receiving river water and afterwards with inverse osmosis salinity reduction of treated waste water, are being installed to compensate for recharge reduction due to increasing land occupation by motorways and railways. Other artificial recharge activities with occasional runoff exist in southern Gran Canaria Island, with potable municipal water in Madrid, and with treated waste water in Mallorca Island and in Dehesas de Guadix (Granada).

Monitoring groundwater levels and quality is an important task, which is compulsory under the WFD and the GWD. Groundwater quantity (piezometric levels) networks, mostly operating wells and some dedicated boreholes are being installed, with irregular densities and patterns. The oldest network, dating back to the mid-1960s is that around Barcelona, later extended to all Catalonia, with recent significant improvements being made (Custodio, 2009; ACA, 2008). A dedicated, especially designed piezometric network was constructed in the 1970s and 1980s in the Doñana area (southwestern Spain), in which up to four point piezometers per site are available (Custodio & Palancar, 1995; Manzano *et al.*, 2009). The Spanish Geological Survey (IGME) operated a Spain-wide quantity and quality groundwater monitoring network between the 1960's and the late 1990's, when the responsibility was assigned to the River Basin Authorities.

4   GROUNDWATER ECONOMICS IN SPAIN

4.1   *General issues*

The few economic studies that did exist before the passing of the WFD were constrained to particular regions or sectors, and for the most part did not differentiate between surface and groundwater sources. An exception of this situation was the research project funded by the Marcelino Botín Foundation (Hernández-Mora & Llamas, 2001; Llamas, 2003).

This situation has started to change with the new obligations that derive from the WFD. The obligation of member states to take into account the full cost recovery principle by 2010 has resulted in the need to undertake an economic analysis of each water use. Consequently the MERMA created the Economics Analysis Group, an internal working group that has been coordinating and guiding the work of the different Basin and Water Management Agencies in the economic aspects of water use. The group has issued internal reports that summarize the content of the work undertaken to comply with WFD Article 5 and Annex II and III reporting obligations (MMA, 2007). Even though the reports have been elaborated with limited and inconsistent economic data, it is possible for the first time to have a more clear understanding of water use economics in Spain. However, the information in the reports mostly fails to distinguish between surface or groundwater sources, so that information specific to the economics of groundwater use is still limited and in some aspects less detailed than in the White Book of Water in Spain (MMA, 2000).

4.2   *Costs of groundwater use*

In Spain and most other semi-arid countries, groundwater is used intensively because the direct benefits that users obtain from a certain level of abstraction greatly outweigh the direct costs of obtaining that water, even when these are high. But the associated indirect or external costs, which could make some levels of abstraction economically or socially inefficient, do not accrue directly to the users. Rather, they are spread over space and time and are borne by other users or by society at large, even by future generations. As a result, the overall costs of intensive groundwater use do not motivate changes toward more economically and socially efficient abstraction regimes.

In order to deal with this discrepancy the WFD requires the calculation of both the direct water service costs as well as environmental and resource costs. In Spain the work done so far has focused on the estimation of the direct service costs. In line with the WFD reporting obligations, the Ministry of the Environment issued in 2003 an internal study evaluating the extraction costs associated with groundwater use for irrigation and urban water supply in Spain. This work calculates groundwater costs as the costs of well-drilling and construction, the replacement value of the infrastructure needed for water abstraction, and the costs of the energy or fuel needed for pumping. No estimation is yet available on environmental or resource costs, which can be significant. What follows is a review of some of the data that can be gathered from these sources.

Service or direct costs. Pumping costs are a function of well yield, terrain characteristics, pumping technology used, depth of water table and energy costs. The 2003 internal study of the Ministry of the Environment estimated the average groundwater abstraction costs to be 0.08 €/m³ for urban water supply and 0.12 €/m³ for irrigation. However, values vary greatly from one hydrogeological unit to another. Costs of groundwater use for urban water supply range from as little as 0.03 €/m³ in some aquifers of the Guadiana river basin to as high as 0.37 €/m³ in the North Basin. In the case of irrigation, values range from 0.04 €/m³, in the Guadiana basin, to as high as 0.74 €/m³ in the Segura river basin. As Llamas & Garrido (2007) point out, this assessment was done without specific field surveys and therefore it should be considered only as a preliminary approach, and refer to the costs of extraction at the well. Other direct service costs, such as the cost of construction and maintenance of irrigation infrastructures and distribution networks have not been calculated specifically for groundwater uses.

Historically, public subsidies have been granted for the conversion of dryland agriculture to irrigation. Regional governments and the national Ministry of Agriculture continue to give economic assistance for the modernization of irrigation infrastructures, mostly using surface water, in order to increase water use efficiency. However, most often, Spanish farmers pay for all direct costs associated with groundwater irrigation, as there is no public support or subsidy for energy use by irrigators.

Existing data indicate that, even where water table levels are very deep, the costs of energy consumption only represent a small portion of farmer's income and therefore are not a deterrent for deeper abstraction levels. In the case of La Mancha Region, in south-central Spain, Llamas *et al.* (2001) estimate that the energy cost of irrigating one hectare by pumping water from a depth of 100 m is about 84 €/year, which only represents about 5% of an average farmer's gross income. Therefore, increasing energy costs resulting from increasing pumping depths will hardly discourage farmers from continuing existing pumping patterns. For instance in the Crevillente aquifer, a small (100 km²) karst aquifer in the Júcar river basin in southeastern Spain, water is pumped from depths of up to 500 m to irrigate highly profitable grapes for export. Pumping costs have increased to 0.29 €/m³, but with crop values ranging between 15,000 to 25,000 €/ha, pumping costs still represent less than 10% of total crop value, as indicated by Garrido *et al.* (2006, p. 346), who point out that as land without water is valueless, farmers will not be deterred by such productivity erosion.

A final set of costs associated with groundwater use for irrigation are the distribution costs. These occur when various users share a well so that pumped water is distributed among them using networks that can be very complex and often inefficient from an economic and a resource use perspective. Shared wells used by irrigators associations are very common in coastal eastern and southeastern Spain, and in the Canary Islands. Pipelines are expanded as new users join the well association and it becomes necessary to service their land. The design of these networks can therefore be very costly (see Chapter 13 to learn how water charges are implemented in these complex organizations). Table 2 shows some results for groundwater costs in the region of Valencia. For comparative purposes, they include the price paid by members in the same area for surface water or a mixture of surface and groundwater. Three significant conclusions can be drawn from the table: (1) there is a great variability in costs; (2) groundwater users pay a higher price for water than surface water users since they pay for all direct costs and also are charged with the added value tax, which is not the case when using surface water; (3) users never pay for environmental or resource costs of groundwater use, except when scarcity requires deepening existing wells or drilling new ones.

Data do not include water quality parameters, mainly water salinity. Salinity may affect crop yield and/or irrigation depth and frequency. Also an inadequate use of relatively saline water may enhance soil salinisation and alkalinisation. This is a serious problem in many irrigated areas in Spain, though no nationwide official report on this issue exists. Except for some analyses conducted in the Ebro delta rice paddies, irrigated with variable quality river water, very little is known about the impact on worsening groundwater quality.

In some cases, farmers have begun to pay attention to these facts and they price water of various salinity and chemical composition differently. In the Canary Islands, farmers have at times selected the best-quality water from their sources for irrigation and sold the worse-quality water for urban and tourist uses. Currently some farmers, aware of the salinity problems, pre-treat irrigation water

Table 2. Average cost of irrigation water in the Valencia region (modified from Carles *et al.*, 2001a).

Groundwater management areas	Source of water	Crop	Average cost[1] (€/m³)
Mijares-Plana de Castelló	Surface	Citrus	0.05
	Groundwater	Citrus	0.15
Palancia-Los Valles	Mixed	Citrus	0.12
	Groundwater	Citrus	0.13
Alarcón-Contreras	Surface	Citrus	0.02
	Mixed	Citrus	0.07
	Groundwater	Citrus	0.10
Serpis	Surface water	Citrus	0.05
	Groundwater	Citrus	0.15
Vinalopó-Alacantí-Vega Baja	Surface water	Various	0.08
	Groundwater[2]	Grapes	0.29
	Groundwater[2]	Various	0.26

[1] Average values of all irrigator associations in each region weighted by surface area.
[2] Groundwater costs in the region in 1999 were 0.51 €/m³, after Rico and Olcina (2001).

by low pressure reverse osmosis, which adds to the direct water cost. The cost of disposing the produced brines is often not considered, and may involve serious environmental costs, or require costly public investments, as in Fuerteventura Island or in the Campo de Cartagena region in the Segura river basin, which are often not accounted for.

Environmental and resource costs. In addition to the direct service costs, groundwater use results in environmental and resource or scarcity costs that need to be evaluated both to comply with WFD requirements and to accurately assess the economic viability and social desirability of different pumping regimes. Intensive and uncontrolled groundwater use can have negative consequences such as aquifer salinisation or contamination, decreased groundwater discharges to dependant aquatic ecosystems (wetlands, rivers and streams), land subsidence, and impact on the rights of other surface or groundwater users.

There are no estimates of the environmental and resource costs associated with groundwater use in Spain. Environmental costs will have to be calculated as the cost of applying the corrective measures to achieve the WFD environmental goals. In any case, it is apparent that if the full cost recovery principle is applied to intensively used aquifers, many existing groundwater uses would not be economically viable. The individualist nature of groundwater abstractions together with the inadequate enforcement of existing rules and regulations, as will be discussed later on, has resulted in the elimination of the scarcity value and made it difficult to guarantee existing rights. It is thus difficult to achieve the goals of the WFD in many intensively used aquifers.

4.3 *Economic issues of groundwater use in Spain*

In spite of data scarcity, Llamas *et al.* (2001) and Hernández-Mora *et al.* (2001) developed a rough estimate of the economic value of groundwater use in Spain, which shows the magnitude of the economic contribution of groundwater. The discussion will be limited to groundwater use for irrigation and public water supply.

Public water supply. In most cases water supply data consider together surface and groundwater. Users pay for the treatment and distribution costs, but do not pay for the resource itself or for external or opportunity costs (Pérez Zabaleta 2001). In 2004 home consumers in Spain paid an average tariff of 1.17 €/m³, with a wide range, from the 0.80 €/m³ paid by home consumers in Castilla y León or Castilla-La Mancha (Duero and Guadiana River Basins respectively), to the 1.72 €/m³ paid in the southeastern Mediterranean region of Murcia, in the Segura River Basin (MMA, 2007 and other internal reports of 2007). Often these tariffs do not cover the investment costs of the necessary infrastructures associated with the service (dams, canals, facilities, etc.), which are usually paid for by general revenue of the State, Region or local government (see Chapter 8 for a more detailed analysis or urban water issues). Current information (MMA, 2007) indicates that while there is some correlation between higher tariffs and lower water consumption rates by domestic users, total household expenditures on water supply services are so low (0.09 €/day for the average 167 liters per day and person of water consumed by domestic users) that water fees are hardly an incentive for lower consumption. However, public education campaigns and careful maintenance of distribution networks have being effective in Madrid and Barcelona in reducing per person water use, keeping a fully satisfactory service quality. In the densely populated areas of Barcelona in-house consumption is currently less than 125 l/day/person.

Irrigation. The primary economic contribution of groundwater in Spain is for irrigation. The recent reports on the economic analysis of water use (MMA, 2007 and internal reports) present updated information on irrigated agriculture economics, but for the most part fail to clearly distinguish between irrigation with surface and groundwater sources. However, some general conclusions are worth highlighting where inferences can be made on groundwater economics in agriculture:

1. Average gross economical productivity of irrigated agriculture is 4.4 times that of rainfed agriculture, although there are significant regional differences. In areas with profitable rainfed crops (olives, grapes, or cherries) the ratio can be as little as 1.1. In the southeastern regions with intensive horticultural production under plastic, which rely heavily or entirely on groundwater sources, net productivities for irrigated agriculture can be as much as 50 times higher than when using surface water, or as high as 12 €/m^3.
2. A large percentage of irrigated agriculture is closely tied to subsidies from the European Common Agricultural Policy (CAP), representing sometimes as much as 50% of farmer's income. In areas which rely on groundwater sources to produce highly profitable horticultural crops, these subsidies represent as little as 1%.
3. Average water services costs to the farmer vary greatly between surface water users (106 €/ha/year), and groundwater users (500 €/ha/year).
4. 58% of water used in agriculture is to produce 5% of gross agricultural production, while 9% of water produces 75% of gross production. This 9% concentrates primarily in river basins and islands that rely heavily on groundwater.

The most comprehensive analysis of the economic contribution of irrigation using groundwater is the Irrigation Inventory for Andalusia (CAPJA, 2003), originally carried out in 1996 and 1997, and updated in 2002. Using data from the original 1997 study, Hernández-Mora *et al.* (2001) show that irrigated agriculture using groundwater is economically over five times more productive and generates almost three times more employment than agriculture using surface water. This can be attributed to several causes: the greater control and supply guarantee provided by groundwater, which allows farmers to introduce more efficient irrigation techniques; the greater dynamism that has characterized farmers that sought out their own water sources and bear the full costs of their water supply; and the fact that the higher financial costs farmers bear motivates them to look for more profitable crops that will allow them to maximize their return on investments.

Using data from the 2002 update, Vives (2003) shows that in Andalusia groundwater supplies 27.3% of all irrigated agricultural land and represents 22.7% of all water used for irrigation. It generates almost 50% of all agricultural output and 50% of employment (Table 3). Irrigation using groundwater is three times more productive than that using surface water, it is at least 20% more efficient in the water use and generates twice as much employment per m^3 used. EU income aid to farmers using groundwater is only 5%, as opposed to 20% for surface water irrigators. In order to compare surface and groundwater productivities, Vives (2003) uses information on the water volume applied in the field, not the volume of water actually pumped or diverted from reservoirs. If the latter data were used (Llamas *et al.*, 2001), the difference between surface and groundwater productivities is even greater since a significant amount of surface water is lost in transportation canals or evaporated in dams and storage basins.

Table 3. Economic indicators in Andalusia for irrigation with ground and surface water, after Vives (2003) and CAPJA (2003).

	Groundwater	Surface water	Total
Irrigated area (km²)	2440	6480	8920
Percentage of irrigated area (%)	27	73	100
Average water consumption (m³/ha)	3900	5000	4700
Total production (10^6 €)	2222	2268	4490
Specific production (€/ha)	9100	3500	5100
Employment generated (jobs/ha)	0.232	0.126	0.154
EU aid to income (% of production value)	5.6	20.8	13.4
Gross water productivity (€/m³)	2.35	0.70	1.08
Total average water price to farmer (€/m³)	7.2	3.3	3.9

The same set of data serves to highlight the economic importance of groundwater, not only for its extractive value, but also for its stabilization value. The availability of groundwater supplies has allowed irrigation agriculture in Andalusia to survive during severe drought periods. Corominas (2000) shows how total agricultural output during dry sequences decreased by only 10%, while 60% of irrigated land received less than 25% of its average surface water allocations. The decrease in water supplies was made up by relying on groundwater sources.

It could be argued that the difference in productivity between surface and groundwater irrigation in Andalusia is largely due to the influence of the Campo de Dalías aquifer region, located in Almería (see Figure 1), where intensive groundwater development for irrigation has fueled a most remarkable economic and social transformation. The combination of ideal climatic conditions, abundant groundwater supplies, and the use of advanced irrigation techniques under greenhouses for the production of highly profitable fruits and vegetables, has allowed the dramatic economic growth of the area since the 1950s, when irrigation begun. Today, irrigation of over 20,000 ha of greenhouses directly or indirectly generates an estimated 1200 M€/year, and it is usual for farmers in the area to have gross revenues of 60,000 €/ha. This has allowed the population in the region to grow from 8000 inhabitants in the 1950s to more than 120,000 in 1999 (Pulido *et al.*, 2000). But the lack of planning or control of these developments by either the water authorities or the users themselves has resulted in social tensions from inadequate integration of the necessary immigrant labor, as well as problems of saline water contamination of soil in some areas, and the need to deepen or relocate some wells.

Data from other regions in Spain serve to underscore the fact that the productive advantage of groundwater irrigation is not only the result of more advantageous climatic conditions. Arrojo (2001) shows that similar advantages are observed in the 8000 ha irrigated in the Alfamén-Cariñena region, an area of intensive groundwater use in Central Ebro river basin, under less favourable climatic conditions. Net water productivity ranges between 0.15 and 0.50 €/m³, depending on the type of fruit crop, while estimated productivities of some large surface water irrigation networks in this semi-arid region, such as Bárdenas or Monegros, are around 0.03 €/m³. Arrojo (2001) estimates that while irrigation with groundwater occupies 30% of the total irrigated area and consumes only 20% of all water used for irrigation in the entire Ebro river basin, it produces almost 50% of the total agricultural output of the basin. The advantage can be attributed to the water supply guarantee provided by groundwater, which allows farmers to invest in more sensitive and water demanding crops that are at the same time more profitable, thus helping them defray the higher costs of searching for and obtaining their own water supplies.

Another regional example of interest in Spain is the Canary Islands. It is significant both for its insularity and the resulting need to be self-sufficient in terms of water resources, as well as for the strategic importance that groundwater, which provides almost 80% of all water resources in the islands. The general scarcity of water resources has resulted in a unique water resources system that is characterized by three factors: the prominent role played by the private sector in the search, extraction and marketing of groundwater resources; the widespread use of water-efficient

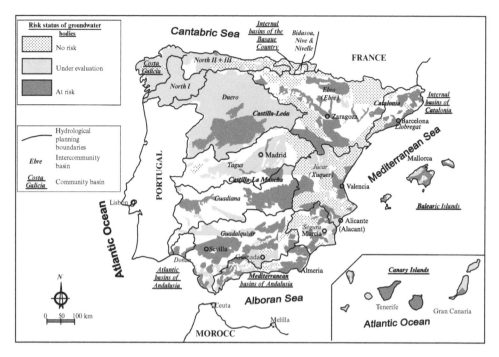

Figure 1.   Areas with groundwater bodies in Spain. White areas inside Spain are devoid of significant aquifers. Shaded areas are the groundwater bodies classifies as no risk, under evaluation and at risk, relative to
the WFD good status to be attained in 2015. The figure is a simplification of maps from the Ministry of the
Environment, of year 2006.

irrigation techniques; and the increasing use of alternative water sources, such as desalinated seawater and brackish groundwater, and treated waste water with salinity reduction. Water resources
have historically been distributed in the islands through largely unregulated and imperfect water
markets where both the resource and the transportation canals are privately owned (Aguilera,
2001), although recently the public sector has invested in regulation ponds and transportation
canals, thus concurring with the private sector. The economic and social importance of agriculture
in the Canary Islands decreased significantly in the second half of the twentieth century, rapidly
losing ground to industrial production and tourism in Gran Canaria. In 1998, agriculture was
responsible for only 3.8% of the islands' total economic output and provided 7.5% of all employment, similarly as what has been observed elsewhere in Spain. In spite of the decline in relative
importance, total agricultural output has remained constant and agriculture continues to be the primary water user in the islands, consuming about 280 Mm^3/year, or 60% of all water used. For the
island of Tenerife, which relies in a larger proportion on agriculture tomatoes under greenhouses is
the most profitable crop, about 4.75 €/m^3, and the traditional banana crop is losing profitability.

5   PRESSURES, IMPACTS AND MEASURES TO ACHIEVE THE GOALS OF THE WATER
FRAMEWORK DIRECTIVE ON GROUNDWATER ISSUES IN SPAIN

5.1   *General considerations*

The Water Framework Directive (WFD, 2000; see chapter 6 for a detailed description of the WFD)
aims to achieve water good ecological status of all water bodies by 2015, with possible negotiated
extensions to 2021 and 2027, if it is demonstrated that the delay is technically, economically, and
socially warranted due to the implementation difficulties or involved costs. The WFD asks for

halting ecological deterioration, reversing negative trends, and carrying out action to achieve a good water quality status.

The application of the WFD to groundwater presented some uncertainties and difficulties and consequently a *daughter* Groundwater Directive (GWD, 2006) has been passed to clarify and develop the WFD concepts. By application of the subsidiarity principle, a guideline of all European Union policies, the WFD and the GWD have to be duly incorporated into European Union Member States water-related acts.

The WFD included, and thus canceled, previous EU Directives on continental and littoral water, except the Nitrates Directives (ND, 1991), which will be in force until the goals of the WFD are achieved. It is aimed at correcting the surface and groundwater increase of nitrates. For groundwater it asks for the definition of nitrate sensitive areas due to agricultural practices, in which control measures have to be undertaken.

The main goal of the WFD is that by 2015 all water bodies (surface and groundwater bodies) achieve a good ecological status (see Chapter 16 for a description of the WFD). In terms of defining good ecological status for groundwater bodies, the WFD focuses primarily on water quality and pollution sources. This outlook may prove challenging to implement when dealing with groundwater in arid or semi-arid Mediterranean EU Member States. The WFD is dominantly influenced by Central and Northern European problems, and consequently may be in conflict with actual situations in other areas, especially in the Mediterranean area, and especifically in Spain, where even natural conditions may not comply with the WFD terms. Exceptions, many of them referring to groundwater bodies, must be fully documented and demonstrated. Excessive groundwater abstraction with its potential impact on water quality degradation and on stream flows and wetlands is a primary concern since, after the WFD, groundwater abstraction should not cause a significant impact on related surface water bodies. If this provision is strictly enforced, many groundwater intensive developments in Spain may have to cease. The social and economic sustainability of such a decision, and its political viability, is problematic (Sahuquillo *et al.*, 2009; Molinero *et al.*, 2008), and if viable, they may need a significant delay to be implemented. Current concerns are also linked to: a) the absence of a long-standing tradition of public participation in policy decision-making; b) the fact that groundwater management results are significantly delayed with respect to surface water, and may only be seen after decades; c) the fact that in the Mediterranean countries the consumptive use of water for irrigation usually represents about 80% of total consumptive use; d) the need to deal with thousands of individual farmers and small communities, which is more complex than dealing with less numerous and usually more organized water supply companies, electric utilities and industries; e) the fact that the natural situation may not comply with the WFD requirements. Further experience and rethinking are needed.

In spite of the above mentioned difficulties, the authors of this chapter consider that the enforcement of the provisions of the WFD in the Mediterranean countries, and especially in Spain, is being, and will be, positive and beneficial from the hydrogeological, economical and social points of view. Even if Spain applies for delays in implementing the goals of the WFD, clear and thorough information on the hydrological, economic and social dimensions of water resource use, accompanied by active public participation will need to be provided. Cost/efficiency analyses of the different measures will need to be undertaken and publicized. Such an exercise of transparent information and public participation will provide a positive outlook on the role of groundwater in water policy and enhance management processes.

5.2 *Pressures on groundwater bodies*

The situation of many intensively used aquifers in Spain can help illustrate some of the difficulties to implement the WFD. As described in Chapter 11, the 1985 Water Act regulated the concept of aquifer overexploitation, giving water authorities broad powers to regulate groundwater use in aquifers that were declared overexploited. The Water Administration identified overexploitation or salinisation problems in 77 hydrogeologic units (MMA-ITGE, 1997) in addition to 15 units in the Canary Islands and the Internal Basins of Catalonia, which have their own water administration.

But the legal declaration of overexploitation was often embedded in intense political and social debate, so that many aquifers subject to intensive use have not been declared overexploited (see Chapter 15). While these declarations should be accompanied by strict regulatory measures, they have most often not been successfully implemented. This explains why exploitation, management and water quality issues can be judged as clearly insufficient and to some extent confusing.

Water management efforts in Spain have focused primarily on quantity, while quality concerns have been secondary. In 1997 there were 60 hydrogeological units where estimated total pumping volumes exceeded estimated natural recharge rates. Although estimates have to be taken cautiously, given the significant data uncertainty and the fact that terms such as natural recharge, overexploitation or water deficit are sometimes misused, they indicate that there is an overall groundwater storage deficit in those aquifers of approximately 665 Mm³/year (MAPA, 2001). Most of them are located in southeastern Spain and in the Balearic and Canary islands. Perhaps the most emblematic case of intensive ground-water use where overdraft is most acute is the Western La Mancha hydrogeological unit, in the Upper Guadiana River basin, where water-table drawdowns have dramatically impacted the wetlands in the "Mancha Húmeda" Biosphere Reserve (Martínez Cortina, 2001, de la Hera, 1998). Other interesting case is that of the Doñana area wetlands, in southern Spain, an emblematic Ramsar site and important European Natural Park, in which groundwater irrigated agriculture established in the late 1970s and early the 1980s has affected groundwater discharges and depth to the water-table (Custodio *et al.*, 2008; Manzano *et al.*, 2009).

Concerning water quality, the primary issue of concern in Spain has historically been aquifer's salinization. This process is usually due to seawater intrusion, dissolution of evaporitic materials or return of excess high salinity irrigation water. Table 4 shows the degree of salinization (chloride and sulphate) observed in stations of the national groundwater monitoring network. Quality limits are naturally attained in some aquifers and springs in arid and semi-arid areas of Spain.

Another water quality concern is nitrate pollution, mostly due to diffuse pollution from agri-cultural and cattle rising activities. Table 4 shows that 20% of the control points yield values in excess of the WFD and water potability norms threshold of 50 mg/LNO$_3$. The Guadiana and Júcar (Xúquer) river basins being the most affected. High nitrate values are found in many aquifers in Catalonia, Murcia, Balearic Islands or the Canary Islands. According to the WFD, values in excess of 25 mg/L oblige to systematic monitoring every four or eight years. Thus, contamination by nitrates is one of the main challenges faced in Spain with regard to groundwater quality, especially because in areas with deep water-tables a large inventory of nitrates is in the unsaturated zone and thus will continue to leach into the aquifers even if new surface sources disappear. Nitrate sensi-tive areas are extensive, but the affected surface area should be still larger for adequate action and protection, although this results in unwanted political problems.

Table 4.   Percentage of groundwater quality monitoring stations classified after chloride, sulphate and nitrate contents, in mg/L, in some of the water districts. Data from MMA of 2003 (http://www.mma.es).

River basin district	Annual average chloride		Annual average sulphate		Maximum 6-month nitrate value		
	0–100	>100	0–150	>150	0–25	25–50	>50
North	100	0	100	0	100	0	0
Douro	87	13	93	7	84	12	4
Guadiana	72	28	65	35	24	41	35
Guadalquivir	72	28	69	31	66	16	18
Mediterranean Basin of Andalusia	60	40	74	26	79	5	16
Segura	40	60	33	67	72	15	13
Júcar	62	38	N.A.	N.A.	27	27	46
Ebro	77	23	61	39	62	18	20

5.3   *Characterization and risk assessment of groundwater bodies*

In order to comply with the WFD requirements, the reports submitted to the European Commission by the Spanish government classify groundwater bodies according to three categories (see Figure 1):

1. *at risk*, which will presumably not attain good status by 2015 and require further characterization.
2. *at risk under evaluation*, in which available information does not allow a clear diagnosis on the possibility of achieving good status, and require additional studies.
3. *no risk*, which according to available data will attain good status by 2015.

The initial characterization identifies two types of pressures: chemical risk as a result of point and non-point (diffuse) pollution, and seawater intrusion; quantitative risk as a result of unsustainable extraction volumes. With few variations, the criteria used in the different river basin districts to evaluate risk has been to apply a matrix that relates pressures and resulting impacts on groundwater bodies. Pressures considered include those resulting from: a) diffuse pollution, primarily nitrate pollution; b) point source pollution (often with limited and insufficient data); c) seawater intrusion; and d) groundwater abstractions (difficult to evaluate because of insufficient information in the terms established by the WFD).

The evaluation of impacts has considered chemical and quantitative aspects separately. In what refers to chemical aspects, several contaminants have been considered with different threshold levels for each river basin district and depending on available information, but risk evaluation has been mainly determined by the 50 mg/L nitrate threshold. In general, several issues have been considered simultaneously for the evaluation of quantitative aspects (MIMAM, 2006): expert opinion; average decrease in piezometric levels; legal declaration of overexploitation; inclusion of the groundwater body in the "Catalogue of aquifers with overexploitation or salinization problems" (MMA-ITGE, 1997); or the impact on associated aquatic ecosystems.

Table 5 presents the summary results of applying this methodology for risk assessment of groundwater bodies (GWB) in all river basin management districts in Spain. Of the 699 GWB characterized, 259 (37%) have been classified as being *at risk* of not achieving good status by 2015; 184 (26%) have been classified as having *no risk*; and the remaining 256 (37% of the total) as being *at risk under evaluation*. The most frequent causes of *risk* are diffuse or non-point source pollution (167 GWB or 24% of those characterized) and quantitative (164 or 23%). In terms of saltwater intrusion, the *risk* results from a deterioration of water quality resulting from inadequate pumping patterns: 71 GWB in all coastal areas except in the North, are at risk for salinization.

For groundwater bodies at risk of not achieving the environmental goals by 2015, the WFD requires an additional characterization that provides information on the hydrogeological and hydrogeochemical aspects and evaluates the impact of human activities on the state of groundwater resources. The Spanish Ministry of the Environment (MMA) and the IGME have developed a methodological guide to support and homogenize the additional characterization work that must be carried out by River Basin Authorities. Only preliminary work on the additional characterization has been done so far, with some results available in the River Basin Authorities' websites.

6   INSTITUTIONS FOR MANAGING INTENSIVELY EXPLOITED GROUNDWATER BODIES IN SPAIN

Spain has experimented with different solutions for groundwater resources management: from the liberal approach that characterized private property of groundwater resources under the 1879 Water Act, to the more government-controlled approach of the 1985 Water Act, responding to intensive groundwater use and transforming the institutional context for groundwater management in Spain. Three innovations are particularly relevant.

First, groundwater was declared a part of the public domain, as surface water resources had been since the first Water Act of 1866 (see Chapters 10 and 11). As a result River Basin Authorities

Table 5.   Risk assessment classification of groundwater bodies (MIMAM, 2006).

River basin district	Number of GWB	Chemical			Quantitative	Total characterized		
		P	D	I	E	At risk	Under evaluation	No risk
Shared River Basins								
North I	6	0	0	0	0	0	6	0
North II and III	34	0	0	0	0	0	12	22
Bidasoa, Nive and Nivelle	2	0	0	0	0	0	1	1
Douro	31	0	3	0	1	3	28	0
Tagus	24	NE	1	0	NE	1	18	5
Guadiana	20	0	9	1	6	11	9	0
Tinto, Odiel and Piedras (1)	4	0	3	1	0	3	1	0
Guadalquivir	71	1	21	1	19	35	29	7
Segura	63	NE	1	2	25	25	33	5
Júcar	79	0	13	8	23	29	26	24
Ebro	105	11	29	0	1	35	7	63
Internal River Basins								
Galicia Coast	18	0	0	0	0	0	15	3
Internal Basins of Basque Country	14	2	0	0	0	2	0	12
Internal Basins of Catalonia	39	23	23	10	10	25	0	14
Mediterranean Basin of Andalusia	67	1	20	11	23	29	23	15
Balearic Islands	90	42	36	30	41	42	35	13
Canary Islands	32	NE	8	8	15	19	13	0
TOTAL*	699	80	167	72	164	259	256	184

P: point source; D: diffuse; I: saline intrusion; E: extraction; GWB: groundwater bodies; NE: not evaluated.
*Some groundwater bodies are at risk for more than one reason. Therefore the sum of each individual risk is larger than the total for both types of risk.
(1) Now within the Atlantic Internal Basin of Andalusia planning district.

acquired, at least on paper, a relevant role in the management of public groundwater resources, and were responsible for granting permits for any uses starting after 1985. The Water Act created a registry system for public water use permits or concessions for both surface and groundwater rights, the Registry of Public Waters. Groundwater uses existing prior to 1986 had the option of remaining in the private property regime by registering the use in the Catalogue of Private Waters. In practice this has been the preferred option by most irrigators.

Second, the Water Act gave Basin Authorities broad powers for the management of aquifers declared overexploited in accordance with the law. When an aquifer is declared legally over-exploited, River Basin Authorities have to draw up a management plan and determine annual pumping regimes. Restrictions apply to users in both the public and private property regimes. No new pumping permits can be granted. All users in the aquifer are required to organize themselves into Groundwater Users Associations, and a General Users Association has to be formed that encompasses all associations existing in a given aquifer system (see Chapters 12 and 15).

Third, the concept of user participation in water management was legally extended and rein-forced. Historically, user participation in Spain was understood as the right of irrigators to organize self-governing institutions for the management of surface water irrigation systems. Since the creation of the River Basin Authorities in the 1920s, representatives of these irrigators associations were part of their governing and management bodies. However, the 1985 Act expanded the concept to groundwater users and representatives of other interests and uses beyond irrigators (See Chapter 12).

The changes introduced by the 1985 Water Act were necessary to deal with the challenges resulting from the intensive use of groundwater resources, although declaring groundwater a public domain is neither unavoidable nor the only possible solution. The implementation of the changes has encountered difficulties which in some ways continue up to present. Two are worth highlighting:

First, River Basin Agencies, without experience in groundwater management, have consistently lacked sufficient human and financial resources to deal with their acquired responsibilities. They have also had difficulties shifting their focus from their traditional water infrastructure development and technical management responsibilities to their new broader water and ecosystem management goals.

Second, the absence of updated groundwater rights records is a significant difficulty, since the Registry of Public Waters and the Catalogue of Private Waters are still quite incomplete. There is no up-to-date record of existing groundwater uses and of total extraction volumes in spite of costly efforts launched by the Ministry of the Environment to try to cover this gap. The White Book on Water in Spain (MMA, 2000) estimated that of the 500,000 operational wells existing in Spain only 50% had been declared and less than 25% had actually been registered. The situation is still not resolved. Moreover, Fornés *et al*. (2005) and Llamas *et al*. (2001) have estimated that the total number of water wells and captured springs in Spain is in the order of two million. If these estimates are accurate, about 90% of all groundwater abstractions in Spain are not registered, and are either illegal or allegal.

6.1   *The reform of the Water Law and the National Hydrologic Plans*

A significant landmark in Spanish water law was the approval in 2001 of the National Hydrologic Plan after more than a decade of intense political debate. The Plan was a legal requirement of the 1985 Water Act and the basic framework to guide water resources management in Spain. It meant to coordinate river basin hydrologic plans and compensate for the uneven geographical distribution of water resources through inter-basin water transfers. Most significantly, the Plan enabled the transfer of 1050 Mm^3/year from the lower Ebro/Ebre river basin to northern and southern areas along the Mediterranean coast, in part to replace excessive pumping in overexploited aquifers along the coast (see Chapter 19). The Plan tried to reinforce the existing groundwater management framework by requiring the declaration of overexploitation of receiving aquifers and the approval of the corresponding management regimes prior to use of inter-basin transferred water (Sánchez, 2003). It required users in the receiving aquifers to be organized in users associations, and established that the user communities would hold the title to the transferred water, making them responsible for reducing pumping rights proportionally to the volume of water received, until total groundwater abstractions were reduced to sustainable levels. In essence the Hydrologic Plans put users for the first time in charge of allocating and limiting their water rights, thus making them responsible for aquifer management decisions together with Basin Authorities.

However, the Ebro river transfer proposal was very controversial and received intense criticism from academics, environmental organizations, other public interest groups nationwide and much of the population in the Ebro basin, for its environmental, economic and social impacts. Massive demonstrations in Zaragoza, Madrid and Brussels were held against it, and also in favor in receiving regions, such as Valencia. In March 2004 the Ebro river transfer plan was cancelled, and the government presented the alternative AGUA Programme (Actions for Water Management and Use), a multifaceted plan aimed to increase available resources in coastal Mediterranean river basins through water efficiency and saving measures, increased use of recycled waters and the construction of new desalination plants along the coastline, in order to address water shortages. The plan was enacted in 2005, modifying the 2001 National Hydrologic Plan. In reality the AGUA programme maintains many of the other controversial dams and public infrastructures that were proposed in the 2001 National Hydrologic Plan with the exception of the Ebro basin transfer. One of them is surface water irrigation scheme for the Garrigues-Segarra area, south of Lleida, in the Ebro basin, a recently completed part of the National Hydrological Plan, with at least a similar impact on the Ebro Delta as the canceled Ebro river transfer.

Furthermore, the AGUA programme proposes the construction of some twenty seawater desalination plants to produce about 600 Mm^3/year for urban water supply and irrigation in coastal Mediterranean basins. However, some authors argue that the acceptance by the farmers of this water is dubious if they have to pay the full cost of that water. It seems that the government is ready to heavily subsidize the cost of desalinated seawater for irrigation (Llamas & López-Gunn, 2007), which is at odds with the full recovery principle that guides the WFD.

6.2   *Organizations for water management: River Basin Authorities and water users associations*

As Chapter 12 shows, Spain has a long-standing tradition of irrigator participation in water management activities. Irrigator associations have existed from as far back as the 11th century. These traditional associations were originally organized around irrigation networks in order to build and maintain the canals, distribute the water among the different members, and resolve water-use related conflicts that could arise between them. Given this tradition, it seemed logical that the 1985 Water Act would encourage a similar participatory management structure for groundwater resources. However, it is questionable whether the system has being transferred successfully, by duly considering aquifer characteristics.

Groundwater user associations in Spain are extremely diverse. According to their goals and objectives they can be grouped into two categories (after Carles *et al.*, 2001b): 1) *associations for the collective management of irrigation networks* whose objective is the common exploitation of a well or group of wells and where members generally pay for all drilling, installation, operation and maintenance costs; and 2) *associations for the collective management of aquifers*, which comprise all or a majority of users within one aquifer; and that, in addition to pursuing their own interests also contribute to the aquifer's management and conservation, a social goal.

In terms of the tenets of the law, associations included in this second group are the ones that can play a significant role in the management of groundwater resources. Of the thousands of existing groundwater users associations, currently only six can be truly included in this last group, and only two (Western Mancha and Campo de Montiel aquifers user associations) were created in response to a legal declaration of overexploitation.

A particularly interesting example of successful groundwater user communities are the Water Users Associations of the Low Llobregat and of the Cubeta de Sant Andreu, both located in Barcelona, Catalonia. These are two linked civil organizations created before the 1985 Water Act, when groundwater was private property. Their members are primarily industrial users and public water supply entities, but also include farmer associations. Affiliation is compulsory for all aquifer users, including those which drain water from structures and buildings. Further to the administrative and legal personnel, they have technical staff that has increased over time. Currently they share a detailed flow and salinity aquifer transport model with the Water Authority of Catalonia in order to facilitate decision-making on aquifer and river water joint use, groundwater recharge operations, and seawater intrusion control. With the exception of the Western Mancha Aquifer User Association, the Spanish Water Administration has so far not succeeded in creating similarly effective organizations in the other aquifers that are *legally* overexploited.

7   CONCLUSIONS

Intensive groundwater development in many regions of Spain, primarily since the 1970s, has brought about significant social and economic benefits. But the unplanned nature of these developments has also resulted in unwanted social and environmental consequences, which in agreement with the WFD, should be corrected by 2015. If this is not considered feasible or possible, the Spanish government has to request for specific exceptions or a deadline extension to 2021 or 2029. However these requests have to be well documented and be subject to public participation processes.

Groundwater is an important economic resource in Spain. Existing data for irrigated agriculture show that groundwater is usually much more productive in economic and employment terms than surface water resources. Some of the reasons that explain this higher productivity are the greater supply guarantee groundwater provides, which allows investment in better irrigation technologies, and the fact that users bear all private costs, thus paying a higher price for water used than irrigators using surface water, and motivating them to look for more profitable crops and using water more efficiently. The Spanish situation should not be extrapolated to any other areas, especially to developing countries, where the crop value is smaller and the cost of abstracting groundwater may be higher (Llamas & Custodio, 2003).

Available data indicate that the economics of intensive groundwater use are such that the direct benefits obtained from a certain level of abstraction greatly exceed the costs of obtaining that water, even when these are very high. This is true even in areas where intensive aquifer use has resulted in dramatic drops in the water table, saltwater intrusion, wetland degradation, or significant social conflict. These environmental and social costs are spread over space and time and do not accrue to the direct user, but to society at large and to future generations. Therefore there is no economic incentive to modify pumping patterns that would be socially and economically inefficient if both direct and indirect benefits and costs were considered. Changes need to be achieved through a transformation of the existing institutional arrangements (both formal and informal) for the management of groundwater resources.

In order to deal with the problems associated with intensive and unplanned use, the 1985 Water Act transformed the institutional context for the management of groundwater resources in Spain. By making groundwater part of the public domain, the Water Act gave River Basin Authorities the power to limit access to the resource and to regulate use. Following a well established Spanish tradition of user participation in water management, the Water Act created the figure of groundwater users associations, giving them a prominent role in the management of groundwater resources. While hundreds of user associations exist throughout the country, a vast majority act as mere water distributors among their members, and very few can be considered true resource managers. The few successful ones have been able to articulate common goals and objectives and to establish mutually accepted rules regarding resource access and use, in order to guarantee the long-term sustainability of the resource and dependant uses. The variety of circumstances under which these successful user associations operate, their ability to bring together thousands of independent users and sometimes manage large and complex aquifer systems, and the way in which some are working cooperatively with-although not subject to-water authorities to establish sustainable management regimes, are all promising developments. The fact that many of these associations were created, not as a result of the statutory requirements of the 1985 Act, but because of a combination of user initiative and administrative support, points to the limitations of a general solution through regulatory means.

The regulatory measures contained in the 1985 Water Act and its reforms have so far proven to be insufficient to solve the problems resulting from intensive groundwater use. It can be expected that the requirements of transparency and participation that derive from the WFD will contribute significantly to improve the situation of groundwater management in Spain. However, considering the still scarce Spanish tradition in transparency and participation, the success needs strong enforcement of the provisions of the WFD from the EU Commission, or groundwater governance has the risk of continuing to be poor. This means that Universities, research institutions, civil organizations and NGOs have to make an important effort in a short time to participate in the discussions, and assume their responsibility instead of blaming the Government for wrongdoing and poor understanding of problems and solutions.

REFERENCES

ACA (2008). *L'Agència, memoria 2007* [Memory of the Agency]. Agència Catalana de l'Aigua. Barcelona: 1–169.

Aguilera, F. (2001). El papel económico de las aguas subterráneas en Canarias [The economic role of groundwater in the Canary Islands]. In: La Economía del Agua Subterránea y su Gestión Colectiva (Hernández-Mora, N. and Llamas, M.R. eds.). Mundi-Prensa/Fundación Marcelino Botín, Madrid: 269–280.

Alcalá, F.J. & Custodio, E. (2007). *Recharge by rainfall to Spanish aquifers through chloride mass balance in the soil.* In: Groundwater and Ecosystems, Proc. XXXV Intern. Assoc. Hydrogeologists Congress, Lisbon. (Ribeiro, L. Chambel, A. and Condesso de Melo, M.T. eds.). In: CD printing. Universidade Técnica de Lisboa, Lisboa: 10 p.

Arrojo, P. (2001). Valoración de las aguas subterráneas en el marco económico general de la gestión de aguas en España [Valuing groundwater in the general economic framework of water management in Spain]. In: *La Economía del Agua Subterránea y su Gestión Colectiva* (Hernández-Mora, N. and Llamas, M.R. eds.). Mundi-Prensa/Fundación Marcelino Botín, Madrid: 3–39.

CAPJA (2003). *Inventario y caracterización de los regadíos de Andalucía* [Inventory and characterization of irrigation in Andalucía]. Departamento de Agricultura y Pesca, Gobierno de Andalucía (CD printing).

Carles, J., García, M. & Avellá, L. (2001a). Aspectos económicos y sociales de la utilización de las aguas subterráneas en la Comunidad Valenciana [Economic and social aspects of groundwater use in the Community of Valencia]. In: *La Economía del Agua Subterránea y su Gestión Colectiva* (Hernández-Mora, N. and Llamas, M.R. eds.). Mundi-Prensa/Fundación Marcelino Botín, Madrid: 153–173.

Carles, J., García, M. & Vega, V. (2001b). Gestión colectiva de las aguas subterráneas en la Comunidad Valenciana [Collective groundwater management in the Community of Valencia]. In: *La Economía del Agua Subterránea y su Gestión Colectiva* (Hernández-Mora, N. and Llamas, M.R. eds.). Mundi-Prensa/Fundación Marcelino Botín, Madrid: 291–321.

Corominas, J. (2000). Las aguas subterráneas en la gestión de sequías en España: Andalucía [Groundwater in drought management in Spain: Andalusia]. *Rev. Real Academia de Ciencias Exactas, Físicas y Naturales* 94(2): 227–286.

Cruces, J. (1999). *Evaluación de los recursos subterráneos en el Libro Blanco del Agua en España* [Estimation of groundwater resources in the White Book of Water in Spain]. In: Jornadas sobre las Aguas Subterráneas en el Libro Blanco del Agua en España. International Association of Hydrogeologists—Spanish Group. Madrid: 11–22.

Custodio, E. & Palancar, M. (1995). *Las aguas subterráneas en Doñana* [Groundwater in Doñana]. Rev. Obras Públicas. Madrid, 142(3340): 31–53.

Custodio, E. (2002). Aquifer overexploitation, what does it mean? *Hydrogeology Journal*, 10(2): 254–277.

Custodio, E. (2009). Utilització de les aigües subterrànies a Catalunya i recuperació d'aqüífers: aspectes bàsics, implicacions econòmiques i integració [Groundwater use in Catalonia and aquifer reclamation: basic aspects, economic implications and integration]. Nota d'Economia, Generalitat de Catalunya, Barcelona. (in press).

Custodio, E., Manzano, M. & Montes, C. (2008). Perspectiva general del papel y gestión de las aguas subterráneas en el área de Doñana, Sudoeste de España [General perspective of groundwater role and management in the Doñana area, southwestern Spain]. Bol. Geológico y Minero, 119(1): 81–92.

de la Hera, A. (1998). Análisis hidrológico de los humedales de la Mancha Húmeda y propuesta de restauración de un humedal ribereño: El Vadancho (Toledo) [The Mancha Húmeda wetland hydrological análisis and proposal for the restoration of a riparian wetland: El Vadancho (Toledo)]. Doctoral Thesis. Universidad Complutense de Madrid: 1–330.

European Commission (EC) (1991). *Directive 91/676/EEC on nitrates from agricultural sources.* Commission of the European Communities. Brussels.

European Comission (EC) (2003). Common implementation strategy for the Water Framework Directive (2000/60/EC): Guidance Document No. 2: Identification of Water Bodies. European Commission, European Communities, Brussels: 1–23.

EEA (1999). Sustainable water use in Europe—Part 1: sectorial use of water. Environmental Assessment Report, no 1. European Environment Agency. Office for Official Publications of the European Communities. Luxembourg: 1–91.

Fornés, J.M., de la Hera, A. & Llamas, M.R. (2005). La propiedad de las aguas subterráneas en España: la situación del Registro y del Catálogo [Groundwater property in Spain: the present situation of the Register and the Catalogue]. *Ingeniería del Agua*, 12(2): 125–136.

Garrido, A., Martínez-Santos, P. & Llamas, M.R. (2006). Groundwater irrigation and its implications for water policy in semiarid countries: the Spanish experience. *Hydrogeology Journal*, 14(3): 340–349.

GWD (2006). Directive 2000/118/EC of the European Parliament and of the Council. Groundwater Directive. Brussels.

Hernández-Mora, N. & Llamas, M.R. (eds.) (2001). *La economía del agua subterránea y su gestión colectiva* [Groundwater economy and its collective management]. Fundación Marcelino Botín/Mundi-Prensa. Madrid: 1–550.

Hernández-Mora, N., Llamas, M.R. & Martínez Cortina, L. (2001). Misconceptions in aquifer over-exploitation: implications for water policy in Southern Europe. In: Agricultural Use of Groundwater: Towards Integration between Agricultural Policy and Water Resources Management (Ed. C. Dosi). Kluwer Academic Publ. Dordrecht, The Netherlands: 107–125.

Hernández-Mora, N., Martínez Cortina, L. & Fornés, J. (2003). Intensive groundwater use in Spain. In: *Intensive use of Groundwater. Challenges and Opportunities* (Llamas, M.R. and Custodio, E. eds.) Balkema, Dordrecht: 387–414.

Hernández-Mora, N., Martínez Cortina, L., Llamas, M.R. & Custodio, E. (2007). Groundwater issues in southern EU member status: Spain country report. Report to the European Academies of Sciences Advisory Council (EASAC)-European Union Water Initiative for the Mediterranean countries. April 2007.

Llamas, M.R. & Custodio, E. (2003). Intensive use of groundwater: a new situation which demands proactive actions. In: Intensive Use of Groundwater: Challenges and Opportunities (Llamas, M.R. and Custodio, E. eds.). Balkema Publishers. Dordrecht, 13–31.

Llamas, M.R. & Garrido, A. (2007). Lessons from intensive groundwater use in Spain: economic and social benefits and conflicts. In: *The Agricultural Groundwater Revolution: Opportunities and Threats to Development* [Giordano, M. and Villholth, K.G. eds], CAB International, Wallingford, UK: 266–295.

Llamas, M.R. & López-Gunn, E. (2007). The role of science and technology in the soft path to solving conflicts about water. Fundation R. Areces Scientific Meeting: *Water Management: Technology, Economics and Environment. Madrid*: 19–20. (http://www.Fundacionareces.Es/Agua_2006_Prog.Htm).

Llamas, M.R. (2003). *El Proyecto Aguas Subterráneas: resumen, resultados y conclusiones* [The Groundwater Project: summary, results and conclusions]. Papeles del Proyecto Aguas Subterráneas no 13. Fundación Marcelino Botín/Mundi-Prensa, Madrid: 1–101.

Llamas, M.R., Fornés, J., Hernández-Mora, N. & Martínez Cortina, L. (2001). *Aguas subterráneas: retos y oportunidades* [Groundwater: challenges and opportunities]. Mundi-Prensa/Fundación Marcelino Botín, Madrid: 1–529.

López Geta, J.A. (2006). La explotación sostenible de las aguas subterráneas: participación de los usuarios [Sustentable groundwater exploitation: users participation]. XI Congreso Nacional de Comunidades de Regantes. Palma de Mallorca: 1–38, (http://www.fenacore.org).

Manzano, M., Custodio, E., Montes, C. & Mediavilla, C. (2009). Groundwater quality and quantity assessment through a dedicated monitoring network: the Doñana aquifer experience (SW Spain). In: *Water Quality Assessment and Monitoring* (Ph. Quevauviller, A-M., Fouillac, J. Grath, R. Ward, eds.). Wiley (in press).

MAPA (2001). *Plan Nacional de Regadíos: horizonte 2008* [National Plan for Irrigation: 2008 horizon]. Ministerio de Agricultura, Pesca y Alimentación, Madrid, Spain: 1–360.

Martínez Cortina, L. (2001). Estimación de la recarga en grandes cuencas sedimentarias mediante modelos de flujo subterráneo: aplicación a la cuenca alta del Guadiana [Recharge estimation in large sedimentary basins by means of groundwater flow models: application to the Guadiana upper basin]. Doctoral Thesis. Universidad de Cantabria: 1–418.

MIMAM (2000). *Libro blanco del agua en España* [White book on water in Spain]. Secretaría de Estado de Aguas y Costas, Direc. Gen. Obras Hidráulicas y Calidad de las Aguas. Madrid: 1–637.

MIMAM (2006). Síntesis de la información remitida por España para dar cumplimiento a los Artículos 5 y 6 de la Directiva Marco del Agua, en materia de aguas subterráneas: Memoria [Synthesis of the information sent by Spain to comply with Articles 5 and 6 of the Water Framework Directive, in groundwater matters: Memory]. Dirección General del Agua, Ministerio de Medio Ambiente. Madrid: 1–85.

MIMAM (2007). Precios y costes de los servicios del agua en España: informe integrado de recuperación de costes de los servicios de agua en España. Articulo 5 y Anejo III de la Directiva Marco de Agua [Water services prices and costs in Spain: integrated report on water services costs in Spain: Article 5 and Annex III of the Water Framework Directive]. Working Documents. Madrid, Ministerio de Medio Ambiente, (http://www.fundacion-biodiversidad.es/opencms/export/fundacion biodiversidad/envios/portal_debate07.html).

MMA-ITGE (1997). Catalogo de acuíferos con problemas de sobreexplotación o salinización [Catalogue of aquifers with overexploitation or salinization problems]. Ministerio de Medio Ambiente/Instituto Tecnológico Geominero de España, Madrid: 1–43 + Appendix.

Molinero, J., Custodio, E., Sahuquillo, A. & Llamas, M.R. (2008). DMA y la gestión del agua subterránea en España [The WFD and groundwater management in Spain]. In: *6° Congreso Ibérico sobre Gestión y Planificación del Agua: Los Nuevos Planes de Gestión de Cuencas, Una Oportunidad para la Recuperación de los Ciclos del Agua.* Fundación Nueva Cultura del Agua, Madrid-Zaragoza. Sesión Paralela V: 9 p.

MOPTMA-MINER (1994). Libro blanco de las aguas subterráneas [White book on groundwater]. Serie Monografías. Ministerio de Obras Públicas, Transportes y Medio Ambiente/Ministerio de Industria y Energía, Madrid: 1–135.

Niñerola, J.M., Queralt, E. & Custodio, E. (2009). Llobregat delta aquifer. In: *Quality Assessment and Monitoring* (Ph. Quevauviller, A.-M. Fouillac. J. Grath and R. Ward, eds.). Wiley (in press).

Pérez Zabaleta, A. (2001). La oferta de aguas subterráneas para abastecimientos urbanos [Groundwater offer for urban supply]. In: *La Economía del Agua Subterránea y su Gestión Colectiva* (Hernández-Mora, N. and Llamas, M.R. eds.). Mundi-Prensa/Fundación Marcelino Botín, Madrid: 75–96.

Pulido, A., Molina, L., Vallejos, A. & Pulido, P. (2000). El Campo de Dalías: paradigma de uso intensivo [The Campo de Dalias: paradigm of intensive use]. *Papeles del Proyecto Aguas Subterráneas. Fundación Marcelino Botín.* Serie A, no. 4, Madrid: 1–54.

Ragone, S., Hernández-Mora, N., de la Hera, A., Bergkamp, G. & McKay, J. (eds.) (2007). The global importance of groundwater in the 21st Century: Proceedings of the International Symposium on Groundwater Sustainability. National Groundwater Association Press, Ohio, USA: 1–382.

Rico, A.M. & Olcina, J. (2001). La gestión colectiva de las aguas subterráneas en tierras alicantinas: algunos ejemplos [The collective groundwater management in the Alacant area: some examples]. In: La Economía del Agua Subterránea y su Gestión Colectiva (Hernández-Mora, N. and Llamas, M.R. eds.). Mundi-Prensa/Fundación Marcelino Botín, Madrid: 475–533.

Sahuquillo, A., Custodio, E. & Llamas, M.R. (2009). La gestión de las aguas subterráneas [Groundwater Management] *Tecnología del Agua.* March and April 2009 issues (in press).

Sánchez, A. (2003). Major challenges to future groundwater policy in Spain. *Water International*, 28(3): 321–325.

Vives, R. (2003). Economic and social profitability of water use for irrigation in Andalucía. *Water International*, 28(3): 326–333.

WFD (2000). Directive 2000/60/EC of the European Parliament and of the Council establishing a framework for Community action in the field of water policy. Water Framework Directive. Brussels.

CHAPTER 15

Making groundwater institutionally visible

Elena López-Gunn

Department of Geodynamics, Complutense University, Madrid, Spain

1 INTRODUCTION

In Spain, and in many other parts of the world, groundwater is an invisible resource. Groundwater is not only 'invisible' physically but also in institutional and governance terms. This chapter will explore this invisibility in institutional and governance structures in Spain. This is crucial because recent history in groundwater use has led to positive and dramatic social and economic changes, yet often with the price tag of negative, external environmental costs. As stated by Molden (2001): *'groundwater has* (had) *tremendous value in rural and urban contexts for poverty alleviations, livelihoods, drought security, agricultural yields, domestic water supply and the environment'* (Molden, 2001, p. 171).

Despite groundwater's importance (Chapter 14), it has only played a minor role in water planning, compared to the dominance of surface water- through reservoirs and water transfers-predominant in Spanish water planning and water paradigms. By and large, groundwater has only generally been an afterthought in Spanish water planning, literally submerged by the dominance of surface water and the hydraulic paradigm that still predominates in the Spanish water administration (del Moral & Sauri, 1999; Lopez-Gunn & Llamas, 2000).

2 INSTITUTIONAL FRAMEWORK

Groundwater is special because the solution to its long-term sustainability and to maximizing its strategic role lies in networks and their self-regulation. That is, groundwater use has boomed mainly due to the private initiative of individuals, mainly farmers and/or small municipalities. In addition, the absence of state planning has led to an impasse, where immediate economic benefits of groundwater use are large. Yet this can be at the expense of extreme social costs in externalities like damage to river flows and wetlands or groundwater pollution. This section will use institutionalist theory to study 'institutions', using two perspectives: as rational institutionalist theory, to analyze the issue of formal water rights, and as historical institutionalism to analyze the key relationship between water users and water authorities (Hall & Taylor, 1996).

2.1 *Groundwater property rights*

According to 'rational choice institutionalism' (Hall & Taylor, 1996:943), institutions solve many of the collective action problems that legislatures usually confront. This type of approach emphasizes the barriers of clarifying property rights, rent-seeking and transaction costs to the development and operation of institutions. In this case, the groundwater use situation in Spain mirrors a tragedy of the commons, when individual farmers acting rationally by abstracting water from their farms, can lead to aquifer over-use. The answer is the creation of institutions to address the problem of collective action. The creation of strong institutions facilitates self-governance and prevents aquifer over-use. A series of variables have been identified that affect the outcome of managing common pool resources sustainably.

As suggested by Ostrom (1992), a key variable identified for self-sustaining, enduring institutions for groundwater management relies on clear property rights. However, in Spain similar to many other major groundwater user countries in the world like India (Kumar, 2000), at present *de jure* rights to groundwater are not clearly defined.

The 1985 Water Act reviewed in Chapters 10 and 11 created a new system, where groundwater used after 1985 was effectively public, and subject to a state permit, and registration in the Registry of Public Water (*Registro de Aguas Públicas*). However, users that had used groundwater before 1985 had a choice, they could either choose to register the permit in the Register, as 'temporary use of groundwater for 50 years', after which groundwater would become public; or they could register their right in the alternative Catalogue of Private Water *(Catalogo de Aguas Privadas)*, which meant in effect groundwater remained private. In the latter case, however, groundwater users are limited in that they cannot change any of the characteristics of the right (like the location or depth of the well, the volume of water pumped, or the beneficiary use that the groundwater is being put to, e.g. type of crop) (Hernandez-Mora *et al.*, 2003).

Yet, the situation in Spain in relation to registering these property rights, an underpinning variable to achieving self-governance is very deficient. For example, the Green Paper on Groundwater in Spain (MIMAM, 2000), estimated 500,000 operational wells, of which only 250,000 were declared, and furthermore only ¼ are registered (Yagüe *et al.*, 2003). However, some authors estimate that the real number of wells in Spain is probably around 2 million (Llamas *et al.*, 2001; Lopez Geta *et al.*, 2001). The lack of definition on groundwater rights centers on two issues.

First is the fact that the registration of many legal wells are still pending. It is an enormous undertaking to register the half a million wells officially estimated. Of these, only 10 to 20% have been registered in the Public Register, the remaining being still un-registered. A number of initiatives have been pursued by the administration to speed up the registration of private and public water rights in the form of investment programmes, like ARYCA (1995–2001), with an initial budget of 42 million euros, rising to 66 million euros, and ALBERCA (2001–2007), with an estimated budget of 153 million euros (Fornés *et al.*, 2004).

Secondly, the administration has to deal with the problem of illegal wells. In Spain there are two types of illegal wells. Those opened after 1985, and not registered as public waters in the Register, and second, originally 'legal wells', listed in the Catalogue but which have changed the characteristics of their private water right, and should therefore be transferred to the Public Register.

The administration will decide in the future, perhaps pressed by the new EU Water Framework Directive, whether to adopt a litigious route or introduce quick regulatory reforms. Meanwhile, a key variable required to establish sound groundwater management—clear property rights—is currently not being met. As Kumar (2000) states: "*In the absence of well-defined property right structure, the increase in groundwater supplies … will lead to increased use by a few, undermining local management efforts. Therefore, unless the communities establish the rights over the resource and regulate the demand, groundwater cannot be sustainably managed*" (Kumar, 2000:284).

The following section uses another branch of institutional analysis, namely historical institutionalism, to seek an explanation of the problem identified above of non-compliance with the Water Act, defined in two ways: first, in the large number of wells operating outside legality and second, in the lack of effective monitoring and sanctioning in groundwater use in Spain, which can partly be held responsible for the opportunity granted to illegal groundwater users to free-ride, and abuse of common pool resource like groundwater.

2.2   *The key relationship between water users and water authorities*

Historical institutionalism operates on the basis that conflict among rival groups lies at the heart of politics. The state therefore is not a neutral broker but is itself engaged in structuring the character and outcomes of group conflict. In this school, institutions are equated with organizations and the rules promulgated by these institutions. Institutions are defined *as the formal or informal procedures, routines, norms and conventions embedded in the organizational structure of the policy or political economy* (Hall & Taylor, 1996:938). In this school, illegal groundwater users are seeking to maximize a number of goals, behaving strategically in the sense that institutions, in this case, organizations (like e.g. the water authorities), provide them with *'greater or lesser degrees of certainty about the present and future behavior of other'* (e.g. groundwater users). More specifically, *'institutions provide information, ... enforcement mechanisms for agreements, penalties for defection ... affect individual action by altering the expectation an actor has about actions that others are likely to take in response or simultaneously with his own actions'* (Hall & Taylor, 1996:939).

In this context water authorities, defined as organizations imbued with an institutional culture and mode of operating, condition the behavior of illegal groundwater users. An analysis based on historical institutionalism can be complemented by the concept of social regulation and regulatory style.

Water laws to address groundwater use can be regarded as social regulations in the sense that regulations deal with protecting and managing a natural resource. The ends pursued by social regulations can be achieved in a number of ways: economic incentives, legal regulation, administrative measures like licensing, or self-regulation and self-audit, where the government seeks to establish principles, yet the industry regulates itself (Hutter, 1997). In Spain social regulation in groundwater is mainly limited to command and control measures, in the form of the water laws discussed above. Even if the possibility of water trading and markets is regulated under the 1999 Water Act, it has not actually materialized, except in areas where there was a previous tradition of water markets (e.g. Canaries and Murcia) and in a few other places (see Chapters 11 and 13). This highlights a systemic reluctance to apply and/or experiment with this new environmental policy instruments (or NEPIs). The approach to environmental protection has been evolving from a regulation driven, adversarial 'government-push' approach to a more pro-active approach involving voluntary and often business led initiatives to self-regulate their 'environmental performance' (Khanna, 2001: 291, as quoted in Kandouri, 2004).

In Spain, the regulatory regime is formal and legalistic, influenced by French traditions, with the aim *'to establish clear legal frameworks, backed up by state agencies and the judiciary'* (Buller, 1998: 70 quoted in Carter, 2001). The culture of the Spanish administration is grounded in administrative law and influenced by the French legal model (Torres & Pina, 2004). This enlarges the role of the judiciary as arbitrators and minimizes the role of regulatory officers, e.g. in water authorities to exercise discretion.

In terms of regulatory style, many water authorities have adopted a highly conflictive, adversarial, regulatory style, which in turn affects the strategic behavior of groundwater users. To dwell deeper on non-compliance with regulation and regulatory style, it is useful to zoom in an aquifer where non-compliance has become the norm and compliance the exception: the case of the Upper Guadiana basin in central Spain.

What can be learnt about the Upper Guadiana case study is that—from a sociological point of view- the regulatory style is fairly conflictive, as opposed to a consensual, accommodative, pluralist style. In this case the perception is that the industry, in the case of the Upper Guadiana farmers who are the main groundwater users, holds a dominant position. This sociological perspective would argue that in a conflictive regulatory style, when law is weakened to the point of ineffectiveness, it might be a case of capture of the regulatory agency. In the Upper Guadiana case almost ten years ago it was estimated that there were 2,500 (Mancha a Mano, 1996), to 6,000 illegal wells (Serna & Gaviria, 1995). It could be argued that the Guadiana Water Authority has been 'captured' by the main water users, the farmers themselves.

Yet this is crucial, the relationship between regulators and the regulated, which lies at the heart of the compliance process. What regulatory theory can help clarify in the case of the Upper Guadiana and possibly many other over-used aquifers is the lack of cooperation, the inherent lack of trust in the regulated-regulator relationship. As Hutton states (1997) *"Compliance is processual, ... a long-term relationship ... (the) relationship between the agency and the regulated is reflexive, each party adapting to the moves and anticipated moves ... enforcement is often social and incremental"* (Hutton, 1997:195).

However, the moves and anticipated moves of the Guadiana Water Authority and groundwater farmers have led to a situation where there is massive non-compliance and the enforcement of regulation is plagued with difficulties.

Adequate monitoring and sanctioning underpin a successful regulatory regime. This sanctioning and monitoring can either be undertaken by the administration itself or by groundwater users, as a devolved system, or a form of social or moral norm (Gezelius, 2002). Set in the context of asymmetric information, groundwater users are in a favorable position compared to regulatory agencies, like water authorities who have to monitor thousands of wells (Carter, 2001). From the point of view of the regulator, maybe the Spanish parliament like the American congress in other environmental programmes *'has underestimated the workload generated by new regulations that produced unrealistic deadlines, excessive administrative rules and virtually unattainable programme objectives'* (Rosenbaum, 1999:172 in Carter, 2001). Yet the ARYCA and ALBERCA projects will likely help the Government to pursue new initiatives based on objectively verifiable data on groundwater use and pumping units.

From the point of view of the relationship of the regulator and the regulated, presently monitoring and sanctioning in groundwater regulation in Spain are the exception, rather than the norm. In most aquifers monitoring, through either water meters or satellite, is only rarely used jointly as a management measure. In cases of conflictive relationships between the regulated and the regulators, there is no communal regulatory regime. The opportunity is there to develop a consensual, accommodative regime based on trust.

The authorities or administration are focused on very traditional, reactive regulatory methods, such as complaints, warnings, notices, prosecutions and formal sanctions, and users respond equally through litigation. Furthermore in Spain the situation is not helped by the fact that the sanctioning regime is highly opaque. The regime does not specify clearly the reasons why a sanction is imposed, therefore halting a crucial process of social learning.

This stands in contrast with a pro-active, preventive style of enforcement, underpinned by strong cooperation. The development of a cooperative style depends on a number of factors like the enforcement style (whether it is flexible or coercive), organizational factors, like resources, issues of trust, the type of business, whether businesses are small or large, whether the relationships are transient or frequent, and the nature of the activity. Often aquifers declared in over-draft are also the scene of coercive enforcement styles. Here trust is limited between regulator and the regulated, and where the administration has to cope with a disabling lack of resources and large number of small farmers.

In effect groundwater management in Spain suffers from the malaise of implementation deficit or ineffective regulation. This conflictive regulatory style often comes with a high price tag in terms of the bad feeling generated between regulator and regulated, which can encourage further flouting of the law (Carter, 2001). The question therefore now turns to one of trust, or social capital. This itself revolves on questions on the state, civil culture and learned attitudes on non-cooperative, non-compliant behavior.

3   TOWARDS GROUNDWATER GOVERNANCE

This brings us back to the question whether groundwater 'institutionally different' to merit separate or special treatment. The answer is yes and no, hydrogeology is a relatively new science, only 30 to 40 years old. Yet 411 hydrogeological units have been identified in Spain, 34% of the country's area, with significant quantities of groundwater, equivalent to 22 km^2/yr, and where use—since hydrogeology started as a science in the 1960s—as stated above- has tripled (Hernandez-Mora *et al.*, 2003; Sanchez, 2003). Groundwater however remains problematic in institutional terms, due to a pervasive problem of non-compliance with the current regulatory system, as described above, and in many cases with difficult, conflictive relationships between groundwater users and the administration.

The example of the Western *Mancha* aquifer in the Upper Guadiana Basin shows how maximizing individual utility can doom collective action. This example of non-compliance will be a particular challenge when faced with both international and European regulatory commitments. For example, the deadline to prepare Integrated Water Resource Management Plans by 2005, or the goal of eliminating groundwater overdraft by 2025 appear unrealistic (Molden, 2001:172). Equally, at European level article 11 of Water Framework Directive states that protection measures have to be fully operational in 2012, which in effect refers to the problem of groundwater over-draft (Sanchez, 2003).

3.1 *Collective management institutions: The case of Groundwater User Associations*

In Spain Irrigation Communities have been traditional institutions for surface water management, some dating back to the 12th century (Gavarro i Castildefort, 1984; see Chapter 12). Yet, ground-water has traditionally been managed in a very individualistic manner. Until the 1985 Water Act, groundwater abstraction was mainly a private initiative, to be undertaken mainly (although not solely) by landowners in the aquifer to be exploited. This means that groundwater has followed a path dependency which is fairly individualistic, with little tradition for collective management. However, the 1985 Water Act updated traditional Irrigation Communities to be User Communities, to make way for two big changes in water management; first, the fact that water users in the late 20th century extended beyond farmers and second, recognize the significance of groundwater use by allowing groundwater users to create their own organizations (Lopez-Gunn, 2003). What is particularly striking when analyzing and comparing Groundwater User Associations in Spain is their range and diversity (see Table 1; see Chapter 12 for an alternative typology of groundwater user groups).

In Spain, as a continental law system, User Communities fall under public law (Codina-Roig, 2003), which are ascribed to the water administration and of which there are three main types; of 1st, 2nd and 3rd order. First order User Communities, are characterized by being composed exclusively of individual users, independent of the main type of use. If however, the main use is irrigation, then they can be called Irrigation Communities, an example is *Comunidad de Usuarios del Campo de Montiel*. Meanwhile, second order User Communities have a federal structure (Subramanian, 1997), with a General Community, and underneath, individual irrigation communities and other uses, like e.g. townhalls- this is the case of the General Community of the Upstream Vinalopó Users (*Comunidad general de Usuarios de Alto Vinalopó*). Third order User Communities are the

Table 1. Typology of groundwater user communities in Spain.

	Public law	Private law	
3rd order	Central Users' Board of the Western Almeria Aquifer	Association	Catalan Association of Users Communities of Groundwater
	(Junta Central de Usuarios del Acuífero del Poniente Almeriense)	(Asociacion)	(Asociació Catalana de Comunitats d'Usuaris d'Aigües Subterranies)
2nd order	General Community of the Upstream Vinalopó Users	Association of wells	Provincial Grouping of Irrigation Wells in Castellon, Association of Irrigation Wells of the Valencian Autonomous Community
	(Comunidad general de Usuarios de Alto Vinalopó)	(Asociaciones de Pozos)	(Agrupación Provincial de Pozos de Riego de Castellón; Asociación de Pozos de Riego de la Comunidad Valenciana)
1st order	General Community of the Montiel Field	Social partnerships	Association of Groundwater users of Castille-La Mancha
	(Comunidad de Usuarios del Acuífero del Campo de Montiel)	(Sociedad civil)	(Asociación de Usuarios de Aguas Subterráneas de Castilla La Mancha)

Source: Author's own elaboration.

most sophisticated, and include both user communities and individual users abstracting water from the same aquifer, like the Central Users' Board of the Western Almeria Aquifer (*Junta Central de Usuarios del Acuífero del Poniente Almeriense*) (Codina Roig, 2003).

However, co-existing with public law institutions are Associations under private law, which groups individual users with their own private wells, financed mainly by private initiative. There is a huge diversity of institutions that had thrived under the Roman Civil Code, which gave pre-eminence to individual rights. For example there are Agrarian Betterment Partnerships (*Sociedades Agrarias de Trasnformacion*, SATs), Wells' Social Partnerships (*Sociedades Civiles* de Pozos), Water Communities (Carles *et al.*, 2001; Rico Amoros & Olcina Cantos, 2001). Key differences to surface water irrigation communities: in that many institutions are regulated under private Act and water is not necessarily tied to the land (Perez Perez, 1988). In our case, there are however two levels within Associations created under Private law; the first are User Associations like the Association of Groundwater users of Castille-La Mancha (*Asociación de Usuarios de Aguas Sub-terráneas de Castilla La Mancha*), which is a *Sociedad Civil* and effectively is like an Irrigation Community, which encompasses farmers with their individual wells, in this case also arranged per village; second, there is an Association of wells, an association that encompasses SATs, Wells Partnerships, Irrigation Communities. Although all these organizations act collectively, the main decision making and management occurs at the level of individual organizations; this is the case of the Catalan Association of Users Communities of Groundwater (*Asociació Catalana de Comu-nitats d'Usuaris d'Aigües Subterranies*), probably the most sophisticated Groundwater User Asso-ciation in Spain, which includes a range of uses (industrial, agricultural public water supply), and a highly sophisticated internal structure in terms of coordination and organizational capacity.

The water authorities have traditionally preferred public bodies to interact with, as devolved agents of the administration, which therefore can issue public acts (e.g. sanctions). In reality, User Communities have a hybrid nature, both in the (public interest) aspect of managing groundwater, and the (private) protection of their private vested interests (Moreu Ballonga, 2003). Yet some argue that inherently groundwater has a different nature to surface water and in groundwater there is a strong tradition of the individuality of managing a well compared to managing a canal. Under Art 79 of the 1985 Water Act the creation of User Organizations was made compulsory in aquifers declared in overdraft (Moreu-Ballonga, 2003), like in the case of Castille-La Mancha and Campo de Montiel.

The potential is enormous for these collective management institutions to fill the gap left by the sclerosis of the administration (in terms of lack of action). This can be seen in actions by groundwater users to monitor their water use through e.g. water meters or traditional figures like the *'regador'*. The *regador* is an employee who works in a local irrigation district and becomes familiar with all the activities in the area; acting by allocating water, providing technical advice whilst also acting informally as a monitor, and in cases, as a witness for official complaints. In other cases, groundwater users like the Association in Catalonia mentioned above have shown the capacity to develop along the whole continuum of issuing an inventory, defining water rights, monitoring and sanctioning. In this case the norm of conflictive relationship between the adminis-tration and groundwater users has been turned around to a collaborative relationship where water users and the administration work closely together, rather than against each other. This synergy has arguably provided a solid foundation for long-term water management. In particular many so called *Convenios* or agreements (as allowed under the 1999 Law reform, Art 97.3) (Garcia-Vizcaino, 2003) have been signed between users and the administration. The best example is Cata-lonia, where certain discrete activities like undertaking the inventory are devolved to groundwater users though these agreements, financed by the administration, yet on the understanding (or trust) that it will be properly undertaken.

In Catalonia the Groundwater User Association has also started to develop reflexive aspects of self-regulation. Groundwater users are increasingly acting to protect the aquifer not only for their own self-interested use, but also for the protection of wider social and environmental aspects. For example, user groups are showing greater awareness on the key importance in long term aquifer management of ensuring careful land-use planning (Codina Roig, 2003). It is the successful relationship between regulator and the regulated that is at the heart of this reflexive evolution; where autonomy, legitimacy and collaboration co-exist in day-to-day groundwater management, and the regulated (groundwater users) become guardians and protectors of their main asset- the aquifer itself.

Yet specific to the Spanish case is the need to make institutions hermetic to politics and slowly generate a shift in institutional culture where groundwater users are perceived as potential co-managers and not solely as users safeguarding their own, private individualistic use. Both groundwater users and the administration are increasingly tied in a symbiotic relationship where groundwater users need a strong and efficient administration, for example in order to provide the security of water rights. Equally the administration needs the back up of users to develop the strong monitoring and sanctioning systems needed to underpin successful groundwater management.

4 CONCLUSION

Groundwater has had a major positive impact in Spain on poverty and livelihoods since the 1960s. However, its challenges in the future are different, since the expectations of Spanish citizens have also risen commensurate with this newfound wealth. New regulatory challenges will be faced internationally, from the pressure to achieve Integrated Water Resource management and from Europe, with the Water Framework Directive. In Spain, evidence points that 'partnership' governance forms between groundwater users and the administration have the potential to break deadlock, conflictive situations over determining groundwater rights, the first step in successful self-governance.

Groundwater has some inherent characteristics that make a key strategic resource, and that also make it stand out in institutional terms. Spain has benefited largely from the intensive use of groundwater resources which has marked a socio-economic revolution. However, institutional development at present is lagging behind and strong frameworks have to be developed to embed groundwater use in its environmental *and* political context. The pending issues in terms of groundwater governance in Spain, and institutionally the priority areas for action are:

1. Use the window of opportunity provided by the Water Framework Directive and the concept of *Full Cost Recovery* to fully internalize the external costs of groundwater use. The central issue is how the main sectorial users of groundwater are going to address this legal requirement. At present, the largest sector benefiting from groundwater use- agriculture- is not paying for the resource itself and this is a difficult question to tackle politically.
2. Address the question over *water rights*, and the lack of information necessary to be able to take informed public policy decisions on current groundwater use and future development scenarios. Clear knowledge and strong regulatory frameworks are needed to fully internalise groundwater's significance in economic and strategic terms into water planning. Data transparency on groundwater use, groundwater rights and potential gaps between use and rights points to a clear weak point in regulation. It is essential to prioritize as a matter of urgency the completion of the water rights registry, by allocating relevant financial and human resources and identify specific geographical areas where allocated water rights exceed the renewable level of resources. Develop potential policy options to deal with the issues of over-allocation, in order to open a dialogue with different stakeholders on a range of different policy choices. In this context, maybe undertake a comparative study on the experience on policy reforms and instruments adopted by other countries like the United Kingdom, Australia and some states in the USA which have started to address the issue of oversubscribed and overallocated aquifers. Engage with society, particularly at regional level, on the issue of illegal wells. Engage and facilitate a dialogue between water authorities to address the root causes of non-implementation and non compliance with the existing regulatory regime. Acknowledge and address the use of water for political gain.
3. Use the *full potential of existing institutions* like participatory management boards discussed in Chapter 13 and collective water management institutions like water user groups, and open them up to newer stakeholders. Encourage and foster aspects of self-reflexivity in these

organizations, where user groups share responsibility to internalize the public good aspects of their water management activities.

4. Strengthen opportunities for *cooperative and collaborative planning*, by allocating sufficient resources (human, organizational) and also sufficient political will and trust. Open up the system in order to avoid potential policy capture by dominant water interests and the use of water for political gain (Zeitoun & Warner, 2006). At present most of the factors that led to collaborative planning are absent, like flexible enforcement, strong organizational structures in terms of resources, or trust.

5. Engage with stakeholders to shift towards a less adversarial regulatory style, particularly in the case of the farming community and invest in the build up of social capital and trust between farmers and water authorities, though undertaking *Convenios* or agreements. These however should be bolstered by simultaneous reforms in organizational structures to open them to other stakeholders from civic society. Again identify potential pilot case studies and initiatives to experiment and innovate in terms of institutional reform. As stated above a key issue is the lack of social capital between the water authorities and the main users of groundwater- farmers. Farmers at present have the information required to manage groundwater. The question is whether farmers are willing to share this information and furthermore whether social capital can be built with water authorities for the joint management of aquifers, where the water authorities in line with new public management steer the overall policy goal but trust water users, like farmers- in the day to day management. Sadly, at present, the adversarial relationship between farmers and water authorities prevent collaborative policy making.

6. Reform the current sanctioning system in Spain to ensure that sanctioning is clear and proportionate with the offence. Open up the system so that the range of sanctioning procedures like complaints, prosecutions, sanctions and liabilities reflect the underpinning social norms, which are developed consensually through debates and discussion with groundwater stakeholders. Start to identify what choices are socially acceptable and which are not, and experiment with a range of methods and tools designed to open up and support decision making like multicriteria decision analyses.

REFERENCES

AEUAS (2004). Referencias que de forma específica, o con mayor relación directa, se hacen a las aguas subterráneas en Plan Hidrológico Nacional (Ley 10/2001 de 5 de julio, modificado por el Real Decreto Ley 2/2004 de 18 de Junio. [Specific references to groundwater in the National Water Plan Law 10/2001 of 5th July, modified by Royal Decree 2/2004 of 18th June]. Madrid, Spain. Asociación Española de Usuarios de Aguas Subterráneas.

Buller, H. (1998). Reflections across the channel: Britain, France and the Europeanisation of National Environmental Policy. In *British Environmental Policy and Europe* edited by Lowe, P.D. and Ward, S. London, UK: Routhledge: 67–83.

Carles, J.C., Garcia Molla, M. & Vega Carrero, V. (2001). La gestión colectiva de las aguas subterráneas en la Comunidad Valenciana. [The Collective management of groundwater in in the Valencia region in *La Economía del agua subterránea y su gestión colectiva*. [The Economics of Groundwater and its collective management] edited by Hernández-Mora, N. and Llamas, R. Madrid Spain: Fundacion Marcelino Botin and Mundiprensa: 291–321.

Carter, N. (2001). Chapter 11: Policy instruments and implementation. In *The Politics of the Environment: Ideas, activism, policy* edited by Carter, N. Cambridge, UK: Cambridge University Press: 284–314.

Codina-Roig, J. (2003). Las comunidades de usuarios de aguas y el marco normativo actual. *Jornadas sobre las Comunidades de Usuarios de Aguas Subterráneas en el marco normativo actual*. [The User communities and the current context in Conference organized on Groundwater User Communities in the current regulatory framework] Prat de Llobregat, Spain, 8–10th May.

del Moral, L. & Sauri, D. (1999). "Changing course: water policy in Spain." *Environment* 41(6): 12–15 and 32–36.

EEA (2007). *Climate Change and Water Adaptation Issues* EEA Technical report No 2/2007 Fornes, J.A., de la Hera, A. & Llamas, R. (2004). El Registro/Catalogo de derechos de aguas subterráneas en España. [The Water rights register/catalogue in Spain] *IV Congreso Ibérico de Gestión y planificación del agua*, [IV Congress on water management and planning] Fundación Nueva Cultura del Agua. Tortosa, 8–12th December.

Garcia-Vizcaino, M.J. (2003). Legislación europea, estatal y autonómica de aguas subterráneas. [European, state and regional legislation on groundwater] in *Jornadas sobre las Comunidades de Usuarios de Aguas Subterráneas en el marco normativo actual* [Conference organized on Groundwater User Communities in the current regulatory framework] Prat de Llobregat, Spain, 8–10th May.

Gavarro i Castellfort, R. (1984). "Las Confederaciones Hidrograficas." [Water Authorities] *El Campo*: 34, 108–116.

Gezelius, S. (2002). "Do norms count? State regulation and compliance in a Norwegian fishing community." *Acta Sociologica* 45: 305–315.

Giordano, M. & Villholth, K.G. (2007). The agricultural groundwater revolution: opportunites and threats to development Comprehensive assessment of Water Management in Agriculture Series, Wallingford, CABI.

Hall, P.A. & Taylor, R. (1996). "Political science and the three new Institutionalisms." *Political Studies* XLIV: 936–957.

Healey, P. (2004). "Creativity and urban governance." *Policy Studies* 25(2): 87–102.

Hernandez-Mora, N., Martinez-Cortina, L. & Fornés, J.A. (2003). Intensive groundwater use in Spain. In *Intensive groundwater use: challenges and opportunities* edited by IPCC R. (2007) Climate Change 2007: Impacts, adaptation and vulnerability-Summary for policy makers http://www.ipcc.ch/SPM13apr07.pdf

Hutter, B. (1997). *Compliance: regulation and environment*. Oxford, UK: Clarendon Press.

Khanna, M. (2001). Non-mandatory approaches to environmental protection. *Journal of Economic Surveys* 15(3): 291–324.

Koundouri, P. (2004). Current Issues in the economics of groundwater resource management- *Journal of Economic Surveys* 18(5): 704–740.

Kumar, D. (2000). Institutional framework for managing groundwater: a case study of community organisations in Gujarat, India. *Water Policy* 2: 423–432.

Llamas, M.R. & Custodio, E. Lisse, The Netherlands: Balkema, 56–78.

Llamas, R. (2003). Lessons learnt from the Impact of the Neglected Role of Groundwater in Spain's water policy. In *Water resources Perspectives: Evaluation, management and Policy*. Edited by Alsharhan, A.S. and Wood, W.W. Amsterdam, The Netherlands: Elsevier Science.

Llamas, R., Fornes, J.A., Hernandez-Mora, N. & Martinez-Cortina, L. (2001). *Aguas Subterráneas: retos y oportunidades*. [Groundwater: challenges and opportunities] Madrid, Spain: Fundación Marcelino Botín and Mundiprensa.

Lopez-Geta, Fornes, J.A., Ramos, G. & Villaroya, F. (2001). *Las aguas Subterráneas: un recurso natural del subsuelo*. [Groundwater: a natural resource in the subsoil] Madrid, Spain: Ministerio de Ciencia y Tecnologia: IGME.

Lopez-Gunn, E. & Llamas, R. (2000). New and Old paradigm's in Spain's water policy. Water Security in the Third Millenium: Mediterranean countries towards a regional vision. Llamas, R.M. *et al.* (eds) UNESCO. Paris, UNESCO. 9: Science for Peace Series: 271–293.

Lopez-Gunn, E. (2003). The role of collective action in water governance: a comparative study of groundwater user associations in La Mancha aquifers (Spain). *Water International* 28(3): 367–378.

Mancha a Mano (1996). Los pozos ilegales enfrentan a los alcaldes con la Confederación [Illegal dwels side the Mayors against the Basin authority]. 10th June 1996.

MIMAM (2000). *El Libro Blanco del Agua en España*. [The Spanish Green Paper on Water] Madrid, Spain: Ministerio de Medio Ambiente; Secretaria de Estado para Aguas y Costas.

Molden, D. (2001). "5.4. Groundwater." *Water Policy* 3: 171–173.

Moreu-Ballonga, J.L. (2003). Evolución y situación actual de la legislación sobre aguas subterráneas en España. [Evolution and current situation of groundwater legislation in Spain] *Jornadas sobre las Comunidades de Usuarios de Aguas Subterráneas en el marco normativo actual*. [Conference organized on Groundwater User Communities in the current regulatory framework] Prat de Llobregat, Spain, 8–10th May.

Ostrom, E. (1993). *Crafting institutions for self-governing irrigation systems*. San Francisco, California: ICS Press.

Perez Perez, F.J. (1988). La constitución de Comunidades de Usuarios de Aguas Subterráneas. [The Creation of groundwater user groups] *Jornadas sobre la aplicación de la Nueva Ley de Aguas en la Aplicación de las Aguas Subterráneas*, [Congress on the implementation of the new water law] October, Zaragoza.

Rico Amoros, A.M. & Olcina Cantos, J. (2001). La gestión colectiva de aguas subterráneas en tierras alicantinas. [The collective management of groundwater in the lands of Alicante]. In *La Economía del agua subterránea y su gestión colectiva*. [The Economics of Groundwater and its collective management] edited by Hernandez Mora, N. and Llamas, R. Madrid, Spain: Fundacion Marcelino Botin and Mundiprensa: 475–533.

Rosenbaum, W. (1999). Escaping the 'Battered Agency Syndrome': EPA's gamble with regulatory re-invention. In *Environmental Policy* edited by Vig, N. and Kraft, M. Washington D.C., CQ Press: 165–189.

Sanchez, A. (2003). Major challenges to future groundwater policy in Spain. *Water Inernational* 28(3): 321–325.

Serna, J. & Gaviria, M. (1995). *La quimera del agua. Presente y futuro de Daimiel y la Mancha Occidental.* [The Water Chimera Present and Future in Daimiel and the Western Mancha]. Daimiel, Spain: Siglo Veintiuno editores.

Subramanian, A., Jagannathan, N.V. & Meinzen-Dick, R. (1997). *User Organizations for Sustainable Development.* Washington D.C.: The World Bank.

Torres, L. & Pina, V. (2004). Reshaping public administration: the Spanish experience compared to the UK. *Public Administration* 82(2): 445–464.

Yague, J., Villaroya, C. & Xucla, R.S. (2003). Proyecto ALBERCA: modernización de los registros de aguas. [ALBERCA Project: modernisation of the Water Registers] Congreso Nacional de Ingenieria Civil, [National Civil Engineering Congress] Madrid 1853–1861.

Zeitoun, M. & Warner, J. (2006). Hydrohegemony: a framework for analysis of transboundary conflict *Water Policy* 8 pp. 435–460.

CHAPTER 16

Facing the challenges of implementing the European Water Directive in Spain

Manuel Menéndez Prieto
Ministry of the Environment, and Rural and Marine Affairs (MERMA), Madrid, Spain

1   INTRODUCTION: BASIC CONTENTS OF THE WATER FRAMEWORK DIRECTIVE

The Water Framework Directive (Directive 2000/60/EC of the European Parliament and of the Council of October, 23, 2000[1] establishing a framework for Community action in the field of water policy) is the most ambitious and complex piece of legislation on environment ever enacted in the European Union.

Mirroring the natural water cycle, the Water Framework Directive (WFD) prescribes the water management activities that are to be carried out within areas based on the natural river catchments. This philosophy, together with its integrated approach to inland, transitional, coastal, groundwater and associated ecosystems, constitutes a revolutionary change in the understanding of water policies on a European scale. One should be warned, however, that the WFD does not replace past EU water legislation, and there are half a dozen directives that are still enforced. The WFD includes the word 'framework' to highlight the need to integrate the objectives of all previous directives under one umbrella to encourage a more integrated set of policies and plans.

1.1   *Main definitions*[2]

The overall environmental goals of the WFD are to achieve a "good water status", prevent "further deterioration" of that status, and enhance the protection and improvement of the aquatic environment through measures "for the progressive reduction of discharges, emissions and losses of priority substances and the cessation or phasing-out of discharges, emissions and losses of the priority hazardous substances"[3].

The concept of water status differs for surface or groundwater. The status of surface water bodies is determined by the worst of its chemical or ecological statuses. Chemical status describes whether or not the concentration of any pollutant exceeds standards that have been set at the European level. Ecological status is primarily a measure of the effects of human activities on water. Generally speaking, the lower the impacts are, the better the ecological status is. The status of groundwater bodies will be determined by the poorest of its chemical and quantitative statuses. Chemical status is defined in the same way as for surface waters. Quantitative status is an expression of the degree to which a body of groundwater is affected by direct and indirect abstractions.

A good status must be achieved for all water bodies, unless the respective Member State applies to the European Commission[4] for derogation. Each water body has to be characterized according

---

[1] Published in the Official Journal of the European Communities on December, 22, 2000.

[2] Readers are referred to the legal text for the definitions of each WFD concept. The focus of this chapter is on meaning rather than on legal definitions.

[3] In the Directive, hazardous substances means the substances that are toxic, persistent and liable to bioaccumulate. Priority substances are listed in one of the annexes of the WFD and include the "priority hazardous substances" identified by the European Commission.

[4] In the WFD jargon 'derogation' means relaxing the fulfillment of any WFD objective, including extending deadlines, and setting up less stringent parameters.

to ecoregion types (which the WFD terms System A) or to the differentiation of water bodies into types using different obligatory and optional factors (System B). This will be the foundation for further activities to establish what "good status" means for each "type".

The next step is to identify, for each type, what the relevant aspects of a water body's characteristics would be like if there were "no or only very minor alterations" resulting from human activities. In the WFD, these nearly undisturbed conditions are called reference conditions.

Reference conditions must be defined for each quality element and each water body type to enable an ecological quality ratio to be calculated and a class determined for each surface water body. They should be selected according to chemical and hydro-morphological characteristics and evaluated more specifically in quantitative terms on the basis of biological parameters. The characterization of surface waters requires that Member States to develop a reference network for each surface water body type. If no reference waters are available, reference conditions could be based on modeling or on expert judgment.

Sometimes it will not be possible to achieve a "nearly undisturbed condition" of a water body because of substantial physical alterations made in order to permit activities like irrigation, drinking water supply, power generation, navigation and so on. The Directive recognizes that in some cases the benefits of such uses need to be retained and, provided a series of criteria are fulfilled, allows their designation as artificial or heavily modified water bodies.

The reference conditions for artificial or heavily modified water bodies of surface water is the "maximum ecological potential", which has to be derived from the water body type that is most similar to the uninfluenced body of surface water.

1.2   *The Directive as guidance for a planning process*

Beyond the specific fulfillment of the above requirements, the WFD defines a complete water planning process. In this respect, the WFD proposes a strategy that ends in the drafting and implementation of a water plan for every river basin district (territorial unit for the application of the Directive, usually grouping more than one natural river catchment).

The WFD defines five main steps in the planning process. The first step is to define the legal and institutional framework, which includes the identification of river basin districts, the establishment of the appropriate administrative arrangements including the transposition of the WFD to national legislations and the designation of competent authorities.

The WFD defines a "river basin district" (*demarcaciones hidrográficas*, in Spanish) as the area of land and sea, made up of one or more neighboring river basins together with their associated groundwaters and coastal waters, which is identified as the main unit for management of river basins. A river basin covering the territory of more than one Member State must be assigned to an international river basin district. Member States shall ensure that the requirements of the WFD and, in particular, all programs of measures are coordinated for the whole of the river basin district (see Chapter 18 to learn about Spanish-Portuguese cooperation with their shared basins). For river basins extending beyond the boundaries of the European Union, Member States should endeavor to ensure appropriate coordination with the relevant non-member states (the Danube is a case in point).

The WFD requires Member States to ensure coordination with the aim of producing a single international river basin management plan, with support from existing structures stemming from international agreements. If such an international river basin management plan is not produced, Member States shall produce river basin management plans covering those parts of the international river basin district lying within their territory.

For every river basin district, the Member states must identify the appropriate competent authority for the application of the rules of the Directive. This authority can be an existing national or international body.

Once the river basin districts have been defined, the WFD establishes the need to develop a series of studies in order to find out the main characteristics of the water bodies and their use, and, then, conduct an assessment of the current status. This is the second step.

The general description of the river basin district has to provide information on the identification of the different water categories—rivers, lakes, transitional waters and coastal waters—and their division into water bodies. Furthermore, water bodies must be differentiated by type and defined by their reference conditions.

The identification of significant current and foreseen anthropogenic pressures and the assessment of their impact is another task that is part of the assessment of current status. Once the main pressures have been identified, an assessment shall be made to predict what impact they are likely to have on the water bodies, i.e. how they influence the achievement of the environmental quality objectives. The susceptibility of the surface water bodies status to the pressures can be obtained using both monitoring data and modeling techniques.

The economic analyses of water uses serve the purpose of evaluating the importance of water in the economic and social development of the river basin. These studies must assess current levels of recovery of the costs of water services. The main elements to be investigated include the status of water services, the extent of the recovery of the costs (financial, environmental and resource costs) of these services, the institutional set-up for cost-recovery and the contribution of key water uses to the costs of water services (see Chapter 13 for a more detailed presentation).

The main aim of the third step is to define environmental objectives. These are the goals and targets for preventing deterioration of the water status, restoring a good water status and implementing the necessary measures with the aim of progressively reducing pollution from priority substances and ceasing or phasing out emissions, discharges and losses of priority hazardous substances. They will serve then as the foundation for decision-making on the program of measures. Goals and targets should be defined with a long-term vision for the river basin district, and be seen as steps to achieve the vision via a sound planning process.

Once the current status of the river basin districts has been assessed and the environmental objectives described above have been set, it will be possible to make a preliminary identification of the water bodies that are at risk of failing to meet the good water status or the non-deterioration principle. This information will be used to design monitoring programs and to define the programs of measures.

As regards the fourth step, the WFD describes three types of monitoring programs to provide information for different purposes. The first is surveillance monitoring that is mainly concerned with improving the assessment of which bodies are at risk of failing to meet the Directive's objectives and which are not. It includes the monitoring of surface water bodies and the chemical status and pollutant trends of groundwater bodies. Operational monitoring focuses exclusively on those water bodies that, on the basis of the risk assessments and the surveillance monitoring programs, are at risk of failing to meet the Directive's environmental objectives (see chapter 14 for a detailed description of the grountwater untis' status). Operational monitoring has to be based on indicators that are sensitive to the identified pressures. This program should also include monitoring of groundwater levels to assess groundwater at risk according to its quantitative status. Investigative monitoring is to be used to ascertain the answer to the question of why a water body is at risk and it should help to design the appropriate management measures.

The fifth step in the implementation of the planning process is the establishment and implementation of the programs of measures and water plans. The WFD explicitly requires Member and Accession States to produce a management plan for each river basin district. The river basin management plan is intended to record the current status of water bodies within the RBD, give a summary of what measures are planned to meet the objectives, and act as the main reporting mechanism to the Commission and the general public. The WFD requires river basin plans to integrate the management of water quality and water resources and surface and groundwater management in order to meet the environmental objectives. A management plan must be produced for each river basin district.

The main deadlines for the tasks described above and the articles of the WFD on which they are based are shown in Figure 1.

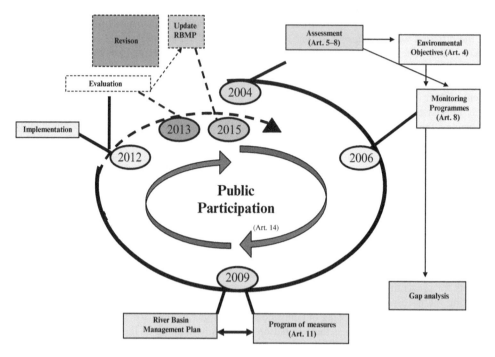

Figure 1.   The implementation process of the Water Framework Directive.

1.3   *The Common Implementation Strategy*

As soon as the WFD came into force, Member States and the European Commission realized the complexity of this piece of legislation, and hence the need to develop a cooperative process for its implementation. This process is called the Common Implementation Strategy (CIS). It aims to assure a common understanding of the different concepts included in the law, providing guidance and technical support for its application. Key principles in this common strategy include sharing information and experiences, developing common methodologies and approaches, involving experts from all the Member States and involving stakeholders from the water community[5].

Under the CIS, a series of working groups and joint activities have been launched to develop and test non-legally binding guidance. A strategic co-ordination group oversees these working groups and reports directly to the water directors of the European Union and Commission that play the role of an overall decision-making body for the CIS.

2   IMPLEMENTATION OF THE WFD: DEFINITION OF RIVER BASIN DISTRICTS AND FIRST REPORTS

2.1   *The Directive transposition into Spanish law*

With regard to the transposition of the Directive, note that European Directives are not regulations to be directly applied at the national level[6]. On the contrary, after their publication in the Official Journal of the European Communities, they must be approved by the respective governments and/or national

---

[5] All guidance documents are accessible from the Water Framework Directive website at: http://forum.europa.eu.int/Public/irc/env/wfd/library
[6] In the EU jargon 'transposition' refers to Member States passing national legislation ensuring the approval of equivalent legal precepts to those established by the 'transposed' EU legislation.

parliaments. The final law at national level is usually a more or less accurate version of the original text and not a mere translation. As the executive body of the European Union, the European Commission oversees the process and ensures that the adaptation to the national legislation is appropriate.

Spain transposed the WFD into national legislation on December, 30, 2003 through Act 62/2003 (referred to here and in other chapters as the 2003 Consolidated-WFD Water Act). The new regulation was accepted by the National Water Council and includes the main contents of the original directive approved by the European Parliament. The philosophy of the transposition was to develop only the key principles, leaving the adaptation of the rest of the WFD for a later regulatory process. In fact, the task of modifying the very large body of Spanish legislation regarding water has only just begun.

2.2   *The issue of defining the river basin districts and their competent authorities*

Spain has a long standing tradition of managing water resources on the river basin scale through the *Confederaciones Hidrograficas* (see Chapter 12). These river basin authorities (RBAs) are known as inter-community authorities if the water runs through more than one Autonomous Community[7] and intra-community authorities if the entire basin territory lies in only one Autonomous Community. In general terms, inter-community RBAs report to Spain's Central Government through the Environment Ministry, whereas intra-community RBAs report to their respective regional administrations.

Although the river basin districts (RBDs), as foreseen in the Directive, do not necessarily have to be defined with the same geographical limits as the basins currently managed by the RBAs, there will be a lot of overlapping. And yet the incorporation of the new elements established in the WFD, like coastal and transitional waters, is certainly one of the major differences. The transposition of the WFD into Spanish legislation placed the Government under the obligation of "entering into a consultation process with the Autonomous Communities" (see Chapter 12 for details on the political setup) before defining the territorial limits of the RBDs. The need for negotiations between both central and regional administrations has become even more crucial after some judicial decisions concerning the present distribution of competences.

At the RBD level, the Spanish transposition of the WFD has introduced the so-called Committee of Competent Authorities designed to ease the cooperation between the administrations with different responsibilities regarding water. Even so, this body respects the actual distribution of competences and takes into account a principle of shared responsibility.

After the transposition of the Directive, another relevant body within the river basin authorities is the River Basin Water Council. According to the new regulations, the Water Council's main aim is to define and submit the river basin water plan to the Spanish Government, promote public participation in the planning process and deal with issues of general interest, including the definition and implementation of measures for water bodies protection and the exploitation of water resources. Members of the Water Council include representatives of related ministries, regional governments, municipalities, water users (no less than one third of total members) and NGOs.

Another difficulty for Spain is to delineate the basin's boundaries to include the transitional and coastal waters. The Spanish Constitution grants the central and the regional governments competencies on the management of coastal areas. However, all regional governments with coastal territory have developed their own programs and passed regional legislation, which will have to be included in the basin's programs of measures. Until the WFD was passed, coastal waters were not managed by the river basin authorities.

2.3   *Article 5 reports*

As described in the last section, one of the first steps in implementing the WFD is to assess the current water status in the river basin districts. Article 5 of the WFD stipulates the need to produce

---

[7]The Spanish Constitution defines the Autonomous Communities as the regional governments.

three main reports: a general description of the river basin district that should include the establishment of reference conditions for surface waters, an identification of significant pressures and assessment of their impact and an economic analysis of water use.

During 2004, the Spanish RBAs began drafting these three reports with the aim of submitting them to the European Commission by the end of March 2005. The first step in the general description of the river basin district was to identify and delineate surface and groundwater bodies. For surface water bodies, a common methodology was used for all the river basin districts. This methodology was based on the application of hydrological and geometrical thresholds that were defined on a national scale after testing different criteria through the application of geographical information system (GIS) modeling tools. The final proposal included a total number of about 3,500 surface water bodies for Spain.

This proposal was refined by the existing RBAs taking into account additional information like the definition of water management units, the results of the pressure and impacts analysis, etc. Groundwater bodies were preliminary delimited using the established hydrogeological units, defined as groups of aquifers designed for efficient water management. Nevertheless, the definition of hydrogeological units is not fully adequate for developing the additional analyses required by the WFD. For that reason, groundwater bodies need to be redefined the Water Directorate.

As described in the previous section, water bodies must be grouped to define the reference conditions in accordance to various types. To define them, RBAs in Spain opted for what the WFD terms "System B", which is based on several obligatory and optional factors. Optional factors were identified through statistical analysis, and 21 different types of surface waters were finally defined in Spain.

The significant pressures were identified and their impacts assessed following a methodology defined by the Water Directorate and based on one of the guidance documents drafted in the CIS process (see previous section).

In general, mathematical modeling was used to assess both the impacts produced by point and diffuse pollution sources and the runoff depletion along water courses due to water abstractions. Monitoring data was used to classify the type of impact as "sure" or "probable". Generally speaking, a "sure" impact takes place when an effect resulting from a pressure is in breach of the legislation in force. A "probable" impact occurs when quality standards and environmental objectives, defined by the WFD or by future environmental legislation, are not being met. Most of the river basin districts have formulated biological and chemical indexes to quantify the impacts on water bodies and, ultimately, to evaluate the risk of failing to meet the environmental objectives.

The economic analysis was carried out in Spain following criteria defined by the Ministry of the Environment, and Marine and Rural Affairs (MERMA) These criteria were based on one of the guidance documents drafted in the CIS process (in this case, the WATECO guidelines). The analysis had two main parts: an economic characterization of water uses and an assessment of the level of cost recovery. The economic characterization of the water uses was mainly made to provide inputs for the analysis of cost-effectiveness of measures to be carried out in the definition of the programs of measures. In general, this analysis took into account four main sectors—agriculture, industry, energy and public supply—, for which the evolution of associated variables over the last years, such as gross added value or employment creation, were studied. Cost recovery studies have shown high rates of recovery for urban, industrial and agricultural uses (above 90%, see MIMAM, 2007).

3   CHALLENGES FOR THE FUTURE: IMPROVEMENT OF THE MONITORING NETWORKS, ADAPTATION OF THE PLANNING PROCESS AND THE IMPORTANCE OF PUBLIC PARTICIPATION

3.1   *Data availability*

Monitoring networks must be adapted with the ultimate goal of confirming the identification of water bodies at risk of not fulfilling the environmental objectives. This task will drive an important effort at modernizing and improving the water monitoring networks currently in operation.

In Spain, there are different types of monitoring networks. Networks controlling water quantity issues, like the Official Network of Gauging Stations or the so-called Automatic Hydrological Information Systems, are usually operated by the river basin authorities. The official gauging network provides information on water levels and discharges at selected sites (about 800) on rivers, channels and reservoirs. The water information systems are real-time monitoring networks designed for floods forecasting and prevention. Neither of them provides data accurate enough on water abstractions, especially for agricultural uses, which usually have to be evaluated through indirect methodologies (assessment of irrigation areas, surveys, etc).

With regard to water quality, the most important network is the ICA (Integrated Water Quality Network) that includes both conventional stations for systematic, periodic sampling and automatic warning stations (part of the automated water quality information system project (SAICA)), which output continuous real-time information. As regards groundwater, the IGME (Spanish Technical Institute of Geology and Mining) has deployed a groundwater quality network with about 2,000 sites.

Generally speaking, the information on water quality is difficult to manage at present due to the heterogeneity of sources and the diversity of standards according to which it is gathered and stored. Sampling frequencies and the number of registered parameters are insufficient in view of the list of priority substances defined by the WFD and the specific needs of information for its implementation. For instance, the quality networks cover only some river reaches or reservoirs where there are declared uses, usually linked to administrative permissions, and do not provide information about (theoretically) non-disturbed areas. The definition of reference conditions is complicated by missing systematic data on biological parameters. Indexes generally used in Spain in this field, like the IBMWP (Iberian Macroinvertebrate Index), were developed at a time when the main objective was to evaluate a water quality standard to be applied to all water bodies, no matter, for instance, what type they belong to. Hence, these indexes have to be adapted to the new concept of ecological status introduced by the Directive.

3.2   *The need to adapt the existing water planning scope*

Another issue that represents a real challenge for the sound implementation of the WFD in Spain is the adaptation of ongoing water planning not only to the new regulations, but also to new concepts. Spain has its own water planning history, which means that water users have their own ways of managing it, including the definition of roles and the allocation of tasks, for instance, between the public and private actors.

The importance of water planning was firstly acknowledged in Spanish legislation in the 1985 Water Act (see Chapters 10 and 11). According to this Act, basin plans are the central instruments for regulating water, to which "all action on the public water domain is subject".

Since 1985, the basin plans have been carried out in two stages: preparation of guidelines and collection of basic data, and drafting of the plans. The stage of laying down guidelines began by preparing the basic documentation of each plan, which had to contain an identification of key issues and the basic information regarding water resources in the basin. The guidelines were then forwarded to ministry departments and regional governments. At the same time, a public consultation process was started. This process took a very long time, and the guidelines for the different basins were not approved until 1992 and 1993. Once the guidelines were approved, the river basin authorities drafted the plans, and the final versions were forwarded to the National Water Council in October 1997. In the end, the plans were approved by Royal Decree 1664/1998 of July, 24, 1998.

Other chapters of this book deal with the making of the National Hydrological Plan and its evolution over the last few years (see Chapter 19). Even so, it is worthwhile recalling here that the National Hydrological Plan is the last planning level in Spain and introduces some elements of coordination between the different basin plans. They include the controversial issue of the conditions of water resource transfers between the territorial areas of different basin plans, and the not-so-well-known identification of shared aquifers or the establishment of common systems to characterize droughts or water stress situations.

This water planning framework in Spain has to be adapted to the requirements of the WFD. A milestone in the river basin planning process (analysis, monitoring, objective setting, and measures for maintaining or improving water status) will be the elaboration of the river basin management plan, which will summarize the relevant planning information for the river basin district.

According to the Directive, the river basin management plans shall in fact include a summary of the results of the analyses; the characteristics of the river basin; a review of the impact of human activity on the status of waters in the basin; estimation of the effect of existing measures and the remaining "gap" to meeting those objectives; and anything else that is required.

In sum, the WFD views a river basin management plan as a strategic planning document and an operational guide to implementing programs of measures that will form the basis for integrated, technically, environmentally and economically sound and sustainable water management within a river basin district.

This concept is very similar to the one applied by the 1985 Water Act to the basin plans. Even so, there are marked differences in the philosophy of the measures to be taken and even in the decision-making process. Regarding the measures to be taken, the existing basin plans mainly target a balanced situation between water resources and demands. Unfortunately, however, the existing plans very often do not deal properly with the ecological requirements as they are defined by the WFD, and, in the end, the current basin plans will have to be redrafted.

3.3 *New planning criteria*

A good example of the need to amend old water planning concepts is shown in new pieces of legislation, like NHP Act 2005, modifying the National Hydrological Plan (hereafter referred by NHP Act 2005).

Not only the legal need to meet the WFD requirements but also a generally accepted feeling of the need to stress measures focused on water demand management moved the new Government to repeal, through NHP Act 2005, some articles of the original National Hydrological Plan, especially those regarding the Ebro-Mediterranean inter-basin water transfer. The cancellation of the Ebro water transfer was not the only aim of NHP Act 2005. On the contrary, it introduced important amendments to the Water Act. They include the need to analyze the feasibility of new water infrastructures and the introduction of new tools for coordination between central and regional administrations concerning regional planning.

The feasibility analysis of the new water infrastructures aims to summarize the technical and environmental aspects taken into account in the project and to clarify the level of cost recovery to be obtained. After approval, the feasibility reports are posted on the Environment Ministry's web site.

Ensuring coordination between regional and water resources planning has become a real challenge in Spain. In the last ten years, there has been a definite population movement from inland to coastal areas in search of better economic prospects. Over the five years leading to the 2008 economic crisis, especially, there has been an extraordinary boom of second residence building, targeting both Spanish and foreign owners. The problem is that urban and land use planning is a regional and municipal competence in Spain, and there are no clear mechanisms for coordination with the water policy developed by the Environment Ministry and its subordinate river basin authorities. In the past, most new urban developments were designed under the understanding that there were sufficient water resources and with the mistaken belief that, if that were not the case, the water authority was under obligation to supply such resources. Under the new NHP Act 2005, this is no longer the case, and water authorities are legally bound to report water resources availability prior to the approval of the new urban developments.

3.4 *The public participation issue*

Besides the different concepts and aims behind the existing basin plans in Spain and the basin management plans defined by the Directive, there are differences in the decision-making process as well, especially in the way they deal with public participation (see Chapter 17 for more details).

Spain's relative inexperience in the design of water plans led to fundamentally technical rather than legal and strategic documents. The consultation process for the approval of the basin plans was very formal and focused more on the characterization of the basin than on the discussion of the different alternatives to the proposed program of measures.

This is not the WFD philosophy. On the contrary, the WFD promotes proactive public participation, especially in the decision-making process (see Chapter 17). In fact, the WFD establishes that Member States shall encourage active involvement and shall ensure consultation. In the first case, Member States have to make a clear effort to promote and facilitate active involvement. In the second case, consultation is an obligation that has to be performed.

Of course, the starting point for embarking on a participatory approach is a commitment at the political level. This commitment has to be based on an understanding and awareness of the new obligations of the WFD and why active involvement is not only beneficial but also crucial in order to implement certain measures.

Nevertheless, public participation will always be a challenge. Some issues can generate conflicts in water resources planning that are not necessarily the result of wrong or illicit approaches. As different people have different goals, perspectives, and values, water resources planning should take into account multiple users, multiple purposes, and multiple objectives. This is especially true in Spain, where water resources are limited and water conflicts between users are an everyday event.

In the implementation of the WFD, planning and public participation will help practitioners to approach complex problems, to organize thinking, and to form the understanding necessary to strike the right balance, bearing in mind that there is no one definition of "public interest" and, if there is, it changes from time to time.

4 SPAIN'S PREPAREDNESS FOR THE 2009–2012 WFD AGENDA

Let's look at what Spain had achieved by 2007 with respect with other Member States. WFD Article 18(3) mandated the Commission to publish a progress report in 2007 making recommendations for the next important milestone: the river basin management plans. The European Commission issued a communication in 2007 summarizing the progress made so far by all Member States (EC, 2007). The report covers four quantitative measures at the Member State level.

The first measure was to answer the question, 'What is the risk of failing to meet WFD environmental objectives based on current data?' (Art. 5 Annex II). Spain's report established that 12% of its surface water bodies were 'at risk' of not meeting the objectives, whereas 23% were not 'at risk' but declared for the remaining 65% that data were 'insufficient'. By way of a comparison, Italy did not submit any data at all, The Netherlands declared 98% of its surface waters to be 'at risk', France declared 30% to be at risk, 25% not to be 'at risk', and the remaining to be undefined due to insufficient data, and Germany reported 55% to be 'at risk', 15% not to be 'at risk' and 30% to be undefined. The second quantitative measure was based on a 'performance indicator of administrative setup' (pursuant to Art. 3). In this regard, Spain ranked 13th among the EU-15, ahead of Greece and Italy. With respect to the third measure, based on Member States' 'reporting performance', Spain ranked 13th as well. The fourth quantitative analysis was a performance indicator regarding the implementation of the environmental and economic analyses (pursuant to Art. 5). Spain ranked sixth among the EU-15.

Spain is not an exception among EU Member States in the struggle to keep up with the timetable and implement the measures. Some of the most intractable problems are related to non-point pollution and the consideration of the changes in rivers' hydromorphology caused by hydropower, navigation and flood control waterworks.

Combating non-point pollution will require the implementation of best agricultural practices, aimed at reducing chemicals use. These measures are not only very costly, but have to be coordinated with the application of the Common Agricultural Policy (CAP). The European Commission negotiated with Member States ways to condition the eligibility for CAP subsidies to the reduction of pollution and to the satisfaction of all environmental and water regulations, including the WFD. So, agricultural and water policies are more coordinated, and incentives have been redefined to work in the right direction.

With respect to the hydromorphology changes, present EU work is aimed at developing agreed methodologies to determine the conditions under which the environmental objectives could be subordinated to the general economic interest. Determining when and how these exceptional measures, which relax the environmental targets, are admissible was one the key CIS activities during 2006.

As has been stressed in this section, Spain is facing both technical and administrative challenges in the implementation of the WFD. From the technical point of view, the monitoring networks are not tailored to the new WFD requirements, leading to problems of data availability, especially regarding biological parameters. In addition, there are a number of knowledge gaps, such as non-state-of-the-art methodologies for evaluating reference conditions and good status, unknown interactions between different water systems and ecosystems, no models for predicting the effects and the combined effects of pressures, and, finally, the unavailability of tools to assess the effect of proposed measures.

From the administrative perspective, there is a need to reach a common understanding of the roles that central and regional administrations must play as competent authorities. Special difficulties arose in the adaptation of the existing water planning practices to the new concepts of the Directive. This will drive changes not only in the contents of the current river basin management plans, but also in the way in which they are developed, especially to improve the public participation process.

REFERENCES

Comisión Europea (2002–2003). Guías de implantación de la Directiva Marco en Europa y otros documentos sobre el proceso común de implantación de la Directiva. [Guidelines for the Implementation of the Water Framework Directive and other Documents for the Common Implementation Process of the Directive] http://forum.europa.eu.int/Public/irc/env/wfd/library

Estrela, T. (Coord.) 2004. Júcar Pilot River Basin, Provisional Article 5 Report Pursuant to the WFD. Confederación Hidrográfica del Jucar. Ministerio de Medio Ambiente. Valencia, Spain.

European Commission (2007). Towards sustainable water management in the European Union- first stage in the implementation of the Water Framework Directive 2000/60/ec. Communication from the Commission to the European Parliament and the Council, Brussels, 22.3.2007. COM(2007) 128 final.

MIMAM (2003). Directiva 2000/60/CE. Análisis de transposición y procedimientos de desarrollo. [Directive 2000/60/CE. Analysis of the Transposition and Development Procedures] http://www.mma.es/rec_hid/dma/index.htm

MIMAM (2007). El Agua en la economía española: Situación y perspectivas. [Water in the Spanish Economy: present and future prospects]. Madrid, Spain: Ministerio de Medio Ambiente. http://www.mma.es/rec_hid/dma/index.htm

CHAPTER 17

Public participation in developing and improving water governance

Ana Barreira

International Institute of Environmental Law (Instituto Internacional de Derecho y Medio Ambiente), Madrid, Spain

1   INTRODUCTION

Democracy is the best system of government, but it is not perfect. It is important to acknowledge that ballot box pressures generally drive politicians, assuring their continuity through the use of short-term policies and strategies. However, in the field of sustainable development and moreover in environmental protection and conservation, temporary frameworks are a great obstacle. The majority of environmental decisions and actions have effects that are only visible in the medium or long term. Similarly, a decision in favour of protecting the environment or that has a negative impact on it does not usually produce immediately visible results. Consequently, I consider that representative democracy is not enough. A truly participatory democratic model is required to achieve sustainable development that does not just depend on politicians. But this implies a good government or governance. The concept of governance includes the relationship between society and its government.

The United Nations Development Programme has defined governance as the employment of economic, political and administrative authority to manage national affairs at all levels, including the mechanisms, processes and institutions through which citizens and groups express their interests, exercise their rights, fulfil their obligations and negotiate their differences. One of the key points of governance is institutional reform, which is necessary in order to return some functions to society and to democratise civil society. There are five essential principles to achieve good governance at all levels of government. These are applicable to environmental governance as well as water governance[1]: *openness, participation, accountability, efficacy and coherence.*

One of the important outcomes of United Nations Conference on Environment and Development (Rio Conference) was the recognition of the need to create mechanisms to improve access to information and public participation in environmental matters (Principle 10 of the Rio Declaration and Agenda 21) such as water policies. This recognition opened a wider debate on the need to reform institutions to make environmental decision making processes more inclusive which in essence goes to the basis of democracy and of the principles of good governance. This process has had its reflection in the European Union of which Spain is a Member State since 1986.

This chapter focuses on the first two principles of water governance: openness[2] and participation[3]. It analyses the legal and institutional framework of openness and participation in developing

---

[1] Water governance refers to a range of political, social, economic and institutional systems that are in place to develop and manage water resources and the delivery of water services at different levels of society (Rogers, P. & Hall, A., 2003).

[2] This principle demonstrates the need for institutions to work in a more open manner. In order to do so, they must develop an active communication on their work and use language accessible and understandable to the general public.

[3] The quality, relevance and effectiveness of policies imply the widespread participation of citizens, at each and every stage of the process, from the policy conception to its implementation. Earlier and more systematic dialogue may lengthen the time taken to prepare a policy proposal, but should improve the quality of regulation and accelerate its implementation.

and implementing water policies in Spain. Firstly, it reviews the international and EU context in particular the European Community (EurComm) environmental law (ie: the Water Framework Directive) which influences the Spanish framework. It follows an analysis of the mechanisms available to Spanish citizens to participate in the development and implementation of water policies. These mechanisms are contained in provisions of the 1978 Spanish Constitution, the 2001 Consolidated Water Act and other administrative provisions. In particular, it will analyse the mechanisms available to citizens to participate in water planning and to some citizens to participate in water management through the Spanish institutional setting of River Basin Authorities (RBA, in Spanish traditionally known as *Confederaciones Hidrográficas*). Finally, it includes a section on conclusions.

2   THE INTERNATIONAL AND EUROPEAN CONTEXTS

At the international level, in addition to the Rio Conference outcomes emphasising the need of transparency and public participation in environmental matters, we find declarations reaffirming this approach not only in the field of environment but also in the field of water policy[4]. Principle 10 of the Rio Declaration has been reflected in an international *hard law* (legally binding) instrument: the United Nations Economic Commission for Europe (UNECE) Convention on Access to Information, Public Participation and Access to Justice in Environmental Matters (the "Aarhus Convention"[5]). In the scope of UNECE, we also find two instruments dealing with the management of international river basins advocating for access to information: the 1992 Helsinki Convention on the Protection of Transboundary Watercourses and International Lakes; and for public participation: the 1999 Protocol on Water and Health to the Helsinki Convention.

In 1986, Spain acceded to the European Economic Community (EEC), today the EurComm. The EurComm is a unique international organization, "it has its own institutions[6], its own personality, its own legal capacity and capacity of representation on the international sphere and, more particularly, real powers stemming from a limitation of sovereignty or a transfer of powers from the States to the Community. The Member States have limited their sovereign rights, albeit within limited fields, and have thus created a body of law which binds both their nationals and themselves"[7]. The EurComm forms the first pillar of what is known as European Union (EU) which it is comprised of another two pillars: the Common Foreign and Security Policy and Cooperation in Justice and Home Affairs.

In 1999, the President of the European Commission stressed the need to carry out a deep reform of EU decision-making processes and the functioning of its institutions. To promote the initiative, the European Commission (2001) adopted the document "European Governance: A White Paper"[8]. In addition, a process to reform the constitutional treaties was opened in 2002, in October 2004 a new Constitution for the EU was signed but never entered into force due to the negative outcome in the French and Dutch referenda. Nevertheless, the reform is still being tried. On December 2007, the Treaty of Lisbon substituting the failed constitution was signed. This Treaty has not entered into force yet since not all Member States have ratified it[9]. The text of the

---

[4] Dublin Statement on Water and Sustainable Development, Chapter 18 of Agenda 21, Ministerial Declaration of Bonn International Conference on Freshwater, the Johannesburg Plan of Implementation and the 3rd World Water Forum Ministerial Declaration.

[5] www.unece.org

[6] The European Commission is a kind of executive power, the Council and the Parliament holding the legislative function and the European Court of Justice as the judiciary.

[7] Costa v. Ente Nationale Per L'Energia Elettrica (ENEL) Case 6/64, [1964] ECR 585.

[8] COM (2001) 428 final of 25.7.2001.

[9] On June 2008, the Irish voted against it. This outcome produced a new institutional crisis at the EU. However, the EU is still committed to this text. Ireland will launch a new consultation.

Lisbon Treaty includes not only the principle of representative democracy[10], but also the principle of participatory democracy[11]. Article 6 of this Treaty provides for the recognition of the rights, freedoms and principles set out in the Charter of Fundamental Rights of the European Union of 7 December 2000, which shall have the same legal value as the Treaties. The Charter includes the right to a good administration[12] and the right of access to documents[13]. It is not clear when the Treaty of Lisbon will be ratified by the remaining Member States[14] to enter into force. The Treaty of the EurComm in force already provides for the right of access to documents of same of the European institutions: Council, Parliament and Commission[15].

Environmental policy, a EurComm policy, which includes water policy it is a shared competence between the EU and Member States. This shared competence implies that the EU establishes minimum standards for environmental protection when it is necessary to intervene at the European level[16] while leaving Member States the freedom to establish stricter protection measures[17]. Member States are obliged to transpose and implement the minimum standards approved at the European level mainly in the form of Directives[18].

In the field of environmental policy there are two paramount instruments at the European Union requiring Member States to act in an openness and participatory manner. Transparency has been promoted through Directive 90/313/EEC on the freedom of access to information on environmental matters[19], which was revoked on the 14th of February 2005, the same date on which the new Directive 2003/4/EC relating to public access to environmental information[20] should have been transposed. This new Directive responds to the actions initiated at the EU level to transpose the Aarhus Convention to Community law. Specifically for developing and implementing water policy, transparency in addition to public participation is promoted by Directive 2000/60/EC establishing a framework for the Community action in the field of water policy[21], known as the Water Framework Directive (see Chapter 16 for a thorough analysis of this Directive).

2.1   *Access to environmental information*

This Directive obliges public administrations of Member States not only to provide information upon request within one month after its receipt but also to take initiatives to disseminate information at their disposal actively via electronic media, meaning that administrations should make efforts to communicate and to be transparent[22]. The Directive provides a wide definition of what

---

[10] Article 8A, 1.

[11] *Every citizen shall have the right to participate in the democratic life of the Union. Decisions shall be taken as openly and as closely as possible to the citizen* (Article 8A, 2).

[12] Art. 41.

[13] Art. 42.

[14] These are the Czech Republic, Germany, Ireland and Poland. In Germany and in Poland although the Treaty has been ratified by the corresponding institutions, their Presidents have to sign the ratification instrument.

[15] Art. 255.

[16] As required by the principle of subsidiarity.

[17] Article 176 of the Treaty establishing the European Community (TEC 2004).

[18] A directive shall be binding, as to the result to be achieved, upon each Member State to which it is addressed, but shall leave to the national authorities the choice of form and methods (Article 249 of the TEC).

[19] OJ L 158 of 23.6.1990, p. 56.

[20] OJ L 41 of 14.2.2003, p. 26.

[21] OJ L 327 of 22.12.2000, p. 1.

[22] The information to disseminate includes: a) texts of international treaties, conventions or agreements, and of Community, national, regional or local legislation, on the environment or relating to it; b) policies, plans and programmes relating to the environment; c) progress reports on the implementation of the items referred to in (a) and (b) when prepared or held in electronic form by public authorities; d) reports on the state of the environment e) data or summaries of data derived from the monitoring of activities affecting, or likely to affect, the environment; f) authorisations with a significant impact on the environment and environmental agreements or a reference to the place; g) environmental impact studies and risk assessments concerning the environmental elements or a reference to the place where the information can be requested or found.

it is understood as environmental information including any information in written, visual, aural, electronic or any other material form about: the state of the elements of the environment such as water; factors such as discharges; measures (including administrative measures), such as policies, legislation, plans, programmes, environmental agreements, and activities affecting or likely to affect the elements and factors; reports on the implementation of environmental legislation; cost-benefit and other economic analyses and assumptions used within the framework of the measures and activities and the state of human health and safety[23].

This new Directive widens the concept of the public authority obliged to provide information, being extended to include entities offering public services related to the environment as it could be a water supply company. The Member States must ensure that:

a. officials are required to support the public in seeking access to information;
b. lists of public authorities are publicly accessible; and
c. the practical arrangements are defined for ensuring that the right of access to environmental information can be effectively exercised, such as:
    – the designation of information officers;
    – the establishment and maintenance of facilities for the examination of the information required,
    – registers or lists of the environmental information held by public authorities or information points, with clear indications of where such information can be found.

The Directive also obliges the Member States to provide for an administrative "appeal", (optional in the Aarhus Convention) which is a procedure that has the advantage of being rapid and free of charge. Spain transposed this Directive into Spanish Law through Law 27/2006, of 18th July, regulating the rights on access to information, public participation and access to justice on environmental matters[24]. The transposition of the Access to Environmental Information Directive into Spanish Law is quite satisfactory. However, its implementation is representing a challenge to the public authorities not used to act transparently and share information. As we will analyze below, its implementation in the field of water policy is even more difficult due to persistent traditions and vested interests.

2.2   *Access to information and public participation in the Water Framework Directive*

At the European Union level, the most important instrument on water management is the Water Framework Directive (WFD) which entered into force on December 22, 2000. This Directive introduces key elements to achieve an effective water governance at the EU level, a coherent and effective legal and institutional framework, water-pricing policies, public participation and an integrated water resources management (IWRM) system (Barreira, 2003).

In reaching its main objective, to achieve a good status of community waters by 2015, this Directive emphasises the importance of guaranteeing information, consultation and public involvement. The main provision on public information and consultation establishes specific requirements for public participation, not only in the development of river basin management plans to be designed to achieve the main objective but also in the whole implementation of the Directive. This implies that citizens should have access to information on, and to mechanisms to participate in, the whole implementation cycle (characterisation of river basin districts, water bodies status classification, elaboration of river basin management plans and of programme of measures).

---

[23] Article 2.
[24] BOE núm. 171 of 19.07.2006.

Article 14- WFD

Public information and consultation

1. Member States shall encourage the active involvement of all interested parties in the implementation of this Directive, in particular in the production, review and updating of the river basin management plans. Member States shall ensure that, for each river basin district, they publish and make available for comments to the public, including users:

   a. a timetable and work programme for the production of the plan, including a statement of the consultation measures to be taken, at least three years before the beginning of the period to which the plan refers;
   b. an interim overview of the significant water management issues identified in the river basin, at least two years before the beginning of the period to which the plan refers;
   c. draft copies of the river basin management plan, at least one year before the beginning of the period to which the plan refers.

   On request, access shall be given to background documents and information used for the development of the draft river basin management plan.
2. Member States shall allow at least six months to comment in writing on those documents in order to allow active involvement and consultation.
3. Paragraphs 1 and 2 shall apply equally to updated river basin management plans.

The first sentence of Article 14(1) encourages active involvement of all interested parties in the whole implementation process of the Directive. "Interested party" can be interpreted as meaning any person, group or organisation with an interest or stake in an issue either because they will be affected or may have some influence on its outcome. The success of this involvement will not be met solely via the three-phased information and consultation procedure pursuant to the second sentence of Article 14(1). With respect to consultation the term 'public' is used. As defined in other EC Directive[25], public means one or more natural or legal persons, and, in accordance with national legislation or practice, their associations, organisations and groups. To ensure transparency and acceptance, public participation has to start as soon as possible.

The participatory aspect has been extensively covered in the WFD Guidance on Public Participation[26] approved within the Common Implementation Strategy (CIS) of the WFD (see Chapter 16 for a more detailed description of the CIS and the WFD). This guidance is horizontal since it is of concern to most activities under the CIS. It provides a common understanding regarding the meaning of public participation in the context of the WFD as a means of improving decision-making, to create awareness of environmental issues and to help increase acceptance and commitment towards intended plans. Public participation for the implementation of the Directive is recommended at any stage in the planning process. This document provides guidance on the forms of public participation covered by the WFD: active involvement, consultation and access to background information.

1. *Active involvement* in all aspects of the implementation of the Directive. While this has particular focus on the production, review and updating of the River Basin Management Plans, the encouragement of active involvement of stakeholders in the wider implementation of the Directive also needs to be considered. Although active involvement has not been defined in the Directive, it implies that stakeholders are invited to contribute actively to the process and thus play a role in advising the competent authorities. Member States must make a clear effort to promote and facilitate active involvement (Barreira, 2004). More intense levels of public

---

[25] Directive on Strategic Environmental Impact Assessment (2001/42/EC).
[26] http://forum.europa.eu.int/Public/irc/wfd/library

participation include: shared decision making to self determination are not required by the WFD but are encouraged;

2. *Consultation.* It aims at learning from comments, perceptions, experiences and ideas of stakeholders. It is a less intensive form of public participation and is only possible after completion of draft plans and other documents. According to Article 14 consultation concerns the following requirements and timetable for consultation:

December 2006 (the latest)	Timetable and work programme for the production of the river basin management plans, including a statement of the consultation measures to be taken
July 2007	Comments in writing
December 2007 (the latest)	Interim overview of the significant water management issues identified in the river basin
July 2008	Comments in writing
December 2008 (the latest)	Draft copies of the river basin management plan available
July 2009	Comments in writing
December 2009 (the latest)	Start implementation of the plan

It is important to emphasise that during the preparation of the river basin management plans, according to Directive 2001/42/EC on the assessment of the effects of certain plans and programmes on the environment[27] it is also necessary to open a public participation process as part of the environmental assessment that must be carried out before a water plan is adopted or before being submitted to the legislative process;

3. *Access to background information.* This itself is a *condition sine qua non* for public participation not a form of public participation and it is an obligatory requirement of the Directive on access to information.

This Directive has already been transposed to Spanish Law as we will see in the following section.

3   THE LEGAL AND INSTITUTIONAL REGIME TO PARTICIPATE IN THE DEVELOPMENT AND IMPLEMENTATION OF WATER POLICIES IN SPAIN

The legal basis for public participation in public affairs is found in the 1978 Spanish Constitution (Articles 9 and 23). In addition, Article 45 provides that citizens have a right to enjoy an adequate environment for their development and a duty to preserve it. To protect this right and to comply with this duty, citizens must have the possibility to contribute to and participate in environmental policymaking and its assessment.

It is important to have in mind that until recently Spanish society has not been conscious of the significance of participating in decision-making process. In addition, the Spanish Administration has not been very opened and has been characterized by a patronizing attitude.

We can identify three phases in the development of participatory models in water policy and management in Spain. The first one, previous to the 20th century, a participatory culture in the management of water at sub-basins and aquifers was well established (Sancho Marco & Parrado Diez, 2004). The second phase started with the establishment in 1926 of the first RBA in the Ebro river. The milestone of this phase was the approval of the 1985 Water Act which introduced a participatory model for the development of water policies and for water management based on the experience of RBA, whose main pillars were water management based on river basin unity and a participatory water administration. However, this participatory model was only opened

---

[27] This Directive was transposed into Spanish Law through Law 9/2006 of 28 April, on the assessment of the effects of certain plans and programmes on the environment.

to water users meaning those holding a water use right or rights to occupy the hydraulic public domain (see Chapter 11 for a description of water rights). The third phase has been launched with the approval of a 2007 Regulation on Hydrological Planning which complies with the requirements of public participation (consultation) provided in Article 14 of the WFD as we will see below.

3.1   *The broad regime of participation in water planning*

The participatory regime provided in the 1985 Water Act has been subsequently developed by different legal instruments. The most relevant are the 2003 Consolidated-WFD Water Act as well as 1988 Regulation on the Water Public Administration and Hydrological Planning and the 2007 Regulation on Hydrological Planning.

In Spain, until July 2007 participation in water planning only took place at the RBAs in charge of water planning and management and at the National Water Council (*Consejo Nacional del Agua*) which is an advisory body. Nevertheless, with the approval of the 2007 Regulation this changed. Participation in water management remains under the scope of RBAs. We must recall that for rivers flowing through more than one than one Autonomous Community (intercommunity river basins), the RBAs are the Confederaciones Hidrográficas (under the State jurisdiction) and that for rivers flowing only through the territory of one Autonomous Community (intracommunity river basins) the RBAs are established by each of those regions and are under their jurisdiction (see Chapters 10 and 11).

The 2007 Regulation on Hydrological Planning transposed the requirements of Article 14 of the WFD into Spanish Law.

According to the 2007 Regulation, RBAs must organize procedures to make public participation effective during the process to adopt river basin management plans[28]. This includes:

a. public information, consultation and active participation procedures organization and timetables
b. coordination of the procedure for the strategic environmental assessment of the river basin plan
c. description of the participation methods and techniques to be employed.

The participation provided in this Regulation goes from access to information, public consultation to active participation. Provisions of the 2006 Law on Access to Information applies this process including **accessing to information** on the preparation of the river basin management plans, especially in what refers to active dissemination by the Administration.

Public consultation take place in three phases where public has six months in each to provide their comments, as required by the WFD. However, the interested parties[29] only counts with

---

[28] Art. 72.
[29] It is important to emphasise that interested party is a broad concept and includes NGOs according to Law 27/2006. Art. 2.2. *Interested persons*:

a. Any natural or legal person complying with any of the requirements provided in Article 31 of Law 30/1992, of 26 November, on the Legal Regime of the Public Administrations and of the Common Administrative Procedure.
b. Any other non-profit legal person satisfying the requirements established by Article 23 of this Law.
   Art. 23 Legal Standing.
   1. Any legal non-profit persons have standing to sue, as regulated in Article 22 if they meet the following requirements:
      a. Include among the goals established in their by-laws the protection of the environment or of any of its elements.
      b. Have been registered at least two years before suing and have executed all necessary activities to reach the goals of their by-laws.
      c. Develop according to their by-laws their activities within a geographical scope affected by the action or administrative omission challenged in the courts.

three months to send their comments on the documents during the second and third phase[30] what is contrary to the WFD. The documents subject to this consultation are:

- first phase (three years before opening the procedure to approve or review the corresponding RBMP): a work programme which includes a timetable, a general study on the river basin district[31];
- second phase (two years before the opening of the above referred procedure): an interim overview of the significant water management issues identified in the river basin;
- third phase (one year before the opening of the above referred procedure) the draft copies of the RBMP including the sustainability environmental report developed as part of the SEA procedure.

The deadlines provided by the WFD have not being complied with, taking into consideration that this Regulation was passed after finalising the first phase provided in that Directive.

In addition to consultation, the 2007 Regulation provides that the RBAs must foster **active participation** which may include the convening of fora and working groups opened to interested parties and recognized professionals.

In the case of RBMP of intercommunity basins, the plan is finally approved by the Government. This Regulation also provides for public consultation during the preparation of the National Hydrological Plan, including the prominent participation of the National Water Council. The approval of this Plan corresponds to the Spanish Legislative Chambers (Congress and Senate Houses).

As a result of the entry into force of this Regulation, the prominent role that the River Basin District Council, one of the bodies of the RBAs, used to have in the preparation of the RBMP has been dissipated, and only is in charge of issuing a report containing its views on the draft plan.

3.2   *The limited regime of participation in water management*

One of the principles applicable to water management is the principle of users' participation[32]. Water users are considered as those users having registered rights (concessions) at the Water Registry. These rights holders are mainly water supply companies, industries and farmers. Therefore, according to Spanish law water users are those only representing the economic use of water.

As detailed in Chapter 12, RBAs are comprised of three kinds of governing or administration bodies: the governing, the management and the planning bodies.

In the *Governing Board* and the management bodies[33] the participatory regime is limited to traditional water users and dominated by representatives of the national administration. Thus, participation in these bodies is restricted to those users having an economic interest.

These bodies perform functions related to aspects of the WFD implementation cycle having social and environmental implications. They take decisions such as building or not a dam which impact on the status of water bodies and can difficult the achievement of the good water status in 2015 as required by the WFD. The ecological services provided by our rivers cannot be left exclusively in the hands of the Administration, which in many occasions do not respect the environmental conditions when carrying out a development project neither of those holding an strict economic interest. According to this Directive, this kind of decisions should be taken in an open manner and not exclusively with the participation of water users, but with the "active involvement" of interested parties. In addition, as seen in other Chapters of this book, the requirements of the WFD go further the traditional way to manage water resources in Spain. It is the public who in many cases is aware of and have valuable knowledge on the environmental and social implications of water management. Other uses as environmental, recreational and social are not represented by members of the public.

---

[30] Art. 79.4 and 80.2.

[31] This study is a collection of the work required by Article 5 of the WFD: Characteristics of the RBD, review of environmental impacts of human activity and economic analysis of water use.

[32] Article 14 Consolidated Water Act.

[33] User's assembly, exploitation boards, dams commission and public works boards.

As seen before, the RBD Council as the planning body has lost its prominent role during the preparation of the RBMPs.

In light of the WFD provisions, it is our view that the principle of users participation in water management limits the participation of ALL interested parties, including users in the planning process as conceived by the CIS.

Without doubt Spain has introduced important reforms to make participation of the public possible during the preparation of RBMPs. However, these reforms missed the opportunity to include reforms in water management and as a result in the minds of those privileged users. These reforms are more complicated, since they go to the core of cultural traditions and vested interests.

As mentioned, there is not a well established practice of public participation in Spain. It will be interesting to analyze the participation processes after the consultations are concluded. While our legal order offers some interesting mechanisms for citizens such as the right of petition or the public legislative initiative, they are hardly used. In addition, the scarce economic resources available to reimburse the expenses originated from participating is another important hindrance to participation in Spain.

4 CONCLUSIONS

Spain was a pioneer in creating river basin authorities and in introducing participation in water planning and management. As a result of the evolution of environmental policy in the EU, Spain had to introduce normative reforms to open participation in the planning process to the public. Nevertheless, the reforms introduced do not respect the terms offered to interested parties to analyse in-depth the different documents prepared as part of the preparation of the RBMPs. But in spite of this failure, the modifications are satisfactory for various reasons. First, they will allow for positive outcomes of involving citizens in the protection of natural resources. Secondly involving citizens may contribute to the accountability, efficacy and coherence of institutions.

The existing participatory mechanisms are in line with some of the principles of good governance as openness and participation. Nevertheless, in the case of water management, they are only opened to some members of the public: those who maintain a special treatment as holders of a water right. In order to achieve a truly opened participatory water governance regime it is advisable to introduce some legal reforms in Spain to open participation of interested parties in the governing and management bodies of RBA. In addition, the challenge is necessary to build capacities in the Administration sector to open its functioning.

But the efforts must not come only from the Administration but also from the privileged actors and from the public in general. It is going to be a difficult path but not an impossible mission. I would like to conclude this article by quoting the EU Sustainable Development Strategy approved in 2001:

> *Many of the changes needed to secure sustainable development can only successfully be undertaken at community level ... In other cases, the action by national, regional and local governments will be more appropriate. However, while public authorities have a key role in providing a clear long-term framework, it is ultimately individual citizens and business who will deliver the changes in consumption and investment patterns needed to achieve sustainable development.*

REFERENCES

Barreira, A. (2003). The Participatory Regime of Water Governance in the Iberian Peninsula. *Water International* 28(3): 350–357.

Barreira, A. (2004). *Dams in Europe: the Water Framework Directive and the World Commission on Dams Recommendation- A Legal and Policy Analysis*. Report prepared for the WWF Dams Initiative. Available at: http://assets.panda.org/downloads/wfddamsineurope.pdf

Embid Irujo, A. (2008). *Ciudadanos y Usuarios en la Gestión del Agua* [Citizens and Users in Water Management]. Ed. Dykinson (Pages 1–520).

European Commission (2001). *White Paper on EU Governance*. Brussels, Belgium: European Commission.

Rogers, P. & Hall, A. (2003). *Dialogue on Effective Water Governance*. Global Water Partnership Technical Committee. TEC Background Papers no. 7, Sweeden. Available at http://www.gwpforum.org/gwp/library/TEC%207.pdf

Sancho, M. & Parrado Diez, L. (2004). Los organismos de cuenca: puntos fuertes y reflexiones para su mejora [Water Basin Authorities: weaknesses and strengths for their improvement]. In Actas del II Congreso Internacional de Ingeniería Civil, Territorio y Medio Ambiente, Vol. II. [Proceedings of the 2nd Internacional Congreso of Civil Engineering, Land and Environment, Vol. II]. Colegio de Ingenieros de Caminos, Canales y Puertos, Madrid: Spain.

Treaty of Lisbon amending the Treaty on European Union and the Treaty establishing the European Community, signed at Lisbon, 13 December 2007 *Official Journal C 306 of 17 December 2007* (Pages 1–231).

Treaty establishing the European Community (consolidated text) *Official Journal C 325 of 24 December 2002* (Pages 1–152).

CHAPTER 18

The Spanish and Portuguese cooperation over their transboundary basins[1]

Alberto Garrido
Department of Agricultural Economics and Social Sciences, Universidad Politécnica de Madrid, Spain

Ana Barreira
International Institute of Environmental Law (Instituto Internacional de Derecho y Medio Ambiente), Madrid, Spain

Shlomi Dinar
Department of Politics and International Relations, Florida International University, USA

Esperanza Luque
Research Centre for the Management of Agricultural and Environmental Risks, Universidad Politécnica de Madrid, Spain

1 INTRODUCTION

Spain and Portugal share four important Iberian rivers, namely from North to South: the Minho-Limia, Douro, Tagus and Guadiana. They all discharge into the Atlantic Ocean. The catchment areas of these basins total 44.8% of the Iberian Peninsula and represent about 46% of the annual surface discharge. These shared rivers account for 740 km of the 1214 km border between the two countries. Very few countries in the world have four shared rivers with four different configurations (see Figure 1), occupying so much of their territory.

The history of Spanish-Portuguese cooperation is landmarked by periods of progress intertwined with moments of standstill and serious strife. Historical distrust between Spain and Portugal dates back to the 16th century, when the Kingdom of Spain annexed the Kingdom of Portugal in 1580 as a result of the dynastic claim of the Spanish King Philip II, who became King Philip I of Portugal. He was succeeded by Philip III who lost Portugal to the Duke of Braganza, who became King John IV of Portugal, after the 1640 uprising. The colonies in America were also a source of conflict, especially during the 16th and 17th centuries. Despite centuries of rivalry, Spain and Portugal have not engaged in military combats, and the Iberian border has not moved significantly since 1640[2].

Spain and Portugal very recently began to enter into the type of integration that other bordering EU states (The Netherlands, Luxemburg and Belgium, being the most telling example) have undergone since the European Economic Community was founded with the Treaty of Rome in 1953. This chapter reviews the history of Portuguese-Spanish cooperation and the sharing of their transboundary rivers, and analyzes what impact the succession of treaties signed by both countries has had on Spanish water policy.

---

[1] This chapter summarises the results of a larger report submitted to the Fundación del Canal de Isabel II (Madrid Water Company), which funded a research project carried in the Universidad Politécnica de Madrid.
[2] There is a territory in Extremadura (Spain), called Olivenza, which Portugal claims, and which Portuguese maps draw as an undefined borderline.

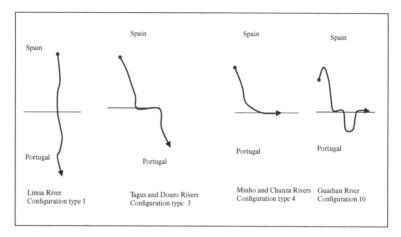

Figure 1.   Geographical configuration of the transboundary rivers (after Dinar, 2008).

The signature of the Albufeira Convention (AC)[3] on November 30, 1998, is the major landmark in the history of cooperation between Spain and Portugal on their shared rivers. With it, both countries embarked upon a process that would map the route for an intense process of cooperation and departed for good from a cooperative process that had been based on specific treaties that dealt with only very partial aspects of the shared rivers.

Very briefly, the position of each country during the years that led up to the AC is as follows. Portugal, the smaller, downstream country with a larger percentage of her territory lying in the shared basins and with fewer dams and water works than Spain, would posit herself as the vulnerable party and would feed the "socio-political perceptions of vulnerability or potential subjugation [existing] in Portugal. This [did] not necessarily have a positive impact on any process for the reconciliation of interest" (Canelas de Castro, 2003, p. 67). Spain would play the role of the more arid, upstream country with fewer per capita renewable resources, always drafting massive water works—some of which would be built and some would not—, and would claim her right to develop irrigation to increase food production and to build numerous dams to compensate for the extreme variability of its semi-arid environment. As Spain has always had far more dams and water works in the shared rivers than Portugal, she would always contend that she provided Portugal with a free flood prevention service, whereas Portugal would claim the right not to be at Spain's mercy, which, in the pursuit of her own concerns, disregarded Portugal's interests.

The AC laid the foundation for a cooperative regime that would thereafter be continuous and irreversible. However, the AC was not meant to solve the problems of the future. Rather it served the valuable goal of ushering in a new era in which the Water Framework Directive objectives would land on both countries' water policy agendas with exactly the same objectives, timetable and accountability obligations in front of the European Commission. It was not by a happenstance that the AC was signed on a date equidistant between the approval by United Nations Convention of the Law of the Non-Navigational Uses of International Watercourses in 1997 and the Water Framework Directive (WFD) in 2000. Undoubtedly, the AC was inspired by the doctrine of the UN Convention and anticipated the challenges posed by the WFD.

---

[3] Convention on the Cooperation for the Protection and Sustainable Use of the Waters of the Spanish and Portuguese Basins (*Convenio sobre Cooperación para la Protección y el Aprovechamiento Sostenible de las Aguas de las Cuencas Hidrográficas Hispano-Portuguesas*).

In the early stages, the AC was negotiated against a backdrop of increasing cooperation and economic integration between both countries, albeit extremely rarefied by the 1993 Spanish National Hydrological Plan (see Chapter 19). This plan envisioned massive water transfers from the Douro to the Tagus (1000 million m^3 per year) and from the Tagus headwaters to the Segura (another 1000 million m^3 per year). In view of these two major inter-basin transfers, which had a serious impact on the discharge of both rivers at the border crossing, Portugal demanded that the two countries should negotiate a comprehensive agreement on the shared rivers that would clarify their future status. Portugal was very concerned that the plan would reduce the hydropower potential of the Douro, from which the country obtains more than a third of her national hydroelectricity.

Spain's Parliament asked the national government to withdraw the entire plan in 1993 because it foresaw unjustifiable water supply increases to enlarge the irrigated area in most basins (see Chapter 19). The withdrawal of the plan unblocked the impasse, and negotiations began in 1994.

Two hypotheses are posited in this chapter. First, it was the WFD that triggered both countries' willingness to take up a serious negotiating position that would lead to the AC. Although the AC preceded the WFD by two years, the WFD had been under negotiation in the EU since 1996, so both processes overlapped significantly. Evidence in favor of this hypothesis is provided by other authors (Ardás *et al.*, 2008; Barreira, 2008; Maia & Vachlos, 2003; Vachlos & Nunes Correia, 2000). The second hypothesis is that the AC is only a small, albeit significant, part of an increasing integration of the two countries' economies, societies and strategic international objectives. The AC released the underlying forces that had, up until its signature, been blocked by the distrust between the two governments, the shared rivers being one of the most contentious issues in the mid 1990s.

However, the WFD foresees clear objectives with respect to the improvement of the ecological status of the water bodies and to the basin plans that are due in December 2009 (see Chapter 16). The rationale of the water plans would be compromised if each Member State were to draft their basin plans for the national part of transboundary rivers independently of the neighboring states. In the Iberian cases, there exist at least four geographical configurations out fourteen possible configurations as defined by Dinar (2008, Ap.B). These configurations are shown in Figure 1 (a fifth one is not shown because it is related to small tributaries).

On February 19, 2008, the second Conference of the Parties, instituted in article 20 of the AC, made great headway regarding the sharing of the transboundary rivers. It was agreed that the flow regimes of the rivers that had been defined in 1998 in the AC would be modified (see below for details), thus responding to a long-standing Portuguese claim that, on top of the annual minimum flows, weekly and quarterly minimum flows should be established at the border. Other examples of the progress made in 2008 were the setting up of a permanent secretariat, which was to be located in one country and alternate every two years (the first turn was for Lisbon), the agreement about the geographical boundaries of the shared water bodies (enabling fully consistent geographical information systems in both countries), and the opening of the bilingual, Spanish and Portuguese, AC web page, offering the general public a wealth of documentation and legal documents. The February 19, 2008 achievements are the latest event reviewed in the chapter.

2 BASIC FEATURES OF THE TRANSBOUNDARY RIVERS

Figure 2 shows the transboundary basins of the five main rivers shared by Spain and Portugal. While all discharge into the Atlantic Ocean, only the Minho and Guadiana form the border at the river mouths. The Guadiana crosses the border, flows into Portugal and then bends to the east to create the border in the last stretch forming an international estuary. As the WFD includes the transition coastal zones in the basin plans, the Guadiana is the river that demands closest cooperation between the two countries.

Table 1 reports the basic quantitative parameters. Note that there are large storage capacity differences between Spain and Portugal in both relative and absolute terms. The Guadiana's storage capacity is more balanced because of the big Alqueva Dam (with more than 4,000 hm^3 of storage).

Figure 2.   Portuguese and Spanish shared rivers.
Source: CADC (http://www.cadc-albufeira.org/).

3   INSTITUTIONAL AND LEGAL COOPERATION FRAMEWORK

3.1   *Prior to the Albufeira Convention—AC (1998)*

Iberian cooperation over its shared rivers has been regulated by a series of treaties and legal instruments, dating back to the end of the 19th century. Until the AC was signed in 1998, the environmental protection of the riverine ecosystems was totally disregarded. The treaties dealt only with aspects that were related to use benefits, where hydropower was by far the most important issue.

The Treaty on Boundaries signed in 1864 established the definitive border, which had not been agreed and expressed in any legally binding text until then. Article 28 defined the international stretches of the rivers, which also form border stretches. Annex I of the Treaty on Boundaries established a regulatory framework for the boundary rivers. Based on the common use of these stretches foreseen in the Treaty, it established the right of free navigation on the Minho, Douro and Tagus, as well as the right to the use of waters that could be beneficial for the people dwelling in both territories.

Later on, this treaty was supplemented by an Exchange of Letters approving the General Act on Border Demarcation (1906), that is, a field demarcation of the borderline between the two countries from the mouth of the Minho River to the confluence of the Guadiana and the Caya. A subsequent Exchange of Letters was agreed in 1912 to define the industrial use of the boundary rivers, limiting the use for hydropower.

The 1927 Convention Regulating the Hydroelectric Development of the International Stretch of the Douro River simply divided the energy potential of this river, defining principles such as equal sharing of the power derived from the principles of common interests and equal rights. Both countries had ambitious hydropower production programs at the time.

The Spanish-Portuguese Convention Regulating the Hydroelectric Development of the International Stretch of the Douro River and its Tributaries (1964) introduced a strict limitation of the withdrawals and flow diversions compared with the 1927 Convention and expanded its scope to the Douro tributaries. Thus, this Convention governed the use of hydroelectric power by allocating each country 50%. The Spanish-Portuguese Convention Regulating Water Use and the International Stretches of the Rivers Minho and Chanza, Limia, Tagus, Guadiana and its tributaries (1968) expanded the cooperation to all rivers shared by the two countries.

Table 1. Basic quantitative characteristics of the shared basins.

	Area (km²)	Population (mill. hab)	Storage capacity[1] (hm³)	Hydropower capacity (MW)	Internal renewable resources (hm³/year)			Water Demands (hm³/year)					
					Surface	Ground-water	Total	Urban	Irrigation	Industry	Cooling	Other	Total
Minho-Limia													
Spain	17,603	0.841	3,040	5,360	12,700	6,400	19,100	82	210	33	40	118	484
Portugal	1,977	0.228	400	700	3,000	300	3,300	11	240	10	n.a.	n.a.	261
Total	19,580	1.07	3,440	6,060	15,700	6,700	22,400	93	450	43	40	118	745
Douro													
Spain	78,859	2.21	7,874	3,370	13,700	3,000	16,700	280	3,800	45	33	453	4,610
Portugal	18,854	1,791	1,080	1,630	9,200	800	10,000	100	1,370	40	n.a.	n.a.	1,510
Total	97,713	4.00	8,954	5,000	22,900	3,800	26,700	380	5,170	85	33	453	6,120
Tagus													
Spain	55,800	7.17	11,140	7,148	10,900	2,400	13,300	1,050	1,900	n.a.	1,250	800	5,000
Portugal	24,800	3.50	2,750	3,266	6,200	2,700	8,900	190	2,020	140	n.a.	n.a.	2,350
Total	80,600	10.67	13,890	10,414	17,100	5,100	22,200	1,240	3,920	140	1,250	800	7,350
Guadiana													
Spain	55,528	1.473	9,220	271	5,500	800	6,300	120	2,280	50	0	0	2,450
Portugal	11,620	0.21	4,610	260	1,900	400	2,300	12	1,280	0	0	0	1,292
Total	67,148	1.68	13,830	531	7,400	1,200	8,600	132	3,560	50	0	0	3,742

Sources: **Minho-Sil:** EPTI Minho-Sil (July, 2008) http://www.chminosil.es; Mendes (2003); Relatório Sintese sobre a Caracterizaçcao das Regioes Hidrográficas-Septembre 2005. **Douro:** EPTI Douro: http://www.cadc-albufeira.org/doc/es/ES9_11.pdf; Mendes (2003); PBH Douro—1999; **Tagus:** EPTI Tajo–Julio 2008; http://www.chtajo.es/pdf_tajo/phc/ETI_TajoProvJulio08.pdf; Plan Bacia Hidrográfico Tejo 1999. **Guadiana:** EPTI Guadiana: http://planhidrologico2009.chguadiana.es, Mendes (2003), Plan Bacia Hidrográfico Guadiana 1999, PNBEPH, nov 2007.

We can summarize the main features of these agreements as follows:

1. The main objective was to ensure the public use of shared river basins, mainly for hydropower, as well as sharing the flow in international stretches.
2. They governed the international and boundary stretches but not the entire river basin.
3. Groundwater resources were outside their scope.
4. They did not provide for the protection of aquatic ecosystems.

3.2   *The Albufeira Convention (AC, 1998)*

The signature of the AC was a milestone in the Spanish-Portuguese relationship, since it introduced significant amendments to respond to the changing needs of the Iberian society at the dawn of the 21st century and Spain and Portugal's obligations as members of the European Union. Consequently many of its provisions are linked to compliance with the WFD and with international law. Thus, the Convention on the Cooperation for the Protection and Sustainable Use of the Waters of the Spanish and Portuguese Basins includes the protection of aquatic ecosystems, pollution of the Spanish-Portuguese watersheds, and its scope extends to all the basins of the Minho-Limia, Douro, Tagus and Guadiana, including groundwater resources.

The objectives of the AC are to coordinate actions to promote and protect a good water status with regard to sustainable use and policies that help to mitigate the effects of floods and droughts. These objectives are very similar to the WFD objectives. To this end, it foresees a cooperation mechanism under the umbrella of the Commission for Development and Implementation of the Convention (CADC, acronym in both Portuguese and Spanish, *Comisión para el Desarrollo y Aplicación del Convenio*) within the Conference of the Parties (CofP), whose main competencies are:

1. Exchange information on, and systematically review, all issues specific to the Convention.
2. Conduct consultations and activities in the bodies established by the Convention.
3. Adopt joint and individual technical, legal, administrative or other measures for the development and implementation of the Convention.

The CADC also has subsidiary bodies, subcommittees, working groups (WG) and forums for public hearings. So far, according to information released on the CADC website, there have been nine plenary meetings of this Committee. The WG and subcommittees of the sixth meeting are:

- WG for the Flow regimes, Drought and Emergency Situations
- WG for the Exchange of Information
- WG for Hydraulic Infrastructure Safety and Flooding
- WG for the WFD and Water Quality
- Subcommittee on Public Participation

The CofP is composed of the representatives appointed by the respective governments and is chaired by a Minister from each State or his/her delegate. So far, the CofP has met twice: the first time in Lisbon on July 27, 2005, and the second, in Madrid on February 19, 2008. On the institutional front, the second CofP decided to establish a Permanent Technical Secretariat composed of delegates from both countries. Table 2 summarizes the main obligations and commitments contained in the AC.

The last CofP, held in February 2008, decided to change the flow regimes that had been negotiated for the AC in 1998. As shown in Table 3, minimum flows for all basins were set up on a weekly basis, except for the Minho basin which only incorporates the quarterly minimum flows.

4   EVALUATION OF THE COOPERATION

The Spanish-Portuguese cooperation has had three distinct phases, as shown in Figure 3. Until 1998, it was developed through partial agreements between the two countries on water use, allocation and distribution. In the run-up to the signing of the AC, the negotiation climate was marked

Table 2. Summary of AC obligations and commitments.

Main chapters	Specific obligations and commitments (only the most relevant)
1. Cooperation among parties	a. Exchange information about the management of the Spanish-Portuguese river basins, and about the activities likely to cause transboundary impacts. b. Provide information to the public about the subjects of this Convention. c. Consultation on projects and activities that may have transboundary impact. d. Establish and/or improve systems for coordinated or joint communication, early warning and emergency. e. Develop programs for infrastructure safety and risk assessment. f. Coordinate management plans and programs of measures established in terms of Community law for each river basin.
2. In matters related to environmental protection and sustainable use	a. Take all appropriate measures to protect water quality, including evaluation, assessment and classification according to EU directives. b. Coordinate procedures for the prevention and control of pollution from point and non-point sources. c. Consider the entire river basins when allocating scarce resources and exchange information about new uses. d. Establish flow regimes that ensure good water status and current and future uses within the terms established by the 1964 and 1968 Conventions.
3. In regard to emergency situations	a. Take measures to prevent accidental pollution incidents and mitigate their consequences. b. Establish mechanisms to minimize the effects of floods (floods). c. Coordinate the actions taken to prevent and monitor drought conditions and shortages, defining criteria and indicators to characterize the situation objectively and to take measures in emergency situations.

Source: CADC (http://www.cadc-albufeira.org/).

by distrust between the two countries and strategic management of the information held by each party. Significant evidence of this point is the great disparity of contributions by Portuguese and Spanish authors in the only two works that have dealt with the joint management of shared basins (Vachlos Nunes & Correia, 2000; and Maia Vachlos, 2003). Both books, ironically published in English, offer very detailed explanations on all areas by the Portuguese authors.

From the signing of the AC to the 2nd CofP meeting on February 19, 2008, half a decade of work, meetings and negotiations took place with the objective of laying the foundations for and building the AC. The entry into force of the WFD in November 2000 forced both countries to step up cooperation in order to comply with Article 13.2. However, it soon became apparent that the coordination of the basin plans required closer cooperation, as well as numerous requirements on models inter-calibration, acceptance of measurement equipment, setting methodologies, and common agreement on the mapping of water bodies.

Thematic working groups were set up and information began to be exchanged intensively in a more trusting climate. The flow regimes established in the AC and the derived commitments led to a stage where the details could be discussed and set more precisely. This second phase was completed at the 2nd CofP February 2008, with agreements that go beyond the scope of the AC. Prospects for the future may be in part based on the steps taken in 2008.

Table 4 summarizes the opinion of the experts consulted on these issues, with specific comments focusing on each basin. The Guadiana basin stands out in this table as having met with

Table 3.   The flow regime of transboundary rivers set by the AC and redefined after the 2nd CofP held in February 2008.

Basin	Before the 2nd CofP Minimum Annual (mill m³/annum)	After the 2nd CofP Annual	Quarterly		Weekly
Minho	3700	3700	440	1 Oct–31 Dec	
			530	1 Jan–31 Mar	
			330	1 Apr–30 Jun	
			180	1 Jul–30 Sep	
Douro	3500	3500	510	1 Oct–31 Dec	10
			630	1 Jan–31 Mar	
			480	1 Apr–30 Jun	
			270	1 Jul–30 Sep	
	–	3500	510	1 Oct–31 Dec	10
			630	1 Jan–31 Marc	
			480	1 Apr–30 Jun	
			270	1 Jul–30 Sep	
	3800	3800	580	1 Oct–31 Dec	15
			720	1 Jan–31 Mar	
			520	1 Apr–30 Jun	
			300	1 Jul–30 Sep	
	5000	5000	770	1 Oct–31 Dec	20
			950	1 Jan–31 Mar	
			690	1 Apr–30 Jun	
			400	1 Jul–30 Sep	
Tagus	2700	2700	295	1 Oct–31 Dec	7
			350	1 Jan–31 Mar	
			220	1 Apr–30 Jun	
			130	1 Jul–30 Sep	
	4000	1300 (1)	150	1 Oct–31 Dec	3
			180	1 Jan–31 Mar	
			110	1 Apr–30 Jun	
			60	1 Jul–30 Sep	
Guadiana	Annual (see (2)) Daily average 2 m³/s	Annual (see (2))			Daily average 2 m³/s
	Daily average 2 m³/s				Daily average 2 m³/s

(2)	Accumulated and average precipitation during the 1 October–1 March period	
Storage (hm³) in the reference dams	>65%	>65%
>4,000	600	400
Between 3,150 and 4,000	500	300
Between 2,650 and 3,150	400	Exceptional circumstances
<2,650	Exceptional circumstances	Exceptional circumstances

Source: CADC (http://www.cadc-albufeira.org/).

greatest agreement among experts in Spain and Portugal in qualifying it as having a major drought and a minor flood risk.

In the other basins, comments reveal few differences regarding the order of priority given to the four categories of problems (droughts, floods, water quality and flow regime). The Minho and

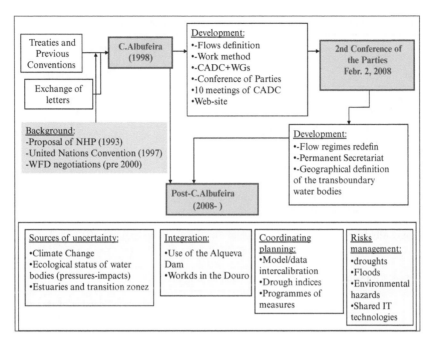

Figure 3.   The stages of Spanish-Portuguese cooperation.
Source: Own data.

Table 4.   Priority issues for the shared management, according to leading experts.

Issue	Experts	Minho and Limia	Douro	Tagus	Guadiana
Drought management	Spanish	Not very important	Important	Very important	Very important
	Portuguese	Not very important	Very important	Very important	Very important
Flood management	Spanish	Not very important	Not very important	Very important	Not very important
	Portuguese	Important	Very important	Very important	Not very important
Water quality	Spanish	Not very important	Important	Important	Very important
	Portuguese	Not very important	Very important	Very important	Very important
Flows regime	Spanish	Not very important	Important	Very important	Very important
	Portuguese	Not very important	Very important	Very important	Very important

Limia basins are seen as less troublesome, and the Tagus and Guadiana show few discrepancies. It is about the Douro where the Spanish and Portuguese experts tend to disagree, where the Spanish experts rate the problems as being less important than the Portuguese, especially as regards flood management.

Another element of analysis was provided by experts attending a workshop held in Lisbon on April 21 and 22, 2008. Ratings on key issues and potential measures were collected, and the 175 attendees were asked to order both the issues and measures by priority (38% response rate). The results are reported in Tables 5 and 6. There is found to be a focus on environmental and conservation issues in the selection of both issues and potential measures. The results suggest that once the flow regimes had finally been settled by means of the agreement on February 19, 2008, experts, managers and stakeholders showed greater concern about the environmental aspects of the shared rivers. The demand for stricter environmental policies and surveillance mechanisms received the highest ratings (Table 6).

Table 5.   Important issues in the transboundary basins (listing top-rated issues only).

Issue	Number of experts mentioning the issue (68 respondents)
Water scarcity	35
Lack of zoning for flood areas	33
Significant alteration of flow regimes	32
Poor and insufficient monitoring	31
Groundwater contamination	30
Eutrophication	28
Coastal zone degradation	24
Contamination with highly contaminant substances	23
Poor irrigation scheme efficiency	23

Source: Albufeira Convention web page: http://www.cadc-albufeira.org/doc/es/ES8_2.pdf

Table 6.   Potential measures (listing top-rated measures only).

Measure	Number of experts mentioning the issue (68 respondents)
Evaluation and control of contaminant spills, especially by the food and meat industries	31
Application of the polluter-pay principle and cost recovery rates	31
Increased and extended surveillance of the water public domain	29
Delimitation of the protection zones for the diversion of surface and ground water	29
Implementation of good farming practices	26
Reuse of treated waste water	25
Construction of water treatment plants equipped with biological treatment for phosphorous and nitrogen removal	23
Drafting of contingent plans for drought periods	22
Modernization of irrigation schemes	22

Source: Albufeira Convention web page: http://www.cadc-albufeira.org/doc/es/ES8_2.pdf

5   DISCUSSION

5.1   *Lessons from the review of treaties and conventions*

The Iberian case has few parallels with other international experiences for two main reasons. First, it is a partnership for four major basins, compared to the more common approach dealing with a single river basin. But, equally or more importantly, both countries are obliged to implement the WFD. This suggests that the AC is only a starting point from which Spain and Portugal must manage to coordinate their planning and management policies. Ultimately it is not about sharing out water, hydropower or navigation conditions. The future calls for much more continuous collaboration and closer integration aimed at achieving the objectives of improving the ecological status of the transboundary rivers and water bodies.

However, the historical roadmap, dotted by treaties and agreements all aimed at solving specific conflicts, cannot be interpreted in the light of the treaties between two sovereign states that share one or more rivers. In no other case is a border crossed by four major rivers, each with a distinctive configuration, so crucially affected by a non-State entity with legal personality, the

European Union, providing both the substance and vision of the future international cooperation on transboundary rivers. The WFD has forced Spain and Portugal to cooperate. However, these two states, their economies and societies are increasingly integrated, and they have taken the opportunity of the WFD to beat out a high-level cooperative agreement on their shared rivers. One can only guess what would have happened over the last decade had the WFD not been approved and in force.

International cooperation on shared rivers is indeed a complex issue. As suggested by Dinar *et al.* (2007, p. 222), "cooperation may be devised around four main factors: providing incentives for cooperation, monitoring and enforcement of the agreement, developing institutional structures for managing potential conflicts, and accounting for (still) likely externality effects resulting from cooperation". Whereas cooperative game theory provides a powerful analytical framework for analyzing the agreements' stability and resilience, there is also the perception that fairness and balance act as equally crucial elements. The analysis of the history, writings, and the analysts' conferences in Spain and Portugal is a key source for assessing the importance they have and have had in the course of their cooperation.

Even though the pre-AC agreements did nothing if not sharpen and detail the terms of cooperation, the process leading to the AC in 1998 was far from easy. Reading the writings by Portuguese and Spaniards before or towards the end of 1998, the profound differences between the two countries' views and perceptions are clear. Agreeing on a minimum annual volume for each basin, with two or three conditions of exceptionality based on average rainfall at two weather stations, could be considered as a simple problem to address. After six years of negotiation, however, neither country had achieved much more in terms of numbers and facts. But they had accomplished something of great value; they had set up a system of cooperation, which goes far beyond just complying with international treaty reporting obligations. Cooperation between Spain and Portugal after November 1998 is not so much a matter of implementing conflict resolution mechanisms but rather an irreversible path towards greater integration. The AC settled the most contentious issues, and provided adequate mechanisms to deal with the unsettled issues.

In short, up to 1998, Spanish-Portuguese cooperation has to be analyzed within the context of the body of international treaties on shared rivers (see Dinar, 2008; and Dinar *et al.*, 2007); not so after that date. The post-1998 era will be about examining cooperation between peers on a number of commitments. Spain and Portugal are working on coordinating river basin management plans by the end of 2009, but all the experts and other analysts, like Ardás *et al.* (2008), believe that drafting common basin plans will soon be a reality.

5.2   *Spain and Portugal's readiness for common planning of their shared river basins*

The river basin management plans must be completed by the planning horizon leading to the 2009 WFD mandate. Both countries have opted to draft separate but coordinated planning documents. Some of the problems and difficulties that could explain this choice are:

- Since the 3rd CADC meeting in 2002, Portugal had been demanding a different AC flow regime. This regime was not agreed and amended until February 2008.
- Until 2008 there was no full inter-calibration of hydrological models that would enable a common treatment of the key quantitative and qualitative parameters of the shared rivers.
- There are still disagreements on the measurement of flows and the measuring equipment used at some points.
- It was not until November 2007 (CADC IX) that both countries agreed on the delineation of the transboundary water bodies.
- There are no agreed quality parameters for shared waters.
- There are no common drought indices that could be applicable to the same basins.

Apart from the above technical grounds, there has not been enough trust and understanding between the two countries so far for them to undertake joint planning an attempt on schedule. On the Spanish side, our survey respondents indicated that Portuguese colleagues were not responding to the agreed commitments on the exchange of information. Another point is that both countries, most notably Portugal, changed their institutional framework in 2007 and 2008. Portugal has just restructured from a highly centralized agency, the National Water Institute (INAG), into hydrographic regions. In Spain, the new Statutes of Autonomy of Catalonia, Andalusia, Aragon and the Valencian Community, which are currently challenged in the Constitutional Court, have expanded their powers over water.

However, experts have predicted that, in the long term, coordination will be as close as if water plans were drafted in common. On the other hand, shared basins have unique characteristics that make it necessary to join both countries' efforts to deal with their planning. Examples of the Tagus and Guadiana are two extremes. In the case of the Tagus, the experts believe that joint planning would not provide any added value to the coordination option because there is a very short common border, and each country can establish its quality objectives and action programs independently. Partly because of its particular geographical setting, the experts considered that, in the case of the Guadiana, a common approach was both appropriate and necessary, and appreciated that sooner or later both countries would have to agree on joint planning.

Based on the likely changes resulting from climate change (see Chapter 7), it will be important to adjust the drought indices to best fit each country's needs. However, national interests can complicate joint planning enormously.

5.3   Prospects for cooperation between countries in a climate change scenario with regard to WFD quality objectives

The projections of climate change impacts published by the Ministry of the Environment (MMA, 2007) suggest that water resources will suffer significant declines in Spain. Suggested temperature increases of 1°C and average precipitation decreases of 5% on the 2030 horizon will cause decreases in average natural water contributions of between 5 and 14%, and further increase demands from irrigated agriculture. The projections also suggest the possibility of increased severity and frequency of extreme events. Regime changes have a direct impact on environmental quality parameters.

The implications of these processes for shared river basins vary considerably. On the one hand, it has been argued that the Tagus River crosses the border with generally good water quality, the result of intense sedimentation in reservoirs and increasing water treatment plants on the Spanish side.

Other basins, such as the Guadiana, and to a lesser extent, the Douro, are subject to environmental pressures with a more sensitive cross-border impact. The WFD places Member States under the obligation to improve the water bodies' ecological status. This necessarily involves setting the conditions for quality objectives to be reached in 2015 in all bodies of water. The main objective of the Environment Ministry's National Strategy for Restoring Rivers is to contribute to the restoration of river ecosystems. Hence, projects that are implemented in shared river basins in Spanish territory undoubtedly have a positive effect in the Spanish-Portuguese area.

5.4   Strengths and weaknesses in the practical development of the contents of the Albufeira Convention

The research conducted in this study suggests that there are areas of concern and hope for the future. We first summarize the strengths and then list the key weaknesses.

First, there is inertia, the cumulative result of more than 12 years of progress, integration and cooperation. There is way of retracing the steps taken so far, which weighs in favor of cooperation along a 'path dependency' that neither country may want to go back on.

Second, the flow regimes that were agreed in February 2008 envision a wide range of contingencies and situations. Any future eventualities outside the formally defined scenarios should be addressed in a spirit of cooperation and mutual loyalty. But, in essence, Spain has already addressed Portugal's demands for greater security and flow regime continuity, which are to be accounted for in their 2009 plans. Furthermore, discrepancies about the calibration gauging have been resolved for the Guadiana, where they are now monitored by means of bi-national measurement teams. In the Douro, the discrepancies at Castro and Miranda will be addressed by a new international measuring station. Once these measurement problems have been solved, water management offers very little room for ambiguity, which will improve enforcement and mutual control.

Third, the examination of records and public documents suggests that the CADC, the Working Groups and the Subcommittee on Public Participation have performed well and worked efficiently.

Fourth, the WFD 'forced' Spain and Portugal to cooperate. While this does not guarantee that such cooperation will not derail, it is certainly a stimulus to the extent that a country can always report breaches by the other party to the EU.

Fifth, although the peninsula's electric market cannot yet be considered to be fully integrated, it is on track to be so in the near future. This implies that the interests of energy generating companies along the Douro will converge. This is likely to reduce the discrepancies in the distribution of hydroelectric power and reservoir operating rules concerning not so much the international section, but upstream flow regulation in Spanish territory and dams.

Sixth, the border economy and increasing integration, materializing in specific aspects such as recent agreements about diversions from the Portuguese Alqueva Dam, the future hydroelectric power building project in the Douro, the taking of water from the Chanza or the construction of bridges and other infrastructure help to create, economic, commercial, institutional and even personal commitments and cross ties.

Seventh, the establishment of the Permanent Technical Secretariat, with offices alternating every two years between Madrid and Lisbon, is a testimony to the willingness to cooperate effectively. Above all, though, it is a vehicle for moving the agenda of cooperation, enhancing trust and transparency and assuring the continuation of joint work.

Eighth, Spain and Portugal have shown loyalty and mutual reciprocity, especially in the management of small incidents, such as certain violations of agreements, the flexibility to accommodate extreme events and impacts arising from construction or actions in the watersheds.

Finally, the implementation of a bilingual website, including all the minutes of CADC meetings, legal texts, CofP agreements, technical documents and texts of cooperation presentations and workshops has also enabled a transparent monitoring of the work open to the general public.

On the downside, there are some weaknesses and areas that require greater effort and commitment by both countries.

First, there has been little involvement of the border regions and their stakeholders in the negotiation process. It is remarkable how marginal the participation of the scientific community has been in the meetings and technical work. Studies required in the negotiation process were conducted by each party separately, thus limiting the possibility of having independent and scientific knowledge and analyses of the shared basins.

Second, the shortage of resources to finance the material and human resources necessary for coordination explains why cooperation has not fostered closer collaboration on planning. Portugal could potentially benefit more from cooperation than Spain if the planning process were performed jointly, reducing the costs of achieving quality targets with more intense action in Spain.

Third, there is some concern over the increasing complexity of the institutional framework and the participation of several administrations (national, regional and local) in the Spanish transboundary river authorities (see Chapter 12). The breakdown of consensus in Spain transformed a monolithic position, which was matched by a similar one in Portugal, and opened up a period of uncertainty.

Fourth and last, the basin plans and models are missing a comprehensive and integrated description. In part this was due to the geographical contention surrounding some border areas. Until 2008 the water bodies were delineated differently in each country. For Spaniards and Portuguese to feel that the basins are shared, they must be able to see maps of the basins and hydrological statistics, environmental and socio-economic including both countries.

REFERENCES

Ardás, I., Consejo, C. López, F. Martín, M. & Octavio de Toledo, F. (2008). La gestión de cuencas hidro-gráficas transfronterizas. El caso de los ríos Hispano-Portugueses. [The management of transboundary hydrogaphic basins. The case of Spanish-Portuguese rivers]. Master in River Management and Sustainable Water Management Thesis. University of Saragossa.

Barreira, A. (2008). "La gestión de las cuencas hispano-portuguesas: El Convenio de Albufeira." [Management of Spanish-Portuguese river basins: The Albufeira Convention] Water Management Scientific and Technical Board. del Moral, L. and Hernández-Mora, N. (Eds.) Seville: Fundación Nueva Cultura del Agua. University of Seville—Ministry of the Environment Agreement.

Canelas de Castro, P. (2003). "New Age in the Luso-Spanish Relations in the Management of Shared Basins? The Challenge of Cooperation in the Protection and Sustainable Utilisation of Waters" In Vachlos, E. & Nunes Correia, F. (Eds). Shared Water Systems and Transboundary Issues. With Special Emphasis on the Iberian Peninsula. Luso-American Foundation, Lisbon, 65–114.

Dinar, S. (2008). International Water Treaties: Negotiation and Cooperation along Transboundary Rivers. London: Routledge.

Dinar, S., Dinar, A. McCaffrey, S. & McKinney, D. (2007). *Bridges Over Water: Understanding Transboundary Water Conflict, Negotiation and Cooperation*. Singapore: World Scientific Publishing.

Maia, R. & Vachlos, E. (Eds) (2003). Implementing Transboundary River Conventions with Emphasis on the Portuguese-Spanish Case: Challenges and Opportunities. Luso-American Foundation, Lisbon.

Mendes, A. (2003). "Luso-Spanish relations" In Maia, R. y Vachlos, E. (Eds) *Implementing Transboundary River Conventions with Emphasis on the Portuguese-Spanish Case: Challenges and Opportunities*. Luso-American Foundation, Lisbon, 311–372.

MMA (2007). "Principales Conclusiones de la Evaluación Preliminar de los Impactos en España por Efecto del Cambio Climático". [Key Conclusions of the Preliminary Evaluation of the Impacts of the Climate Change Effect on Spain]. Secretary General's Office for Pollution and Climate Change Prevention—Spanish Climate Change Office, Madrid.

Vachlos, E. & Nunes Correia, F. (Eds) (2000). Shared Water Systems and Transboundary Issues. With Special Emphasis on the Iberian Peninsula. Luso-American Foundation, Lisbon.

CHAPTER 19

The end of large-scale water projects

Pedro Arrojo Agudo
Department of Economic Theory, University of Zaragoza, Spain

1 HISTORICAL BACKGROUND

The origins of water management in Spain are rooted in its Mediterranean culture. It was at the beginning of the 20th century, however, when the basics of modern hydraulics were laid down. At that time, modern hydraulic engineering offered the possibility of making water a key element of socio-economic development. The idea of redeeming Spain from its backwardness in comparison with the rest of Europe, adopting hydraulic policy as a powerful lever for action, was promoted by the *Regenerationist* movement (Harrison & Hoyle, 2000; see also Chapter 9).

Earlier, the large investments and long pay-back periods involved had led to the failure of private initiative in fomenting major hydraulic works, which by then were technically feasible. The *Regenerationist* movement contributed with the final ingredients needed to undertake these works which were to transform the country, inspired by a *political will and the stimulus of public financing* through the vigorous action of a modern and *regenerated* state. The combination of these three factors; cultural tradition, modern hydraulic technology and decisive financial and managerial involvement by government, gave birth to a new management model, *Hydraulic Structuralism*. New institutional and legal frameworks that were revolutionary at the time, provided an impulse to traditional supply-side strategies through massive public subsidies. This policy presided over the management of surface water during the 20th century.

Spain's conflicting socio-political situation at the end of the 19th century, along with its particular geoclimatic conditions, make it difficult to find references comparable to Spanish *Regenerationism* in the rest of Europe. However, in the western US, albeit with different cultural traditions and historical circumstances, under the leadership of J.W. Powell, developments took place which were surprisingly similar to those promoted by Joaquín Costa's *Regenerationism* in Spain (Arrojo & Naredo, 1997).

In spite of the fact that water was regarded as a key economic resource, it was never treated as a good to be managed on the basis of purely economic criteria. The North American hydraulic experience, without being exemplary as regards economic management, at least took on board the tradition of cost-benefit analysis. In Spain, social consensus over the intrinsic benefits of building dams, canals and irrigation provided the opportunity to apply *economic analysis* at levels as irrelevant as those applied to literacy or health plans. In this context, hydraulic works have gradually been shrouded in a sort of *productivist* mystery which has made them more of an end in themselves than a means.

During a lengthy initial phase, the socio-economic success of *regenerationist* strategies led to broad social consensus over the development of major hydraulic works. The existing socio-economic structure, based upon the primary sector, low costs and pressing socio-economic needs led, at that moment, to positive cost-benefit balances. Soft loans and long-term financial arrangements were widely granted. Through this consensus, in the course of the 20th century Spain developed an impressive hydraulic infrastructure. Nonetheless, during the final quarter of the 20th century, this initial socio-economic rationale of major hydraulic works entered a crisis (Diaz Marta, 1999).

The concept of *general interest* which provided a rationale for public subsidies, gradually subsided. Powerful pressure groups with interests in hydraulic policy began to question the supposed *general interest*. What in the US came to be termed *"pork barrel politics"* accurately reflects the spirit in which hydraulic power lobbies handled the federal budget in the US. In Spain, things were not much different under Franco's *hydropopulism*. Yet, the lack of democracy precluded any possibility of debate and rationalisation, even on technical and economic aspects.

In the US, it was the economic debate on the profitability of major hydraulic works that triggered the crisis of this model at the start of the 70s, with subsequent critical reinforcement provided by the ecological movement. In Spain, economic, social and environmental criticism came later, at the start of the 90s, to give birth to the drafting of the *Water Plans for River Basins* and the series of *National Hydrological Plans* (NHPs) (Arrojo & Naredo, 1997).

2   THE PRELIMINARY NATIONAL HYDROLOGICAL PLAN (PNHP)

In 1985, under the Socialist government, the drafting of a new Water Act, replacing the one in force since 1879, represented an effort at modernisation. However, the 1985 Water Act did not sufficiently emphasise demand and conservation-based management strategies (see Chapters 10 and 11).

The 1985 Act gave backing to hydrological planning, both at river basin and national level, envisaging an ambitious research and documentation plan. Eight years later, in February 1993 a document entitled *"Proyecto de Bases de Directrices" (Basic Guideline Project)* was made public by the Spanish Government. It candidly recognised a number of blind spots as regards knowledge of the problems, which the National Plan aimed to tackle. The sizeable budgets directed at specific studies had not succeeded in creating the necessary databases envisaged in the 1985 Water Act. Data-gathering, publication of yearbooks, and the hydrometric and hydrochemical networks were not making sufficient progress. Nevertheless, political urgency imposed itself, prompting the presentation of the Preliminary National Hydrological Plan (PNHP).

Traditional *supply-side* approaches, considering water as a simple *resource*, divorced from any *ecosystem-based approach*, were made explicit in the PNHP's *Discussion of Reasons*. Under this approach certain suppositions were taken as axioms that predetermined the diagnosis: the territory of Spain was considered as *hydraulically unbalanced*; it was assumed that some basins have *surplus flows*, which are *thrown away, unused and lost* to the sea; while others have *structural deficits*; the country was assumed to be marked by profound *hydrological and rainfall imbalances*.

In order to amend these supposed *"historical shortfalls"* the drafters of the PNHP proposed *balancing the country hydraulically* by appropriate *transfers between different river basins*. For this purpose a complex system of works was to be embarked upon through the SIEHNA: *Sistema Integrado de Explotación Hidrológico Nacional* (the *National Integrated System for Water Use*). Briefly, the proposal was to transfer 2,800 Mm³/yr of *"excess water"* through a *"hydraulic combing system"* from North to South and from West to East.

A number of serious flaws and omissions can be identified in the PNHP. First, no serious studies of the potential of integrated surface and underground resources were carried out. Secondly, little attention was paid to demand management strategies or to reuse and desalination technologies. Thus, the potential for offering incentives for water conservation in the 'deficit' areas was given secondary priority. In the third place, the PNHP did not include feasibility studies on the projects or their alternatives. Fourthly, criteria to determine minimum environmentally responsible flows were not reported. In short, the claimed water shortfalls or imbalances could not be deeply analysed.

Nevertheless, the most significant mistakes and deficiencies came from the PNHP's general approach, based on the *structuralist model* and *supply-side strategies*. The concept of water demand was seriously misused. Demand mirrored resource requirements, assuming massive subsidies and almost zero prices, thus ignoring any user liability for costs.

Leakage rates in urban water supply networks stood at around 35% (Cabrera, 2000; see also Chapter 8 in this volume), which offered ample opportunity for savings through adequate modernisation processes. On the other hand, while demographic projections at the time offered prospects of stability, the PNHP assumed an annual growth in urban demand of 2000 Mm³. The increase in urban-industrial demand was estimated at 40%.

For agricultural uses, a shortfall of 3000 Mm^3/yr was estimated. Moreover, 600,000 new hectares of irrigated land were planned, which required additional supplies of 4,500 Mm^3/yr. Savings through modernisation of irrigated land were calculated at only 1,000 Mm^3/yr. The increase in irrigation demand was assumed to be 6,500 Mm^3/yr, an increase of 31% from 1992 farm consumption.

To summarise, the PNHP planned to provide a growth in supply of nearly 10,000 Mm^3/yr over 20 years, equivalent to 30% of the baseline estimates in 1992. To achieve this, it planned the reuse of 600 Mm^3 of used urban water, obtaining almost 200 Mm^3/yr through desalination and an increase in groundwater abstraction of around 1,100 Mm^3/yr. Groundwater, following the tradition, was left aside, thus ignoring the potential of surface and groundwater integrative approaches (Llamas, 2001).

But above all, the plan included the construction of 200 new dams, which would provide almost 8,000 Mm^3/yr of new regulated or controlled water flow, of which 5,000 Mm^3 would be used in original river basins and around 3,000 Mm^3 would result from inter-basin transfers (Arrojo, 2003a). We now turn our attention to the PNHP's diversion plans.

3   THE PNHP INTER-BASIN TRANSFERS

The PNHP described the SIEHNA as the Plan's masterpiece. Two rivers, the Douro and the Ebro, were identified as the principal sources for the planned transfers, while the Internal Basins of Catalonia, Júcar, Segura, Guadiana and Guadalquivir river basins were to be the recipients (see detailed maps in Chapter 2).

The tradition of the *National Hydraulic Works Plan* of 1933 was revived with SIEHNA. Since its proposal as long ago as 1933, the inter-basin *Tagus-Segura* transfer had been finished in 1980. The project for the *Ebro-Eastern Pyrenees* diversion, proposed in 1973, was based on the *"urgent need"* to transfer 1,400 Mm^3/yr from the Ebro to Barcelona, in order to avoid the anticipated collapse of Catalonian industry in the 80s. The diversion fortunately did not take place, but the collapse did not happen. By the 90s these demands planned for the 80s had diminished to 900 Mm^3/yr.

According to the PNHP, a new network of major infrastructures was to connect the North II river basin with the Douro; the Douro with the Ebro and the Tagus; the Ebro with the Catalan basins and the Júcar; the Tagus with the Segura and the Guadiana; the Guadiana with the Guadalquivir; the Júcar with the Segura; the Segura with the southern river basins and these with the Guadalquivir.

In a first phase it was planned to transfer 1,505 Mm^3, basically from the Ebro. The second phase, with 1,310 Mm^3/yr, was centred on the grand diversion from the Douro (930 Mm^3), although the Ebro was once again to make a considerable contribution (380 Mm^3).

One of the controversial points which has emerged in the process of processing the PNHP was its arguable relationship with the River Basin Water Plans (*Planes Hidrológicos de Cuenca*), being drafted at that time, which we next review.

4   THE RIVER BASIN WATER PLANS (PLANES HIDROLÓGICOS DE CUENCA)

Since the end of the 80s, the Spanish government has taken on the task of water planning at river basin levels, pursuant to the provisions of the 1985 Water Act.

Although there was a commitment to open a broad debate all over the country through the River basin Councils, the participatory process was to be frustrated. The preponderance of short-term political interests and the influence of hydroelectric-construction and irrigating communities lobbies eventually prevailed. This could happen because of their high representation profiles within the Councils, as shown in Chapter 12, and because of opaque procedures for public participation, as detailed in Chapter 17.

5   THE NATIONAL IRRIGATION PLANS (NIP)

As the drafting of water planning moved forward, the need to first clarify the future of irrigation in Spain became increasingly obvious, as this is responsible for 80% of total demand. The government itself ended up embracing this argument 1993, delaying the processing of the PNHP, in order to first approve the National Irrigation Plan (NIP).

NIP explicitly set out a number of serious contradictions within the government's water plans. As opposed to the provision of transforming 1,200,000 new hectares of irrigated land, as an overall balance for the different River Basin Plans (or even the 600,000 new hectares projected in the PNHP), the preliminary NIP document (MAPA, 1995) envisaged developing only about 200,000 hectares. The new priority was now to concentrate on modernising existing irrigation systems.

The emergence of explicit contradictions between the River Basin Plans, the PNHP and the NIP was to lead the government to put off the final passage of the National Hydrological Plan through Parliament, a situation that continued to the end of its mandate without the Plan being approved.

A few years later, when the Popular Party (*Partido Popular*, PP) entered office in 1996, similar contradictions between hydrological and irrigation planning were again made explicit. In the course of the past decade, such contradictions have made quite clear the strict surveillance which the European Commission maintains over agricultural plans, due to the fact that the EU's Common Agricultural Policy absorbs about 50% of the EU's budget. This new NIP limited the extension of new irrigated lands to 242,791 ha (MAPA, 2001). However, as water planning legislation has the rank of an Act of Parliament (see Chapter 11), it has tended to be imposed in an irrational manner, overriding more rational planning of irrigation schemes.

6   THE WHITE PAPER ON WATER

After the 1996 elections, the PP's new term in office broke up the main impasse in the process of water planning. The tough opposition to the PNHP, which the PP had kept up, opened the way to expectations that water management would be significantly modernised. The reform of the 1985 Water Act and the drafting of a new White Paper (*Libro Blanco del Agua*), which was to serve as the basis for water planning, were the commitments of the new government in this respect.

The reform of the 1985 Water Act, which was presided over by liberalisation of the concessionary rights markets, was more of a political reaction guided by ideological muscle-flexing than the expression of a new water management strategy, clearly decided on by the new government (see a more detailed analysis in Chapters 10 and 11).

By then, the *1st Iberian Congress on Water Planning and Management*, held in 1998, had generated a powerful set of arguments and alternatives that emanated from a broadly based and prestigious sector of the scientific community (Arrojo & Martínez Gil, 1998). This wealth of arguments was to give strength and cohesion to civic groups in favour of a *New Water Culture*, which were to take unprecedented action not much later. But it was, above all, the decision of the European Commission to draft a new Directive—the Water Framework Directive (WFD; see Chapter 16 in this volume), clearly in harmony with these critical arguments, which inexorably marked that times were changing.

With the White Paper (MIMAM, 2000), the government provided a positive opportunity for reflection. This White Paper may be considered the fruit of a time of transition. A notable effort, unprecedented in Spain, was made to bring together and publish data and studies which had been produced since the start of the 90s and which provided a sufficient wealth of information to draft a relevant diagnosis of Spain's complex water scenario. This was in spite of the fact that the information, often scattered and piece-meal, led to occasionally glaring contradictions (Llamas & Samper, 1999).

Unfortunately, however, over and above the wealth of arguments and data, the final diagnosis was a political issue. In fact, it was to be the start of a NHP, the essential lines of which, by that time (the end of 1999) had already been decided by the Ministry of Environment.

The White Paper's analysis of financial policy, one of the most important and sensitive aspects of Spanish water policy, may serve as an example (MIMAM, 1998: 539–560). Despite the candid recognition of the complete failure of all economic aspects of water policy in Spain (as detailed in Chapter 13), the White Paper concluded: "*Nevertheless, the justified criticism of our administrative system as regards the economic handling of water resources does not necessarily have to lead to the need for profound legislative reforms, as if this were an inevitable consequence ...*" MIMAM (2000: 559).

In the end, the overall diagnosis and conclusions established that the main problem concerns "*hydrological imbalances*" between different river basins, requiring major works which enable "*surplus flows*" to be transferred to "*structurally deficient*" river basins.

In this context, the reform of the 1985 Water Act, which gave birth to the 1999 Water Act, previously approved, was left wide open to the greatest possible perversion of the *free market*. It was accepted that the future National Hydrology Plan (NHP) should prioritise *supply-side* strategies through massive subsidies, but at the same time it envisaged the management of these subsidised concessionary rights via the *free-market* dynamics laid down in the new provision of the 1999 Water Act (Arrojo, 2003a; Chapter 10).

With regard to the diagnosis and the alternatives proposed in the White Paper, authors such as Llamas once again underlined the lack of attention paid to groundwater management, and denounced this sort of "hydroschizophrenia" that has traditionally affected water planning in Spain (Llamas, 2001; Chapter 14).

7   THE NATIONAL HYDROLOGICAL PLAN (NHP) OF 2001

The final consistent approach of the NHP that was proposed in 1998 was eventually to re-publish the diagnosis and solutions proposed by the PNHP that had been tabled by the previous government. Once again the backbone of the Plan became a pro-transfer policy. Nevertheless, the consistent approach of the WFD, passed in 2000, along with growing citizen-led mobilisation and the critical pressure of the scientific community, was to force unprecedented changes.

The NHP (pursuant to the NHP Act 2001) ruled out the Douro basin as a possible source of diverted water in order to avoid conflicts with Portugal, which would have blocked any European financing (see Chapter 18). The number of major dams to be constructed was reduced, as were the volumes to be diverted. In an effort to smooth out the *structuralist* essence of the Plan, projects to modernise and improve efficiency were emphasized. The term *sustainable development*, which had scarcely appeared until then, now became omnipresent in documents as a means of justifying the Plan to the European Commission in Brussels in pursuit of European financing. The NHP was finally to become an absolute hodgepodge of measures, admitting everything, provided that the priority of pro-diversion projects was maintained.

The 2001 NHP Act envisaged the construction of more than 100 major new dams which, along with the Ebro diversions, had an estimated cost of €6.5 billion. This new wave of major dams was mainly justified by the development of more than one million new hectares of irrigated land envisaged in the *River Basin Plans* that had been approved in 1998. If these irrigation schemes became reality, another €20 billion would have to be added to the budget. In fact, in view of the misgivings of the European Commission, the government justified the excessive scale of this growth in irrigation by arguing that it should be regarded as a catalogue of possibilities rather than a programme of commitments to be undertaken.

As regards the modernisation of irrigation schemes and improvements in water supply, sanitation and purification, the NHP 2001 finally provided for significant investments of €6.2 billion and €5.4 billion respectively. However, in contrast to the inter-basin transfer and some of the major reservoirs planned, commitments to develop these were not unequivocally made. All of the above echoes the fact that arguments in favour of a change in the model were beginning to be heeded. On the other hand, these projects now provide the core which would enable this Plan to be opened up to models that are more consistent with the WFD.

After the elections of 2000, the PP extended its political authority with an absolute majority in Parliament. At that time the process of negotiating the WFD in Brussels was entering its final stages. In this context there were two options: to modify the process of passing the NHP in order to be consistent with the WFD, or to accelerate its passage by applying a policy of *fait accompli* before the Directive was approved and put into effect. The government, relying on its absolute majority, chose the second option.

8   THE EBRO TRANSFER: THE REACTIVATION OF OLD "SUPPLY-SIDE" STRATEGIES

The so-called Ebro Inter-basin Transfer, planned in the NHP 2001, comprised two transfers: the North Diversion, 180 km in length, which was planned to transport 189 Mm^3/yr to the metropolitan area of Barcelona for urban-industrial use; and the South Diversion which was planned to transport 861 Mm^3/yr over a distance of 850 km down the Mediterranean coast, carrying water to Castellón, Valencia, Alicante and Murcia, to as far south as Almería (see Figure 1 in chapter 14). Of this flow, 275 Mm^3 would be for urban-industrial uses and 586 Mm^3 for irrigation. The two diversions totalled 1,050 Mm^3/yr, which planned to divert from the lower Ebro near its delta.

In order to justify these diversions, the so-called *"structural deficits"* created by the rising demand along the Mediterranean coast, were presented as an inexorable reality of the present and future. At no time were the causes of this unsustainable growth in demand diagnosed. Any option to deter this growth through suitable demand management strategies was marginalised. Furthermore, the NHP argued in favour of the surplus nature of the Ebro River basin, avoiding consideration of the new environmental objectives in the WFD, and minimising the current impacts of climate change.

It is significant that the government had refused to analyse the most significant experience of diversions in Spain: the Inter-basin Tagus-Segura Transfer, which had been in operation since 1980. In fact, this diversion was aimed at satisfying the same "structural deficits" in the same south-eastern zones as the Ebro's *South Diversion* proposed by the NHP. The Tagus-Segura Transfer was planned and constructed to divert 1,000 Mm^3/yr, but practical circumstances have led to it diverting an average of only 330 Mm^3/yr.

The development of 50,880 new hectares of irrigated land had been planned with the water to be transferred, but twenty years later the surface area of new, irrigated, recognised and legalised lands was in fact 87,825 ha (CHS, 1997). That is to say, in spite of a drop of 70% in the available water flow, the authorities approved an increase in irrigated land 70% greater than the surface area originally planned. If we add the thousands of hectares transformed illegally in the heat of the expectations created by the diversion, we find ourselves with a new *"structural deficit"* which, far from being resolved, mushroomed. This negative balance between expectations and availability has gradually been resolved, in many cases, by the abusive and illegal use of aquifers. (Martínez, 2000; see also Chapters 14 and 15).

9   UNDERSTANDING THE DEBATE ABOUT NHP 2001 AND ITS FAILURE

9.1   *The availability of water flows of the Ebro River*

The estimate of current average flows (13,400 Mm^3) was erroneous, as it did not take into account the existing recessive trend. Rigorous statistical analysis shows that the resulting level is only 10,000 Mm^3/yr. When explaining this recessive phenomenon, growing afforestation in headwater regions (Gallart, 2001) with a consequent increase in evapotranspiration must be taken into account, as must the impact of *climate change* (see Chapter 7), in consistency with the application of the precautionary principle. The NHP assumed a relatively optimistic scenario in this respect. Even so, insofar as the supposed *surpluses* of the Ebro were expected to disappear in the course of the next 50 years, the authorities opted to restrict the horizon to only 20 years.

9.2 *The poor quality of divertible water*

As regards the quality of the divertible water, the average salinity level of the Lower Ebro is 1200 μS/cm, already above the maximum recommended by the EU for urban water supply (1,000 μS/cm), and it is estimated that this level will go beyond 1,500 μS/cm in the coming decades (Arrojo, 2003b).

9.3 *Environmental problems and the sustainability of the Ebro delta*

The concept of *"ecological flows"* adopted by the NHP was that of the River Basin Plans, the bases of which had been drawn up in the 80s, quite divorced from any consideration of the sustainability of ecosystems. Logically, the contradictions with the ecosystem approach of the WFD were quite striking.

The Ebro delta, recognised by both the EU's *Natura 2000 Network* and the *Ramsar Convention* as one of the most valuable natural spaces to be protected in Spain, is currently an endangered ecosystem as a result of salinisation and subsidence. The Ebro River Basin Plan recognised that the *"ecological flow"*, set for the delta at 100 m³/second (some 3,000 Mm³/yr), was an arbitrary *administrative* decision that would have to be reviewed on the basis of sound scientific evidence. In spite of this, the NHP's diagnosis of *"surplus river basin"* was, in fact, based on this arbitrary decision, and evaded the objective of recovering the good ecological status of water bodies (including deltas, estuaries and coastal platforms) required by the WFD.

The studies commissioned by the Ebro River Basin Authority (Prat, 2001), which in the end were ignored, argued the need to guarantee a minimum environmental regime that would require some 10,000 Mm³/yr and left out any qualification of excess flow.

9.4 *Groundwater—a subject still pending*

According to the NHP's diagnosis, the processes of abusive exploitation of Mediterranean aquifers and the thousands of existing illegal wells were justified, incorporating their impact within the sphere of *structural deficits* to be compensated for (Chapters 14 and 15). The challenge of organising user communities and encouraging integrated management of surface and groundwater continued to be relegated very much to second place. Finally, once again the Plan avoided analysing the practical experience of cost recovery for groundwater irrigation, as a background to the viability of this principle, demanded by the WFD (Llamas, 2001).

9.5 *Energy costs*

The net energy cost of the transfers, discounting possible turbine systems on downhill stretches, reached values of over 3.5 Kwh/m³ in Alicante and over 4 Kwh/m³ in Almería (Valero *et al.*, 2001), which is equivalent to the energy needed to pump flows from a depth of 1,200 m. Current desalination technology consumes 3.5 Kwh/m³, with a downward trend to 3 Kwh/m³ (Arrojo, 2003a).

9.6 *Economic and financial inconsistencies*

In the economic and financial sphere there were evident contradictions between the subsidy-based *supply-side* strategies and the new economic rationale based on cost recovery, as envisioned in the WFD. The government tried to justify the economic rationale of the Ebro diversions via an economic cost-benefit study (MIMAM, 2000). However, according to calculations made by Arrojo (2003b), the cost-benefit analysis was severely flawed.

After correction of these errors, average costs rose from €0.31/m³ to €0.72/m³ (Arrojo, 2003b). The government intentionally avoided identifying the costs which could be allotted to each specific stretch of diversion, and which varied from €0.2/m³ in Castellón to €1.5/m³ in Almería. In contrast to these costs, according to the government's own data, expected demand would plummet from €0.36/m^3 in Alicante-Murcia-Almería, €0.24/m^3 in Castellón and €0.12/m^3 in Valencia. Therefore, only Castellón, closest to the diversion point, would be able to pay the costs attributable to it (Arrojo, 2003a).

As regards the expected benefits, the government avoided calculating the *opportunity value* of water in the different recipient basins. The study simplistically and erroneously assumed desalination to be the most economic alternative, at a disproportionate and outdated cost of €0.81/m^3, far higher than the current cost of €0.45/m^3. The value of the transferable water flow was calculated from the opportunity value curves of water in the different receiver basins. Arrojo (2003a) found that the average opportunity value for the whole project was €0.14/m^3, ranging from €0.04/m^3 for the 315 Mm^3 planned to be transfer to the Júcar basin to €0.19/m^3 for the 546 allocated to the Segura-Almería basins. After correcting these errors, the cost-benefit economic balance is negative, with a Net Present Value of minus €3.5 billion (Arrojo, 2003a).

As regards financial aspects, after a prolonged silence, the government finally published the document *Study on the usage and tariff system* (MIMAM, 2003), which specified the financial strategy of the diversions as follows: 30% of financing would come from non-returnable European funds, 30% interest-free public funds with repayment after 50 years and 40% from the capital market at 4% interest.

Since no interest was assigned nor any inflation considered for this 30% of national public funds, the net subsidy was actually being raised to 60% (Arrojo & Sánchez, 2004), in contrast to the *cost recovery* principle recommended by the WFD.

Since these costs far exceeded the payment capacity ceilings of the users estimated by the government itself, financial bankruptcy loomed on the horizon; bankruptcy such as had been experienced by the Central Valley (California) and the Central Arizona Projects (Hanemann, 2003), which had been managed in a similar way. The government itself, in its non-public debates with the European Commission, finally had to admit that the refusal of European funds would produce a deficit of more than €3.1 billion, a figure close to the €3.5 billion that we had calculated (Arrojo, 2003a).

9.7   *The repeal of the diversions: New prospects and new challenges*

After the victory of the Socialist Party in the 2004 election, the new government repealed the NHP diversions in accordance with its electoral promises. This was done along with the passing in Parliament of the 2004 NHP Act. As alternative measures, they moved to prioritise desalination and the reuse and modernisation of urban water networks and irrigation systems. This decisive move was to be a *volte face* in the history of water management in Spain, as had been the presidential veto of the so-called *Hit List* in the US (Reisner, 1993).

Nevertheless, adopting desalination as the main alternative to the inter-basin transfers has created controversy. It is a fact that today, the contrast between long-distance diversions and modern desalination technologies by reverse osmosis provide a balance which is increasingly favourable to these technologies in a number of aspects.

The essential challenge, however, is not so much a question of producing new supply sources but rather of designing and implementing *demand management* strategies. The application of the cost-recovery criterion in this regard is left as a prospect for the future, with foreseeable problems of opposition from the users (see Chapter 13).

However, such difficulties cannot excuse any delay in public clarification of the costs arising from the supply of flows in each case, nor in the rigorous application of the cost-recovery criterion to future projects. This simple line of action will produce significant changes in the immediate future.

Another important challenge is to stop the mismanagement of groundwater, with the River Basin Authorities accepting responsibility in cases of abusive exploitation, and strongly promoting communities of groundwater users, making them responsible for aquifer management.

In the end, the pressure placed by the European policy on the Spanish government, and its multiple ramifications, altered national priorities. Complying with the WFD would eventually become an intense and resource-demanding task. Water planning was to focus primarily on environmental problems, and water uses would need to pass tests of financial and economic soundness. Besides meeting the most urgent urban needs and bottlenecks, most resources and administrative capacity would be devoted to meeting the demanding agenda of the WFD.

10   FINAL REMARKS

In the course of the 20th century, Spain has gained broad experience in promoting hydraulic works and water management on the basis of a dual model: first, public management of surface water based on supply-side strategies through massive subsidies; and secondly, management of a private nature and individualistic style, with groundwater costs covered by the user. This dual model, based on a *productivist* philosophy, has entered a crisis for reasons which are not merely economic but also environmental and social, and which have converged to produce serious problems of unsustainability.

After a new Water Act was passed in 1985, an active process of water planning began at both river-basin and national level. This process, based on an essentially *continuist* approach, has been questioned by a broad citizen-led movement. At the same time it motivated a significant scientific debate which, without any doubt, offers guidelines for the European Mediterranean region. The National Hydrological Plan, which was due to be finally approved in 2000, was structured once again around the promotion of major public works (more than one hundred large new reservoirs and major diversions of the Ebro).

The NHP was based on traditional supply-side strategies through massive public subsidies for surface water, while it continued to ignore the problems of misgovernment and unsustainability in Spain's main aquifers. The approval of the EU Water Framework Directive in 2000 was to confirm the need for a drastic change from traditional supply-side strategies and *resource management* approaches to new ecosystem-based approaches through demand management and conservation strategies. After the last elections in 2004, this new consistent approach is apparently now being definitively imposed. The transfers envisaged under the NHP were repealed, at the same time as an ambitious process of implementing the Directive was embarked upon, including wide-ranging institutional reform. It is therefore clear that times of change are approaching which go far beyond political challenges; changes that reveal two essential aspects: the *recovery of the good state of ecosystems* and the *rational economic use* of water management consistent with the new WFD.

REFERENCES

Arrojo, P. & Naredo, J.M. (1997). *La gestión del agua en España y California* [Water Management in Spain and California]. Bilbao, Spain: Bakeaz.

Arrojo, P. & Martínez Gil, J. (1998). El Agua a Debate desde la Universidad: Hacia una Nueva Cultura del Agua [Water Policy Debate from the Academic World: Towards a New Water Culture]. Proceedings of *First Iberian Congress on Water Planning and Management*. Zaragoza, Spain: Fundación ?? Fernando el Católico-CSIC.

Arrojo, P. (2003a). *El Plan Hidrológico Nacional: una cita frustrada con la historia.* [The National Hydrologic Plan: a frustrated historical attempt]. Barcelona, Spain: RBA.

―――. (2003b). Spanish National Hydrological Plan: reasons for its failure and arguments for the future. *Water International* 28(3): 295–303.

Arrojo, P. & Sánchez, L. (2004). Valoración económica y financiera de los trasvases previstos en el PHN Español [Economic and Financial Evaluation of the Inter-basin Transfers Envisioned in the Spanish NHP]. Working document, Review of the Faculty of Economics, University of Zaragoza, Spain.

Cabrera, E. (2000). Estado general de los abastecimientos de agua en España [General situation of the urban water supply systems in Spain]. In *La eficiencia del agua en las ciudades* [Efficiency of water use in cities], edited by Estevan, A. & Viñuales, V. Bilbao, Spain: Bakeaz, 53–93.

CHS (Confederación Hidrográfica del Segura) (1997). *Plan Hidrológico de la Cuenca del Segura* [Segura Basin Hydrological Plan]. Murcia, Spain: MIMAM-CHS.

Diaz Marta, M. (1999). Evolución de las políticas hidráulicas españolas desde la Ilustración hasta nuestros días [Evolution of Spanish water policies since the Illustration until the present]. In *El Agua a Debate desde la Universidad: por una Nueva Cultura del Agua* [Water Policy Debate from the Academic world: Towards a New Water Culture], edited by Arrojo, P. & Martínez Gil, J. Zaragoza-Spain: Fundación Fernando el Católico-CSIC, 67–79.

Gallart, F. (2001). La estimación de los recursos hídricos en el PHN: insuficiencias del método empleado ante los cambios de uso y cubiertas del suelo en las cabeceras de las cuencas [The evaluation of water resources in the NHP: methodological incompleteness resulting from changes in land use and vegetation in the basin headwaters]. In *El Plan Hidrológico Nacional a Debate*, edited by Arrojo, P. Bilbao, Spain: Bakeaz, 201–211.

Hanneman, M. (2003). Appendix C- Economic findings and recommendations. In *A technical review of the Spanish National Hydrological Plan (Ebro River out-of-basin diversion)*. Murcia, Spain: Fundación Universidad Politécnica de Cartagena.

Harrison, J. & Hoyle, A. (2000). *The Spanish Crisis of 1898. Regenerationism, Modernism, Post-Colonialism*. Manchester, UK: Manchester University Press.

Llamas, R. & Samper, J. (1999). *Las aguas subterráneas en el Libro Blanco del Agua en España* [Grounwater resources in the Spanish WaterWhite Paper]. Madrid, Spain: AIH-GE.

Llamas, R. (2001). La inserción de las aguas subterráneas en los sistemas de gestión integrada [The inclusion of groundwater in integrated resource management systems]. In *Una cita europea con la Nueva Cultura del Agua: perspectivas en España y Portugal* [An European encounter with the New Water Culture: Perspectives in Spain and Portugal], edited by Grande, N., Arrojo, P. & Martínez Gil, J. Zaragoza, Spain: Fundación Fernando el Católico, CSIC, 233–254.

MAPA (Ministerio de Agricultura, Pesca y Alimentación) (1995). *Avance del Plan Nacional de Regadíos* [Advances in the National Irrigation Plan]. Madrid, Spain.

MAPA (Ministerio de Agricultura, Pesca y Alimentación) (2001). *Plan Nacional de Regadíos. Horizonte 2008* [National Irrigation Plan for 2008]. Dirección General de Desarrollo Rural. Madrid, Spain.

Martínez, J. (2000). Modelos de simulación dinámica en el estudio de las externalidades ambientales del regadío en sistemas áridos y semiáridos del Sudeste Ibérico [Dynamic simulation models for the analysis of environmental externalities in south-eastern Iberian arid and semi-arid systems]. Doctoral Thesis. University of Murcia, Spain.

MIMAM (Ministerio de Medio Ambiente) (2000). *Libro Blanco del Agua en España* [Water White Book in Spain], Madrid, Spain: Ministerio de Medio Ambiente.

MIMAM (2000). *Análisis Económicos* [Economic evaluations], Appendix to the National Water Plan. Madrid, Spain: Ministerio de Medio Ambiente.

MIMAM (2003). *Estudio sobre régimen de utilización y tarifas.* [Review of water use regimes and tariffs]. Document n.6 of the Environmental Assessment of the Ebro Diversions planned in the PHN. Madrid, Spain: Ministerio de Medio Ambiente.

Prat, N. (2001). Afecciones al Bajo Ebro derivadas del PHN, alternativas y necesidad de un nuevo sistema de gestión del agua [Impacts of the NHP on the Lower Ebro; alternatives and new management systems]. In *El Plan Hidrológico Nacional a Debate* [The debate on the National Hydrological Plan], edited by Arrojo, P. Bilbao, Spain: Bakeaz, 413–426.

Reisner, M. (1993). Cadillac Desert: the American West and its disappearing water. Penguin Books, New York, USA.

Valero, A., Uche, J. & Serra, L. (2001). *La desalación como alternativa al PHN* [Desalination as an alternative to the NHP]. Study by CIRCE (Centro de Investigación de Recursos y Consumos Energéticos). Universidad de Zaragoza, Centro Politécnico Superior, Government of Aragón (DGA), Zaragoza-Spain.

V
Conclusions

CHAPTER 20

Meeting future water challenges: Spain's strengths and weaknesses

Alberto Garrido
Department of Agricultural and Resource Economics, Universidad Politécnica de Madrid, Madrid, Spain

M. Ramón Llamas
Department of Geodynamics, Complutense University, Madrid, Spain

1 THE STARTING POINT: A MIXTURE OF HISTORY AND INTERNATIONAL DEVELOPMENTS

Spain has a long and enlightened history of water resources management. In the late 19th century *regenerationist* politics and politicians saw the potential of harnessing Spanish rivers and transforming the semiarid rural landscape (Del Moral, **Chapter 9**, and Arrojo **Chapter 19**). To pull Spain out of poverty and illiteracy, there was no choice but to transform the country's barren hinterlands. Engineers, political leaders and lawyers were convinced that the development of ambitious water planning and works was a national objective that would have to be achieved by the efforts of the government.

Spain passed its first Water Act 1866, and amended it to produce the 1879 Act, which all experts agree on recognising as a monument of fine legal drafting (see **Chapter 10** by Ariño and Sastre, and Embid Irujo's **Chapter 11**). However, during the 20th century, several waves of policy initiatives responded to changing paradigms: the Gasset plan in 1902, the creation of the Ebro Basin Water Authority in 1926 and finally the draft of the first National Water Plan in 1933. These paradigms—the unquestionable usefulness of controlling the river systems with dams to mitigate floods and to provide urban water, irrigation water and hydropower—only materialised in the 1950s during the Franco regime. During the 50s and 60s about twenty new large dams were built every year (Arrojo, **Chapter 19**). Spain doubled its irrigated surface in fifty years, reaching 2.5 million hectares of surface water-irrigated land at the end of the century.

Gradually, almost another million hectares that drew on groundwater resources were added to Spanish irrigated acreage. Custodio *et al.*, in **Chapter 14**, discuss this 'silent revolution' that began in the late 60s and early 70s with the use of groundwater. Ironically the most productive agricultural water uses were those initiated by private individuals using groundwater. As a matter of fact, in the most water stressed basins, Segura and Jucar, the expansion of groundwater irrigation has reduced significantly the basins' average surface reservoir levels, and the generation of hydropower, leaving a significant part of the water works' capacity severely underutilised (Gómez, 2009).

Spain turned its back for good on a history of political unrest and isolation in 1978, when the Spanish Constitution was enacted, and in 1986, when it became part of the European Economic Community (now the European Union). By the time the 1985 Water Act was passed, as Embid Irujo points out in **Chapter 11**, the Spanish Constitution provided sufficient support to governments to administer the hydraulic public domain (all surface and ground waters) and to intervene in cases of groundwater 'overexploitation'. The leadership of the public agencies and legislators was reinforced in theory by the 1985 Water Act, but got weakened by three widely different National Hydrological Plans approved between 1993 and 2005, none of which got beyond their initial stages. Each of them seemed to lag behind the evolution of the Spanish economy and society, soon becoming obsolete and being discarded by new governments entering into office.

The 1993 attempt created serious tensions with Portugal because significant transfers out of two shared rivers were planned, completely disregarding the neighbour country's interests (see Garrido *et al.*, in **Chapter 18**). And yet this opportunity, and the anticipated mandates of the WFD, put the Iberian countries in a negotiation path which lead to the Albufeira Convention in 1998, which can be considered a historical landmark for two countries with five centuries of 'peacefully' conflicting relationship.

Garrido and Calatrava, in **Chapter 13**, review the economics underlying this whole sequence of government initiatives. In times of shortage, water allocation was governed by engineering constraints and, administrative regulations. Constrained by water legislation that was very generous to water-rights holders, the cost of water never reflected actual supply costs. By the beginning of the 90s, fuelled by an extreme three-year drought that had serious consequences for irrigators and millions of consumers, the economics of water resources reached the point of no return. Exchanges of water-rights were legally authorised for the first time in 1999 (see **Chapters 10** and **11**). Although water markets have never been very liquid, several administrations have recently opened public tenders to buyout water rights for irrigation at prices that would have been unimaginable even as recently as the early 90s (up to 0.19 €/m^3). The EU's Water Framework Directive (WFD) went even farther in mandating that Member States shall take account of the principle of recovery of the costs of water services, including environmental and resource costs. Providing cheap water and supporting water-works with weak financial support could well leave Spain in breach of European legislation after the enactment of the WFD in 2000. The entire financial design of the water sector of the 1900s became not only outdated but will become outlawed in 2010 if Spain wants to show the best records in enforcing the WFD.

The demise in the economic aspects of the water sector resulting from WFD was just one inflexion point. But there were other, as for most of the 20th century, water policy was drafted and implemented by a close-knit community of hydraulic and agricultural engineers (Del Moral, **Chapter 9**). The academic community scarcely participated in the process, with the notable exception of civil and hydraulic engineers. Dooge (1999) describes the progressive interdisciplinary analysis of hydroprojects worldwide, with the successive integration of economists, ecologists and sociologists into the process. This process arrived in Spain in the 1990s, and with the arrival came the serious questioning of the engineering approach.

At the turn of the 21st century, Spanish water conflicts ran deeply along regional, political and environmental axes. The WFD set out a demanding agenda, enshrining ideas that were foreign to most traditional tenets of water management. Spain's long water-policy history was forced to change course when the WFD was enacted by the European Union in December 2000. A more balanced mix of objectives replaced the traditional emphasis on supply policies, which had little concern about the quality of water and ecosystems. But the regional conflicts embittered as the political negotiations of the new Autonomous Statues for Aragón, Catalonia, Andalusia, Valencia (already approved in 2006 and 2007), and Castille-la Mancha (still under discussion in April 2009) took place with the NHPs of 2004 and 2005 fresh in the minds of voters and of Members of National and Regional legislatives. In local, regional and national elections, water issues became political ammunition, and the ensuing debate took place with very little scientific base. Ironically, during the last decade Spanish scientific production in all fields reached, including water resources, put in the country among the ten most advanced countries in the world.

We may question whether the WFD's ambitious goals would have been set out in a truly national water policy. The WFD has been qualified as a 'too Northern European', posing serious enforcement difficulties in Mediterranean contexts. Given the lessons drawn by Del Moral and Arrojo, it is irrefutable that Spain would never have passed national legislation of the sort of the WFD.

2   MAIN WATER POLICY ISSUES IN TODAY'S SPAIN

Spain's per capita income in 2006 was 97% of the European Union of 25 Member States. Its agricultural sector contributes a meagre 3% to the GDP, but uses between 70 and 75% of all the water in Spain. As Maestu and Gómez (**Chapter 5**) and Aldaya *et al.* (**Chapter 6**) show, Spanish water economy is markedly dual. A small percentage of water use is very valuable, whereas the remainder is used for very low-productivity purposes. Urban consumption grew at an annual rate of 4 to 5% between 2001 and 2007; tourist resorts, urban expansion and golf courses experienced rapid growth. The irrigation sector is not supplying the resources needed to meet this demand, as neither the liberalisation of the 1999 Water Act nor national irrigation plans enabled the farm sector to free up the necessary resources. The increasing scarcity of water to meet current and future needs, including environmental ones, is being accompanied and reinforced by a number of inter-related processes.

First, recurrent droughts and long-term climate projections suggest that Spain's drought contingency planning may be hindered by reduced runoff and precipitation and ever stricter constraints on the development of new infrastructure (Iglesias *et al.*, in **Chapter 7**). Spain will be among the EU countries which will suffer the most from climate change. All models predict that run-off in all basins will diminish and extreme events will become more frequent (Bates *et al.*, 2008).

At the urban level, water shortages were faced in the early 90s with water supply interruptions of various degrees of severity. Cabrera *et al.* (**Chapter 8**) provide a sombre view of the present state of most Spanish urban supply systems. Their conclusion is that the country is not ready to provide reliable, healthy and sustained drinking water to one third of the population. Such unreliable service may be due in general to low urban water rates (Garrido & Calatrava, **Chapter 13**), insufficient private/public investment resulting from excessive bureaucracy (Sastre & Ariño, **Chapter 10**) or the lack of stewardship of strategic groundwater resources (Custodio *et al.*, **Chapter 14**; Lopez-Gunn, **Chapter 15**).

Secondly, as Schmidt & de Stefano (**Chapter 4**) show, Spanish water ecosystems have been undergoing severe processes of deterioration as the country has raised its level of water usage, together with harmful impacts and pressures. Insufficient wastewater treatment of spills both from urban and industrial users remains an unsolved problem which will take years to be tackled to the extent the most environmentally-conscious EU member states have accomplished. In the view of Garcia Novo *et al.* (**Chapter 3**), most indicators of ecosystem quality have deteriorated within the period of time they have been recorded and analysed. The irreversible loss of wetlands, biodiversity and habitats is a disaster for some, and is unquestioned by all experts. The restoration of water bodies in response to the requirements of the WFD will be a gigantic endeavour, and is one that river basin plans, due in 2009 according to the WFD timetable, will need to address with initiatives capable of withstanding the scrutiny of society through serious public participation processes (see Menéndez-Prieto, **Chapter 16**, and Barreira, **Chapter 17**). At the same time, the most pressing environmental problems seem to be worsening, and will likely be aggravated by the effects of climate change (Iglesias *et al.*, **Chapter 7**).

Thirdly, Spain shows the remarkable growth of desalination capacity installed in Spain before the coming into office of the new government in 2004. Its major water policy decision has been to halt the huge Ebro transfer, instead aiming to double the country's desalination capacity in six years. However, this plan is facing more problems than expected, as few farmers and water retailers are signing long-term contracts to become customers. In total only one third of the capacity that should be installed in 2009 became operative. Custodio *et al.* (in **Chapter 14**) offer evidence that most users prefer cheaper groundwater resources, despite the drawdown limitations and deterioration in quality of most coastal aquifers. There are clear and convincing signs to the effect that patterns of groundwater use are not being effectively controlled by government agencies. Lopez-Gunn, in **Chapter 15**, offers a compelling hypothesis about why enforcement of the 1985 Water Act provisions regarding groundwater resources failed.[1]

---

[1] In view of the inadequate treatment of groundwater resources by the WFD, the EU enacted the Directive 2006/118/EC of the European Parliament and of the Council of 12 December 2006 on the protection of groundwater against pollution and deterioration.

3   STRENGTHS AND WEAKNESSES IN MEETING THE WFD POLICY OBJECTIVES

3.1   *Strengths*

a.   Institutional framework

Spain has 75-year old basin authorities that have been working on water planning according to widely accepted rules and paradigms until very recently. Water laws and statutes are also rich and flexible enough to accommodate the set of policy instruments required. However, these agencies will need to be reorganised to fit with the administration set-up required by the WFD. And yet, Spain has decades of experience in running the water institutions, and all stakeholder, users and government agencies consider their role as essential and effective (Varela-Ortega & Hernández-Mora, **Chapter 12**).

b.   Learning from past mistakes

Spain's recent history of water planning failures shows that, in the future, governments will need to draft their policies in a much open and participatory way (see del Moral and Arrojo, **Chapters 9** and **19**). Spanish society is perhaps tired of, and confused by, the inconsistencies and discontinuities of government action during the past twenty years. Data, experience and information are now readily available that would allow for broader and more rigorous discussions to be held. The new planning guidelines[2] issued by the government represent a huge step in terms of policy thinking and issues integration.

c.   The emergence of alternative forms of dispute resolution

Negotiations and new strategies to solve water disputes have been ever more widely used in the last decade, breaking deadlocks and stalemates that had been insurmountable until recently. While still timid, the effects of these diverse initiatives are accumulating in a rich social capital, which will likely pave the way for settlements on a larger scale. At the national and regional levels, the disputes have not subsided, with some ACs requesting that more inter-basin transfers should be built, and some others claiming full rights on all resources running within their territory. The best examples of conflicts resolution are found at local or even regional levels.

d.   A dynamic urban water supply sector

In spite of the deficiencies identified by Cabrera *et al.* (**Chapter 8**), Spanish urban suppliers have made significant changes in the way they charge for their services, the quality of wastewater treatment and sewage collection. Most analyses suggest that consumers' willingness to pay for a reliable water service lies above current water rates (see **Chapter 13**). This paves the way for improving waste water collection and treatment, contributing in turn to healthier bodies of water. While not to the extent it was planned, the rapid increase in the desalination capacity that has been installed in several coastal regions shows that the private sector is also responding to the rise in demand for water, primarily rooted in urban growth and the development of tourism. Per capita urban water consumption has been stabilised if not diminished in many cities, after growing between 1997 and 2001 by 4% on average.

e.   Greater water productivity and investment in rural areas

As Custodio *et al.* (**Chapter 14**) demonstrate, the agricultural sector attests to the relative strength of the rural economy of many areas. It seems highly likely that new assessments will show that the well-known situation in Andalusia and the Mediterranean regions could be extended to all the semiarid regions of Spain. This would mean that groundwater-irrigated agriculture that consumes about five cubic kilometers of water per year produces more value and jobs than surface water-irrigated agriculture that uses about 20 cubic kilometres per year. Although the direction of the causality cannot be established to explain these differences in efficiency, it does show that there is ample scope for increasing irrigation with surface water. The National Irrigation Plan is among the least contested planning efforts in Spanish history (Arrojo, **Chapter 19**). About 1.3 million

---

[2] *Instrucción de Planificación Hidrológica* (http://boe.es/boe/dias/2008/09/22/pdfs/A38472-38582.pdf).

hectares of irrigated land have been intensively rehabilitated, implementing modern water and soil nutrient control systems.

Another enabling factor is the change in EU Common Agricultural Policy (CAP), which was last reformed in 2003. The change in policy means that farmers receive support payments that are fully decoupled from production. Decoupled support measures brought to a halt farm policy incentives to produce more, and to use irrigation water for that purpose. All analyses show that the reform of the CAP has been followed by increases in land and water productivity (Gil *et al.*, 2009; Garrido & Iglesias, 2009), especially in the mainland provinces where farmers' revenue is more dependent on CAP's subsidies (Varela-Ortega and Hernández-Mora, **Chapter 12**). However, the preferential agreements of the EU with North African countries and the possible integration of Turkey in the EU are going to have a very serious impact on the Mediterranean agriculture of Spain in the long term, which will come on top of serious labour market constraints for most non-mechanised farming operations.

f.   The role of European Union policies and other international agreements

All the chapters in this volume have mentioned the profound impact of the EU's WFD. While the domestic water policy agenda is clearly a national and regional matter, all initiatives must meet several important criteria laid down in the WFD. In the case of the Iberian transboundary basins, the WFD gave the co-operation between Spain and Portugal a significant push (Garrido *et al.*, **Chapter 18**), which led to the Albufeira Convention signed in 1998 in anticipation of what the WFD would request Member states about internationally-shared river basins. This opened a period of fruitful co-operation among the two countries, leaving behind decades of controversies.

Also, the agreements related to the food trade to be achieved by the World Trade Organization will also have an impact on agricultural activities, mainly in irrigated agriculture. The traditional concepts of water and food security will change in the near future, while the related concept of "virtual water" is becoming more familiar to water decision-makers. Aldaya *et al.* (**Chapter 6**) report recent evaluations of the Spanish footprint and virtual water trade. In light of the significant amounts of water virtually traded through farm trade, the water stress indicators of Spain should at least put widely accepted notions of scarcity under severe scrutiny and be thought in a more global context. Most water is used to produce low-value commodities, like cereals, which Spain is increasingly importing to feed a growing and competitive livestock sector. This process has been reinforced by the reform of the CAP, by decoupling farmers' aid regime from their farming decisions.

g.   Innovative politics by regional and local governments

Spain's highly decentralised state is a source of concern, but also a hotbed of experiments and initiatives. For instance, both Garrido & Calatrava, and Embid Irujo highlight the importance of regional initiatives in the area of water pricing and of environmental policies. Water policy in Spain has been devolved to the regional governments, some of them acting as frontrunners of environmental policies and taxation. Since 2004, the Central (Spanish) government has taken coordination role, and the Spanish parliament has not passed any significant piece of legislation on water affairs, which stands in sharp contrast by the numerous initiatives of the Autonomous Communities.

h.   The remarkable increase in research, outreach and dissemination efforts in all water-related disciplines

Based on standard scientific indices (ISI Web of Knowledge),[3] the scientific community has improved its productivity, making Spain number 6 in world science rankings in agricultural sciences, 8th in chemistry, 9th in mathematics, and 10th in engineering.

---

[3] Thomson Reuters.

3.2   *Weaknesses*

a.   Financial prospects of supporting the programmes of measures

The Spanish economy was relatively healthy and has managed to maintain growth rates between 2.5% and 4% for the 1995–2007 period, but dropped to negative growth rates in 2009, together with most world economies. The crisis is expected to hit Spain harder, though, because of its large contribution of the construction sector during the 2000–09 decade, which has collapsed after the explosion of the house market. Unemployment rate is expected to reach 20% by 2010, topping the ranking of all OECD countries. These prospects have lowered the environmental objectives in the list of priorities, and will further delay the water price increases for households, industries and farmers that, according to the WFD, should be implemented in 2010. The cost of complying with the WFD in full has been evaluated at €30 billion (2.4% of GDP), so it is very likely that Spain will request numerous derogations in the enforcement of the WFD (see **Chapter 16**).

b.   The marked swinging character of water policies

Since 1993, Spain has attempted and failed three times to have a national hydrological plan widely accepted by the main national parties. The most recent Water Policy initiative, passed on April 22nd 2005, was backed in Parliament only by the Socialist MPs and their allies. While the initiatives taken after the 2004 election show a clear willingness to get away from outdated concepts of water planning, by 2008 it was clear that the programme AGUA was seriously delayed, especially in the number of desalination plants in operation. In 2008 too, government resumed talks to discuss a new transfer from the low Tagus to the Segura basin.

c.   The chaos of groundwater governance

With a few notable exceptions, Spain's records on groundwater sustainability and government are discouraging (see Custodio *et al.*, **Chapter 14**, and Lopez-Gunn, **Chapter 15**). All efforts made since the passing of the 1985 Water Act have failed to tame the 'silent revolution' occurring in the groundwater sector since the early 70s. Twenty years of failed attempts to tackle the most serious problems of groundwater overexploitation and pollution leave a bad record. And yet, examples like the high Jucar (Mancha oriental, **Chapter 15**) or the low Llobregat (Catalonia, as reported in **Chapter 14**) provide hope for agreeing on more sustainable paths in the high Guadiana and most Southeastern aquifers.

d.   The failure of water statutes to solve the most intractable problems

The leadership of the public authorities in brokering or imposing solutions to the most intractable problems has been significantly eroded after two decades of failure. Citizens and companies still regard public agencies as adversarial and inquisitive. These in turn do not inspire respect and scarcely posses the moral authority to convene opposing parties and build a sufficient level of trust (**Chapters 12** and **15**). Citizens still do not trust public agencies not grant them the required leadership to revert intractable problems.

e.   The strength of the agricultural sector

Never in the course of Spanish history have irrigators paid tariffs that reflect the full direct costs of supplying surface water. This does not imply that water is cheap for all of them, as a significant proportion of irrigators pay the highest prices among OECD standards. Yet, irrigators, who take about 70–75% of all water used, have succeeded in maintaining their relatively favourable status as opposed to other water users, who pay significantly higher charges. An example of this is the regional water pricing policy of Catalonia (see Garrido & Calatrava, in **Chapter 13**). In this Autonomous Community, whose water prices are among the highest in Spain for urban and industrial users, farmers are exempt from paying the regional 'canon del'aigua'. In 2008 the Ministries of the Environment and Agriculture were merged to create the Ministry of Environment, and Rural and Marine Affairs, a move that was interpreted by the Foundation for a New Water Culture and most environmental organisations as the defeat of environmental camp to the agricultural interests.

f. The worsening of water pollution and growing environmental problems

Spain's water bodies have been undergoing a serious process of deterioration, as witnessed by the loss of natural habitats (García Novo *et al.*, **Chapter 3**) and by man-made impacts (Schmidt & de Stefano, **Chapter 4**). Both of these chapters offer sombre views of these processes, and do not provide much room for short-term hope. Important basins, like the Segura and Júcar basins, have deteriorated so much that it will be extremely difficult to restore them to good ecological status. At the root of many environmental problems in most water bodies are two factors which will drive the agenda for the 2010–2015 term of the WFD. First, the need to upgrade and improve the urban water treatments of virtually all urban areas will necessarily imply increasing the households' and commercial water tariffs, which are about half or one third of those paid by French, German, or Danish consumers (see **Chapter 8**). Second, tourist and urban development in the last 15 years, coupled with very intensive agricultural use, has brought the water systems beyond the carrying capacity in most Mediterranean regions. Reversing this trend will be a daunting task that will require financial resources and a strong alliance between all users, environmentalists and government agencies.

g. The breakdown of the Constitutional design for the (regional) inter-community basins

The discussions and debate about the 2001 NHP gave rise to another equally important breakdown of consensus. In this case, regional disputes over transboundary rivers became explicit and turned into political ammunition. Although the management of inter-community water resources is, according to the Spanish Constitution, a national jurisdiction, some Autonomous Communities claimed area-of-origin rights in order to question the grand Ebro transfer scheme. Its beneficiary regions, in turn, claimed that inter-community basins were a national jurisdiction and inter-basin transfers were strategic projects for the whole country. While the 2001 NHP was repealed soon after the Socialist Administration came into office in 2004, the conflicts subsided but did not disappear. For one thing, the region of Castille-La Mancha demanded that the Tagus–Segura transfer should eventually be phased out, on the basis that the region itself needs the water resources that are transferred annually to the Segura basin. Furthermore, the 2004–08 political term opened a period of political discussions in Catalonia, Andalusia, Valencia, Castille-La Mancha, Aragón and Basque Country among others, to draft and approve new Autonomous Statutes. These statutes represent the cornerstone of the political autonomy of the Autonomous Communities (ACs) and mark the dividing line between the competencies of the Central administration and those of each AC. The Catalonian Autonomous Statute was the first to be established, but it was soon followed by a number of other ACs. The implications of the redefinition of the Autonomies' regimes for water and the management of inter-community river basins are doubtful. On the one hand, all new Statutes define to a larger or smaller extent new competencies over inter-community basins; the Andalusian being as deep as to declare in article 51 that the region 'has exclusive competencies over the Guadalquivir resources that flow within its territory and do not affect other Autonomous Community', adding that '[those competencies] should not affect the National Planning of the hydrological cycle, ... nor be in breach with article 149 of the Constitution', which establishes the exclusive competencies of inter-basin river basins. On the other hand, the Andalusian Statute has been brought to the Constitutional Court on the grounds, among others, that the Guadalquivir provisions of her Statute breach the constitutional principles. While the Court has yet not settled this issue (in April 2009) the Andalusian regional government has already been given competencies on the Guadalquivir and setup a regional office to manage it, which co-exists in Seville with the previous one, now in charge only of planning issues and managing the waters running through the neighbour regions. Aragón's statute mandates explicitly that its regional government and representatives should see to prevent water transfers from the rivers flowing within its territory. While it is still too soon to ascertain the impacts of this process of devolution, a prudent judgement would indicate that the role of the Central government in inter-community basins has been diminished. Water policy is increasingly a regional policy, and regions, with the eventual support of their Autonomous Statutes, will surely develop their own legislative initiatives.

Although Spain has always praised itself for not having the type of State-run water policies that Australia and the US have, the last significant policy changes with which this book ends would seem to suggest that the Spanish model has been changed in a bottom-up process to make it closer to a federal one. Garrido *et al.* (**Chapter 18**) conclude that it may be easier for Spain to co-operate with Portugal in dealing with the shared Iberian rivers than for the Autonomous communities themselves to solve their water disputes. While the WFD provides the foundation, spirit and a timetable to manage the Spanish-Portuguese co-operation, it does help very little to solve domestic water issues.

REFERENCES

Bates, B.C., Kundzewicz, Z.W., Wu, S. & Palutikof, J.P. (Eds) (2008). *Climate Change and Water.* Technical Paper of the Intergovernmental Panel on Climate Change, IPCC Secretariat, Geneva, 210 p.

Dooge, J.C.I. (1999). Hydrological science and social problems. *Arbor* **646** (October), 191–202.

Garrido, A. & Iglesias, A. (2009). Lessons for Spain: a critical assessment of the role of science and society. In Garrido, A. & Ingram, H. (Eds). *Water for Food in a Changing World.* Routledge, London, in press.

Gil, M., Garrido, A. & Gómez-Ramos, A. (2009). Análisis de la productividad de la tierra y del agua en el regadío español [Analysis of land and water productivity of Spanish irrigation]. Paper presented at the *Ecoriego Workshop*, Granada, Febr. 16–17.

Gómez, C.M. (2009). La eficiencia en la asignación del agua: principios básicos y hechos estilizados en España [Water allocation efficieny: basic principles and stylized facts about Spain]. *Información Comercial Española* 847: 23–40.

Author Index

Subject index